COMPREHENSIVE BIOCHEMISTRY

COMPREHENSIVE BIOCHEMISTRY

SECTION I (VOLUMES 1-4)
PHYSICO-CHEMICAL AND ORGANIC ASPECTS OF BIOCHEMISTRY

SECTION II (VOLUMES 5-11)
CHEMISTRY OF BIOLOGICAL COMPOUNDS

SECTION III (VOLUMES 12-16)
BIOCHEMICAL REACTION MECHANISMS

SECTION IV (VOLUMES 17-21)
METABOLISM

SECTION V (VOLUMES 22-29)
CHEMICAL BIOLOGY

SECTION VI (VOLUMES 30-39)
A HISTORY OF BIOCHEMISTRY

COMPREHENSIVE BIOCHEMISTRY

Series Editor:

GIORGIO SEMENZA

Laboratorium für Biochemie, ETH-Zentrum, CH-8092 Zürich
(Switzerland)

VOLUME 40

SELECTED TOPICS IN THE HISTORY
OF BIOCHEMISTRY
PERSONAL RECOLLECTIONS. V.

Volume Editors:

RAINER JAENICKE
Institut für Biophysik und Physikalische Biochemie,
Universität Regensburg, D-93040 Regensburg (Germany)

GIORGIO SEMENZA
Laboratorium für Biochemie, ETH-Zentrum, CH-8092 Zürich (Switzerland)

ELSEVIER
AMSTERDAM·LAUSANNE·NEW YORK·OXFORD·SHANNON·TOKYO
1997

ELSEVIER SCIENCE B.V.
Sara Burgerhartstraat 25
P.O. Box 211, 1000 AE Amsterdam, The Netherlands

ISBN 0 444 82658 0 (Volume)
ISBN 0 444 80151 0 (Series)

PREFACE TO VOLUME 40

As in previous volumes (Vols. 35–38) in the Comprehensive Biochemistry series encompassing Selected Topics in the History of Biochemistry, the chapters in this volume complement The History of Biochemistry in Vols. 30–33 by M. Florkin, Vol. 34A by P. Laszlo and Vol. 39 by A. Kleinzeller. The aim of the editors was to invite selected authors who had participated in or observed the explosive development of biochemistry and molecular biology particularly in the second half of this century to record their personal recollections of the times and circumstances in which they did their work. The authors were given a completely free rein with respect to both content and style and the editors have made no attempt to impose any sort of uniformity in the chapters. Each reflects the flavour of the personality of the author.

The series was started some time ago by one of us (G.S.) who was struck by the fact that the breathtaking progress in biochemistry, molecular biology and related sciences had led to the almost unique situation that these fields had become of age at a time when their founding fathers, or their scientific children, were alive and well. In the intervening years, time has taken its toll and sadly many leading figures have died in the last few years.

The contributors to this volume encompass a wide variety of experiences in many different countries and in very different fields of biochemistry. Some have worked close to the laboratory bench throughout their scientific life and are continuing to do so. Others have been closely engaged in organisational matters, both nationally and internationally. All mention incidents in their own career or have observed those in others that will be of interest to future historians who will record and assess the period in which our contributors have lived and worked. It was an extremely exciting time for the life sciences. It was also a period of major and often tragic historical events that deeply affected the life and work of the

generation to which our contributors belong. The editors
wish to express their gratitude to all those who made this
series possible, especially the authors.

University of Regensburg
Regensburg, 1996   *R. Jaenicke*

Swiss Institute of Technology
Zürich, 1996   *G. Semenza*

CONTRIBUTORS TO THIS VOLUME

A.A. KRASNOVSKY
*Laboratory of Photobiochemistry, A.N. Bakh Institute of
Biochemistry, Lenynsky prospect, 33, Moscow 117071,
Russian Federation*

S. LIFSON
Weizmann Institute of Science, Rehovot 76100, Israel

B.G. MALMSTRÖM
*Department of Biochemistry and Biophysics, Göteborg University,
Medicinaregatan 9C, S-413 90 Göteborg, Sweden*

S.V. PERRY
*Department of Physiology, School of Medicine,
University of Birmingham, UK*

M.F. PERUTZ
MRC Laboratory of Molecular Biology, Cambridge CB2 2QH, UK

G. SCHATZ
*Biozentrum der Univ. Basel, Dept. Biochemistry,
Klingelbergstrasse 70, CH-4056 Basel, Switzerland*

R. SINGLETON, JR.
*Department of Biology, University of Delaware,
Newark, DE 19716-2590, USA*

E.C. SLATER
*Department of Biochemistry, University of Southampton,
Southampton, UK*

CONTENTS

VOLUME 40

A HISTORY OF BIOCHEMISTRY

Selected Topics in the History of Biochemistry
Personal Recollections. V

Chapter 4. A Lifetime Journey with Photosynthesis by A.A. KRASNOVSKY (Ed.: R.J.) 205

Chapter 5. Efraim Racker: 28 June 1913 to 9 September 1991
by G. SCHATZ (Ed.: R.J.) 253

Chapter 6. A Life with the Metals of Life
by B.G. MALMSTRÖM (Ed.: G.S.) 277

Chapter 7. Harland Goff Wood: An American Biochemist
by R. SINGLETON, JR. (Ed.: R.J.) 333

Chapter 8. Fate has Smiled Kindly
by S.V. PERRY (Ed.: G.S.)....................................... 383

G. Semenza and R. Jaenicke (Eds.)
Selected Topics in the History of Biochemistry: Personal Recollections, V
(Comprehensive Biochemistry Vol. 40) © 1997 Elsevier Science B.V.

Chapter 1

Wandering in the Fields of Science*

SHNEIOR LIFSON

Weizmann Institute of Science, Rehovot 76100, Israel

There are people who love to talk, love to listen, love to tell stories, but when it comes to writing, they find it extremely difficult to choose the right words. I am one of those. As I start to tell my scientific life story, I glance at the calendar. It is 6 March 1996. In 12 days, so says my passport, I shall be 82 years old. I ask myself, is the story worth telling? How long will it take me to tell it? Will I ever finish it? If I ever will, it is only thanks to a dear friend of mine whom I promised, in a moment of weakness, to try and dig out of my personal memories something of more general interest. The dear friend is Rainer Jaenicke, and it is to you, Rainer, that this story is dedicated.

I was born in Tel-Aviv, which was in those days a tiny suburb of Jaffa, a small fishermen town and harbor in one of the many districts of the Turkish Ottoman Empire. My parents were born in two neighboring villages, in one of the many districts of the Russian Empire, in the middle of the second

---

* Based on a lecture to a symposium at the Weizmann Institute of Science, 1979, whose proceedings were published as Molecular Structure and Dynmics, Balaban International Services, 1980.

half of the 19th century. When my father was a teenager, he was recruited to the Czarist army, where he was assigned the job of a tailor, and where he devoted all his free time to study, on his own, the Bible, the Talmud, the Hebrew language, and whatever general knowledge he could get hold of in those circumstances. In 1905, he was released from the army and moved with my mother and their two children, to St. Petersburg, and there he founded the first Hebrew school. He was an ardent Zionist and a devoted lover of the Hebrew language. Therefore, he chose to live and bring up his children in 'Eretz-Israel' (the land of Israel, the Hebrew name for Palestine), where Hebrew started to be the spoken language. In 1909, my family moved from St. Petersburg to Jaffa.

My father affected, enormously, my intellectual and spiritual development. He possessed the thirst for knowledge of a true intellectual, on the one hand, and the religiosity of a true devotee, on the other hand, and he did everything he could to endow upon me both trends. However, even as a child I felt torn between them, until one succumbed to the other. I remember vividly in this connection, several episodes which left a strong mark on me as a boy, and which may, perhaps, interest the reader.

I remember myself as a little child, sitting at the table with a plate of porridge in front of me, dreaming as usual instead of eating. My elder brother was just released from the Turkish army, and was amused with his little brother, whom he hardly knew. He urged me to finish my plate. If I won't, he warned, an angel will take the plate away; and indeed the plate disappeared. I was very excited. Here was a real chance to see an angel. I asked my brother: 'where is the angel?'. 'He flew away', said my brother. 'But the door is closed?', I asked. 'Well, an angel can fly even through the ceiling'. This made some sense to me. Obviously, angels can do odd things. Yet, I was still puzzled: 'But what about the plate?', I asked. 'When the plate is held by an angel it becomes like an angel and can pass through the ceiling together with the angel'. I became suspicious, it didn't make sense to me that a plate can pass through a ceiling. 'If you promise to be a good boy and to close your eyes, the angel will return the plate', said my brother. I promised, and indeed the plate reappeared. This intrigued me to try again. Apparently, I wanted to observe the disappearance of the plate. Is it really an angel or is my brother pulling my leg? I refused to eat, and the angel stole the plate again by surprise. On the third round, I caught my brother snapping the plate. My unexpected reaction took him by surprise. I jumped on him with my little fists, crying and sobbing bitterly. He could not have realized how deeply I felt betrayed, cheated, humiliated. I was told beforehand, by my sister and others, what angels can do to you. Now it appeared that all these stories were lies that grown-ups invented to cheat me.

This episode haunted me for a long time, and was linked in my memory with another episode which happened a few years later. I was sitting in the synagogue near my father, attending the Sabbath services. The 'Cohanim' (the plural of 'Cohen' which is, in the Jewish tradition, a descendent of the Jewish priests of biblical times) were gathering on the front stand to perform the ceremony of Blessing the Congregation. I asked my father why do they cover their faces during the ceremony. 'You asked a very wise question, my boy', said my father. 'When the Cohanim bless the congregation, God rests on their faces, and it says in the holy book: 'Man shall not see my face and live', therefore they have to cover their faces'. In my imagination, I saw God surrounded by a mysterious aura that can kill anybody who is exposed to it. So I asked my father how come that the Cohanim themselves stayed alive when God rested on their faces. 'This is also a very wise question', said my father. He always encouraged me to be inquisitive, as he himself was. 'It is written in the holy books: 'Messengers of righteousness are not harmed'. Since blessing the congregation is a 'mitzvah' (meaning a righteous deed) they are protected'. Being a good boy and very naive, I accepted his explanation literally. However, something inside me nagged, is it really so? What will happen if I shall risk my life and go behind the front-stand of the Cohanim to peep and have a short glance at God? Will I immediately drop dead? Perhaps if I close one eye and peep with the other eye very very quickly I may risk only one eye? I sneaked out of my seat, and stealthily approached the stand from behind. My heart was beating wildly as I came closer and closer. Finally, I stood behind these good innocent people. There was no aura, no God, only dull faces and murmur of the prayer. I sneaked back to my seat and tried to continue reading in the prayer-book, but the experience was too traumatic. A compulsive yawning beset me. I yawned every few seconds and could not stop. My father got upset. 'Go out and have some fresh air, and come back only when it is over'. I went out and the compulsive yawning stopped, but it reappeared when I returned. I never told my father what happened. I felt disillusioned, on the one hand, and a sinner, on the other hand. Now there were no angels, and there was no God. There was only my father, who believed in them and who would never forgive me if he knew what I had done.

However, even without God, the world was full of mysteries. Heaven without god, the skies with their sun and moon and all these stars, were an even bigger mystery. I was very excited when my father told me the little he knew about the fixed stars, the planets, the solar system. But my curiosity was not satisfied, and there was nobody else around to answer my questions. I remember when electricity was first introduced in Tel-Aviv. It was in the early 1920s, and I was younger than teenage. Watching the workers as they stretched the copper wires to be hung on the high poles, I was sure that these should be hollow pipes, because how else could the electric current flow. I was astounded to discover that they were definitely full wires, and there was nobody who could solve this mysterious puzzle for me. Moreover, electric light was so different from that of the kerosene lamps to which I was used. You press a button and the room is filled with light. What would happen if the whole room would be made of mirrors? Would the room become brighter and brighter as light accumulates? These are only few examples of how my world was full of puzzles, of mysteries, of questions which begged for answers.

As I grew up, I discovered gradually that there are solutions to puzzles, that the mystery is not in the nature of things but in our feelings about them. Things stop to seem mysterious if we can explain them. However, new explanations generate new mysteries. I felt that uncovering the secrets of nature is a most exciting game, a source of greatest happiness. I started to dream about becoming an astronomer or physicist, and my parents were encouraging and supportive. After matriculation, they would send me to a good European university, if only they could afford the financial burden. I looked forward eagerly to such an opportunity.

Events took a sharp turn, however, in a totally different direction. Toward the end of my high-school years, I became more and more involved in the problems of the world around me. These were the times of the big world depression of the late 1920s, the times of growing anti-Semitism in Eastern Europe and Germany, and of growing tension between the Jews, the Arabs and the British authorities in Palestine. These were also the days of naive beliefs in political ideologies. Zionism, socialism, communism, fascism; these words triggered high emotions in my environment, particularly among the young. This was also the time of large expansion of the Kibbutz movement. A stream of Jewish youngsters from Eastern Europe was flowing to Palestine. They had not a penny in their pockets, but their hearts were over-flowing with idealistic enthusiasm to make their Zionist and socialist dreams come true. I was swept by this enthusiasm.

Serving the causes of my people and of humanity should take precedence, I felt, over serving my personal dreams. I joined the youth movement 'Hashomer Hatza'yr', and soon after matriculation, together with a score of class-mates and friends, we founded a new Kibbutz of the 'Hashomer Hatza'yr'. Like the 'flower children' of the late 1960s, we exuberated in our togetherness and in our desire to change the world like nobody else ever did. However, in contrast with the 'flower children', we were persevering and practical. Our Kibbutz soon matured and settled in a beautiful southeastern corner of the Jezreel valley, where it flourished economically and as a particular way of communal life. It became an exemplary symbol for the younger 'Tzabres' (the nickname for Jewish boys born in Palestine, or later in Israel) who established more Kibbutzim of the 'Hashomer Hatza'yr'. Even today, three generations later, my Kibbutz is still flourishing, economically and otherwise.

I remember the Kibbutz period as one of the most exciting periods in my life. I loved and cherished the communal way of life, and still miss it. Also, in a more personal way, I enjoyed my role in the Kibbutz as one of the active members who molded its unique character. There was, however, one source of growing frustration. The sanctity of manual labor and its moral preference over intellectual professions was one of the principles of the Kibbutz ideology in those days (not any more). I did comply with it in practice, but could not accept it as a dogma. The Kibbutz, I thought, could and should benefit from the brains of some of its more gifted members, more than from their muscles. Personally, I was eager to combine my yearning for science and intellectual matters with the needs of the Kibbutz.

After much soul-searching, I decided to aim at being a high-school teacher, although the children in my Kibbutz were still very young. In the meantime, I would prepare myself for the task. There existed a newly established boarding-school in one of the nearby veteran Kibbutzim, Mishmar-Ha'Emek. It was named the 'Educational Institute', and it served as a central high-school for all the Kibbutzim of the 'Hashomer Hatza'yr'. It was organized as a 'youth-community', and was renowned for its innovative methods in education and teaching. I sought the advice of Milek Golan, the leader and director of this school. I told Milek that I wished to join the school, but had neither the qualification, nor the experience, which I believed were essential in such a vanguard school, and asked him what should I do in order to qualify as a teacher. His response was totally unexpected. As he knew me personally, he said, he had no doubt about my natural qualifications, and experience is gained only by time and hard work. I should start teaching at the school immediately, accumulate experience for, say, three years, then complement my education with some academic studies to be financed by the school.

Thus, I plunged into the stormy waters of the 'youth-community', trying to learn to swim the hard way. Being a teacher in the Educational Institute was an around-the-clock, non-stop, total commitment. I had to acquaint myself with the innovating teaching methods, as well as with the subject matters which I was supposed to teach. Sometimes I learned, in the late hours of the night, subjects which I taught next morning. Like my students, I lived in the youth-community, and served day and night as their mentor and guide. Never in my life did I work so hard. However, I loved what I did. I learned a lot in a whole variety of fields, firstly, because the multidisciplinary nature of our teaching methods forced it upon me, and secondly, because I had at my disposal the school-vacation periods. I got myself acquainted for the first time with Darwin's theory of evolution, with Mendel's theory of heredity, with wave-particle duality, with space-time relativity, with some branches of high mathematics, and so on. Accumulation of knowledge in such a wide range of subjects was very exciting and gave me an enormous satisfaction, but it was necessarily superficial, and wetted my appetite for orderly academic studies.

Three years passed by, and I was waiting impatiently to spend the coming year at the Hebrew University in Jerusalem with the promised support of the Educational Institute. However, reality wished it otherwise. The World War II was raging and Hitler was still victorious on all fronts. The German forces in north Africa exhibited a frightening superiority over the British forces ever since the German general, Rommel, landed in Tripoli, at the beginning of 1941. They pushed persistently eastward, and within a few months, they were close to the Egyptian border. It seemed that nothing could stop them, and it was clear that if they overran Egypt, the British forces would abandon Palestine as well. I remembered Hitler's vow in one of his chilling speeches broadcasted worldwide: 'What the Jews have built in Palestine in 25 years, I shall destroy in 25 hours'. I took his threat seriously, and joined the 'Palmach' in the middle of 1942, about the time when Rommel penetrated deep into Egyptian territory, reaching El-Alamein, 90 km from Alexandria. The Palmach was a small, well-trained underground force organized by the Kibbutz movement. It was the advance force of the Hagana, the defense organization of the Jewish community in Palestine. It was illegal under the British Mandate over Palestine. Nevertheless, the British military supported and even trained the Palmach for a short period, as a potential guerrilla force against the Germans.

A few years later, before and during Israel's war of independence, 1947–1948, the Palmach made history. It saved the Jewish community in Palestine from extinction, being the forerunner and kernel of the Defense Army of the newly-born State of Israel. However, toward the year 1943, the Palmach seemed to have lost its immediate role, since the danger of German invasion of Palestine was over. In October 1942, Rommel was defeated at El-Alamein, and by May

1943, the German African Corpus was eliminated. I asked to be released from the Palmach, but my request was refused. Instead, I was given a kind of 'leave of absence', against my commitment to go on any mission abroad when required. In the meantime, I was allowed to study at the Hebrew University in Jerusalem.

By this time, I was already 30 years old, and it seemed to me that my dreams to become a scientist had to be given up. First, I thought it was too late; second, it was incompatible with Kibbutz life to which I felt committed; and third, I doubted whether my modest talents were sufficient to make me a good scientist. Mediocre scientists are needless, I thought, while teachers are always needed. However, my desire to study modern science was unabated.

My mission abroad, to which I was faithfully committed, never materialized. In the meantime, the war was over, Nazi Germany was defeated, so my commitment somehow evaporated. While waiting to be called, I continued to study my favorite subjects, physics and mathematics. After graduation, I returned to my position as a teacher at the Educational Institute in Mishmar Ha'Emek. My plans were now to gain more experience as a teacher, then move, if accepted, to the Kibbutzim's teachers college, where I would devote myself to the development of innovative technologies in science teaching.

Reality, however, again wished it otherwise. No sooner did I start teaching in Mishmar Ha'Emek, than the War of Independence broke out. Iraqi forces attacked Mishmar Ha'Emek in April 1948. Their artillery destroyed the campus of the Educational Institute, the students were evacuated and dispersed to their respective Kibbutzim, and I was soon recruited to the newly established Military Research Unit of the Israeli Defense Army, where I was assigned to work on some primitive military physics, improvising booby-traps, hollow-charges and the like. There I met Aharon Katchalsky, one of the founders and leaders of the 'Research Unit'. He was my age, and we became friends. Aharon was always bursting with new ideas and new problems, involving everybody around him in their solutions. I enjoyed our discussions very much, and once in a while helped him in various problems, such as to calculate burning rates of propellants, to solve diffusion equations, etc. One evening he invited me for a walk outside the military labs, located in the cellar of the still unfinished first building of the Weizmann Institute. As we strolled leisurely, he suddenly said: 'How would you like to join the Weizmann Institute's Polymer Department when the war ends? I need somebody to do light scattering of polyelectrolytes'. I was completely taken by surprise, but quickly recovered. I said: 'I'll consider it if you tell me what polyelectrolytes are'. In fact I had great doubts about being suitable for the job. At the university I had not aimed for a scientific career. Moreover, my interests had always focussed on physics and mathematics, not chemistry; the infinite details of chemical reactions and chemical compounds bored me. As a Kibbutz teacher, I needed no diploma, and managed to postpone my final chemistry exams for the indefinite future. Now, unexpectedly, I was invited to join the Polymer Department of the Weizmann Institute, to work with the legendary Aharon Katchalsky. This was a temptation that even a devoted Kibbutznik couldn't resist. I considered it an adventure rather than the start of a career. If I failed, or if I didn't like it, I could always go back to my Kibbutz, I told myself.

Aharon was very persuasive in his scientific philosophy and research program. He told me a lot about the young science of polymers. He considered synthetic polyelectrolytes to be models for biological polymers. Proteins, he explained, are essentially polyelectrolytes; once we understand the laws of behaviour of polyelectrolyte solutions, the road will be open to understanding proteins, indeed to deciphering the secrets of life. He maintained that experimental facts are meaningless unless correlated by a theoretical analysis. He believed in cooperative research, where experimentalists and theoreticians, biologists, chemists and physicists join together to achieve results which none could reach alone. I was attracted by this idealistic attitude. It suited my social ideals and values as a Kibbutznik. It also gave me the comfortable feeling of participating in a great research program without having to initiate independent reasearch in a field in which I had no academic background.

Light scattering of dilute solutions of polyelectrolytes appeared a hard nut to crack. Polymethacrylic acid was evasive. The light scattered was irreproducible, and proved to be due to dust particles. I spent months struggling with experimental difficulties, but the pure solution refused to scatter light, defying our naive expectations to determine the high molecular weight of the polyelectrolyte according to the Debye theory. Then one day I had a 'brain wave', and the riddle was solved. Light scattering is the result of fluctuations in the polymer concentration. The polyelectrolyte molecule carries thousands of fixed charges and is surrounded by an equal number of counter-ions. Just as the osmotic pressure of the solution is orders of magnitude larger than that expected from the polymer concentration, so the light scattering must be orders of magnitude smaller. In order to measure molecular weight, one should eliminate the effect of the counter-ions. Indeed, when I added a 0.1 M concentration of a simple salt such as NaCl, the polyelectrolyte solution scattered light normally, according to the Debye formula. I was delighted to experience my first 'scientific discovery', the agony of so many useless efforts, and the joy of finding that there is such a simple and clear solution of a problem that, only a while ago, seemed hopelessly difficult. For some reason this work was published without my name.

Subsequently, I shifted my attention to theoretical problems of polyelectrolytes. How can one quantitatively derive osmotic pressure, light scattering and other properties of polyelectrolyte solutions, with and without added salt? Soon I found myself involved in Aharon's efforts to calculate the free energy of dilute polyelectrolyte solutions. It was an extension of his studies with Werner Kuhn in Basel, a few years earlier, that originated 'the science of polyelectrolytes'. Following Debye–Hückel's theory of electrolytes, we assumed the ionized groups of the polyelectrolyte molecules to possess a pairwise repulsive energy, $u_{ij} = \varepsilon^2 \exp(-\kappa r_{ij})/Dr_{ij}$, where ε is the proton charge, κ the Debye–Hückel reciprocal radius of ionic atmosphere, r_{ij} the distance between the charges, and D the dielectric constant. The free energy was assumed to be the sum over all ij pairs, averaging over the distances r_{ij}. The averaging was obtained using the probability distribution $W(r_{ij};h)$ of non-ionic polymers derived by Kuhn, Künzle and Katchalsky [1].

The theory led to a simple closed expression for the 'electrostatic free energy' [2]. It successfully fitted the experimental curves of potentiometric titration and activity coefficients of polyelectrolyte solutions, and it made a great impression in international conferences. It failed, however, to give the right end-to-end distance, predicting full extension already at very low degrees of ionization; the good agreement with titration experiments was obtained, only provided we used the theoretical, overestimated end-to-end distance, not the experimental one. Furthermore, the basic assumption of the theory, the superposition of the u_{ij} terms, and the use of the non-ionic polymer distribution $W(r_{ij};h)$ had an ad hoc character.

One day, as I was pondering about these problems, suddenly I was struck by an idea which set me on to a new track. A polyelectrolyte chain dissolved in a highly polar solvent like water, and surrounded by an ionic atmosphere, is analogous to a hot metal filament in a vacuum surrounded by an electron cloud, such as the filament in an incandescent lamp. Why not use the theory of incandescent filaments as a basis for deriving the free energy of polyelectrolyte solutions? Aharon was very enthusiastic about my idea, and when, some time later, Professor R.M. Fuoss from Yale came to Rehovot for a few months on his sabbatical leave, he joined our work on this model. Two papers, one on 'The potential of an infinite rod-like molecule' [3], and the other on 'The electrostatic free energy of fully streched macromolecules' [4], summarized the work.

This theory was limited to pure polyelectrolyte solutions without added salt, and it ignored all aspects related to the flexibility of the macromolecule. It was concerned merely with the interaction of the fixed charges on the polyelectrolyte and the counter-ions, supposed to surround the chain by cylindrically symmetric atmosphere. Despite these limitations, this model is still used and quoted often, surprisingly, even now, after more than 40 years.

Another subject in which I became deeply involved during that period, was the theory of mechanochemical processes, or mechanochemistry. Acid polyelectrolyte gels expand at high pH and contract at low pH, reflecting in a macroscopic way, the microscopic expansion and contraction of the polyelectrolyte upon ionization and deionization, respectively. If a fiber of such a gel is made to contract while a weight hangs at its end, it will lift the weight, thus converting chemical energy directly to mechanical work. I analyzed the basic mechanochemical processes by drawing an analogy with the Carnot reversible cycle of thermal engines. Consider a fiber streched reversibly in two steps. First, it is immersed and stretched in a bath of constant chemical potential, e.g. constant pH. In the second step, it is taken out of the bath and further stretched. The first step, which we called 'isopotential', must be accompanied by a change in degree of ionization, while the second step, the 'isophoric', must be accompanied by a change of pH. To form a mechanochemical cycle, we now immerse the fiber in a solution of the newly acquired pH and contract it reversibly, first isopotentially and then isophorically, so as to return the fiber to its original state. The net work performed by such a cycle is derived from the reversible transfer of H^+ ions from one pH to another. This work was also published, regrettably, without my name.

Mechanochemistry was among Aharon's dearest research projects. The mechanochemical cycle was expected to be a breakthrough on the road to the understanding of mechano-chemical processes in biology. After all, what is a muscle if not a highly efficient mechanochemical engine? [5]. Admittedly, our mechanochemical gels, like osmotic cells, obtained mechanical work from dilution processes rather than from chemical reactions. However, the general principles of mechanochemistry hold equally for dilutions and reactions, namely the reversible conversion of chemical potential differences into mechanical work [6].

My role in the Polymer Department gradually developed into that of a theoretical adviser. I assisted Aharon on all theoretical problems, but had no definite research program of my own. As I once put it, I was the Court Physicist in the Katchalsky Kingdom. This rather peculiar state of affairs came to an end, however, just as suddenly, as accidentally and unexpectedly as my joining the Weizmann Institute, five years earlier. One Friday morning my good friend, Pnina Elson, came to my office. 'Do you realize that today is the deadline for applying for a Weizmann Fellowship?' I knew of the newly established fellowships, but didn't consider myself a suitable candidate. 'Don't be a fool', she insisted. 'It's time for you to change and to be on your own'. I thought for a minute, then quickly filled out the application and submitted it, just on time, to the Academic Secretary. That very day, at noon, the Institute's Scientific Committee met, and in the afternoon, Aharon returned to the Department rather agitated. 'You got the fellowship', he told me. 'I had a big fight with the Committee. I told them how important your contri-

butions to science and to the Department as a whole are, but
David Rittenberg insisted that you had no record of independent reasearch. 'If he is so smart why ain't he rich', he
argued'. Aharon was very critical of this 'narrow careeristic'
attitude but for me, this incident was an eye-opener. I told
Aharon that while I was grateful for his support, David was
right, and that in a way I was grateful to him also.

I spent most of the year of the Weizmann Fellowship with
Peter Debye in Cornell, and the rest with J.J. Hermans in
Leiden. Upon my arrival at Cornell, I tried to get Debye interested in the theory of conductance of polyelectrolytes. I
presented to him my thoughts on the subject and the difficulties I had encountered. He considered the topic with interest, and after about an hour of discussion said: 'It is
rather complicated, are you sure it is worthwhile to work on
it?' It was the first of many lessons I learned from this great
man. Using sound judgement in choosing a problem is as
important as using talent in solving a problem. Striking the
right balance between the two is essential for good science.
Debye did not interfere at all in my work, but was very
genereous in advice and judgement. When I brought up the
question of whether a polymer molecule in a solution, flowing through fine capillaries, experiences a radial centripetal
force, he guessed (rightly) that the answer should be negative, but encouraged me to calculate the effect of inhomogeneity of the field of flow on the forces acting on a random
polymer chain in solution. Indeed, I found no radial forces,
but the theory predicted an increment of viscosity propor-
tional to the square of the ratio between the dimensional av-
erages of the polymer molecule, and the radius of the capil-
lary [7]. When I showed Debye the exact solution of the ensuing equations, which were some Bessel functions of an
imaginary argument, he was critical. An approximate solution whose consequences are easy to grasp, he said, is better
than an exact but obscure solution. Since the inhomogeneity

of the field of flow was only a perturbation superimposed on the homogeneous flow, a perturbation method should yield an excellent simple approximation. Indeed it did.

Debye was at that time interested in the effect of viscosity on the hydrodynamic instability of fluid jets. He wrote an excellent preliminary report on the subject, and tried to obtain an instability condition from the Stokes–Navier equations. I thought that the effect of viscosity on instability should be negligible, at least to the first order. Viscosity slows down the growth of perturbations of flow leading to instabilities, but it also slows down the decay of such perturbations due to stabilizing forces. To examine this problem to its first order, I chose an analogous problem, the hydrodynamic instability of flow of viscoelastic gels. These obey the equations of elasticity, in which the elasticity coefficient is complex, with the imaginary part being responsible for energy dissipation, thus playing the same role as viscosity. I found that only the real part of the elasticity coefficient contributes to the limiting velocity of the jet, above which instability sets in [8]. To my delight, Debye considered this result surprising.

While in Leiden, I became more closely acquainted with the Dutch school of polyelectrolytes. I thought it desirable and feasible to make a comparative analysis of the theories of polyelectrolytes that were then availabe. The theories of Hermans and Overbeek (HO) [9], Kimball et al. (KCS) [10] and Katchalsky and Lifson (KL) [2], had all tried, each one in its own way, to link conformational statistics and electrostatic interactions. I observed that the theories of HO and KL, though widely differing in many ways, had one shortcoming in common: both yielded too large expansions of the macromolecule even at very low degrees of ionization. The source of this weakness was traced to the assumption of additivity of the electrostatic and conformational contributions to the free energy. The KCS theory had much in common with HO, but assumed the additivity of conformational and electrostatic potential, rather than free energies. This seemingly trivial difference was, however, the reason for the puzzling result of the KCS theory, that the maximum expansion factor of the polyelectrolyte by ionization was calculated to be $\sqrt{2}$, i.e., much too small. This study [11] gave me the satisfaction of finally obtaining a broad, objective view of the 'state of the art' in the field of theoretical polyelectrolyte research. It also helped those who entered the field later, and was quoted often in the literature.

When I returned from abroad I looked for a new approach to polyelectrolytes, one that would be free of ad hoc assumptions such as continuous charge distribution, etc. It should take into account 'more important' properties of such systems, and – by neglecting the 'less important' properties – should offer a simple model, amenable to a quantitative analysis. One such obviously important property was the regular arrangement of the ionizable groups along the molecular chain. Each group has its close neighbours in well ordered positions, and the neighbouring groups interact with an energy which depends on their state of ionization. If the chain is long enough and the number of ionizable groups large, common tools of statistical mechanics are applicable. The polyelectrolyte chain is an open system, since equilibrium is maintained between the ionized and unionized groups, say COO^- and $COOH$ respectively, and the surrounding H^+ ions. The resepective chemical potentials μ_- and μ_0 fulfil the condition of equilibrium

$$\mu_- + \mu_{H^+} = \mu_0$$

The statistical mechanical tool appropriate for such a system is the semi-grand canonical partition function ('semi' for N fixed, ν variable)

$$g.\,p.\,f. = \sum_{\nu=0}^{N} \sum_{\{s\}} \exp[-E_s / kT + \nu\mu_- / kT + (N-\nu)\mu_o / kT] \quad (1a)$$

where N is the fixed number of ionizable groups (monomers) comprising polyelectrolyte molecules; ν is the dynamically variable number of ionized groups; E_s is the energy of interaction between the groups, when each group is in a well defined 'microscopic state', i.e., either ionized or unionized; and $\{s\}$ represents the set of all microscopic states of the molecule. The great simplicity of this system is due to the fact that the molecule is essentially a one-dimensional system. Further simplicity arises when we confine the energy E_s to be composed of neighbour interactions only, neglecting interactions between groups situated far from each other along the chain. If we consider, for example, only first neighbours then

$$E_s = \sum_{i=1}^{N-1} E_{i,i+1}$$

where $E_{i,i+1}$ can have just 4 values, E_{00}, E_{0-}, E_{-0} and E_{--}, and we commonly assume that only E_{--} is non-zero, although formally this is not required.

Those familiar with the one-dimensional Ising model of ferromagnetism will recognize immediately, the similarity of the two models. By applying mathematical techniques developed for this model, with some appropriate extensions and modifications, the following results are obtained, remarkable in their simplicity and elegance. We define two matrices, the activity matrix

$$\mathbf{A} = \begin{pmatrix} a_- & 0 \\ 0 & a_0 \end{pmatrix}$$

where $a_- = \exp(\mu_-/kT)$, $a_0 = \exp(\mu_0/kT)$ are the activities of the ionized and unionized components, respectively, and the neighbor interaction matrix

$$\mathbf{U} = \begin{pmatrix} u_{--} & u_{-0} \\ u_{0-} & u_{00} \end{pmatrix}$$

where $u_- = \exp(-E_-/kT)$ is the statistical weight for first neighbour interactions of ionized groups and similarly for the other elements. Then the g.p.f. defined in Eq. (1a) is obtained precisely as

$$\text{g.p.f.} = (1,1)\mathbf{A}(\mathbf{UA})^{N-1}\begin{pmatrix} 1 \\ 1 \end{pmatrix} \tag{1b}$$

In this matrix product, each factor UA belongs to one ionizable group, representing the set of statistical weights contributed by this group to the g.p.f..

It is not appropriate here to enter into technical details of how the secular equation, derived from the matrix **UA**, is used to obtain the degree of ionization as a function of pH, or any other observable property. What fascinated me while working on the Ising model of polyelectrolytes was the feeling that the theory recognized the 'more important' characteristics of the polyelectrolyte in solution over the 'less important' ones [12]. Furthermore, because of its simplicity, the theory could be generalized in various directions. For example, it was easy to obtain partition functions for polyampholytes (polymers with alternating acidic and basic groups) [12], or to include second and higher neighbour interactions [13].

The subject of binding of mono-valent ions such as Na+ to polyelectrolytes evoked great interest in the mid-1950s, an interest prompted largely by Wall and co-workers [14] who performed transference experiments using radioactive Na^+ tracer. They observed that part of the Na^+ counterions were dragged along with the polyacid molecule towards the anode. This implied that the counter-ions were associated with the macromolecule for very long times, because had the association and the dissociation followed each other in rapid succession, all Na^+ ions would have acquired the same average transference number and moved in the same direction.

Years later, this observation was shown to be an artifact, caused by mixing of the solution across the porous wall separating the anode and cathode compartments. However, there were authors who accepted these experiments as solid facts, and consequently produced theories agreeing with them. I was rather skeptical about such theories, as well as the experiments that prompted them, but thought it worthwhile to put the subject to the test of thorough analysis. The Ising model appeared an excellent tool to examine the possibility of association between COO^- and Na^+ in polyacids. All that was needed was to assume that each ionizable group can have three states: ionized (COO^-), unionized ($COOH$) and associated ($COONa$). The semi-grand partition function was still given by the matrix product in Eq. (1b), except that the matrices A and U were of order 3. The conclusion of this analysis [12, 15] was that neighbour interactions among the carboxylic groups could not possibly account for Wall's experiments.

There remained another possible explanation for Wall's observation: that the electrostatic potential attracts the Na^+ ions to the region at the 'core' of the molecule where the (negative) potential is largest, and that it takes the ion a long time to diffuse away from the core against the electrostatic force which constantly drags it back. Pondering over this problem I recalled studying with Aharon, a few years earlier, a Russian book on statistical mechanics in which the following problem was discussed, originally solved by Pontrjagin. Consider a volume V bounded by surface S. A particle, located initially at a point $\mathbf{r}$ inside S, performs a Brownian motion in the presence of a field of force $-\varepsilon\nabla\Psi$. What is the average time $\bar{t}$, it will take to reach, for the first time, some point on the surface? The solution satisfies the differential equation

$$\nabla \cdot (e^{\phi}\nabla\bar{t}) = -e^{\phi} / D \tag{2}$$

where $\phi = \varepsilon\Psi/kT$ and D is the diffusion coefficient. This equation was obviously applicable to the problem of the escape time of an ion from the field exerted on it by the polyelectrolyte. The answer indicated that Wall's experiments were also incompatible with the 'escape model'.

When Julius Jackson visited Rehovot, we further developed the escape problem [16] and derived the effective diffusion constant D_{eff} of ions in a polyelectrolyte solution

$$D_{\text{eff}} = D / \left\langle e^{\phi} \right\rangle\left\langle e^{-\phi} \right\rangle \tag{3}$$

Julius continued his interest in the subject, which led to a series of important papers on ionic conductance in polyelectrolyte solutions.

Thus, the problem that Debye had thought too difficult to bother about, seemed, nevertheless, to insist on being solved, as if it had a strong will of its own.

Once the use of matrix algebra was found to be so appropriate to the study of the electrostatic properties of linear chains by generalizations of the Ising model, it was tempting to take the method another step forward and to calculate the statistical mechanics of conformational changes, such as the average end-to-end distance and its dependence on neighbour interactions of the groups along the chains.

Here the task was mathematically more difficult. One needed, first, an expression for the end-to-end distance h of a polymer chain in any particular microscopic state. For this, one needed the vector **h** as a function of all the bond lengths, bond angles and torsional angles of rotation along the backbone of the polymeric chain. I scanned the literature to find a clue to the solution of this problem, but failed. Professor Giulio Racah, my Master Thesis advisor at the Hebrew University used to say: 'It is often easier to solve a problem than to find its solution in the literature; and it is more fun'. As I pondered over the problem one evening, as if trying to follow Racah's advice, it occurred to me that in my student years, I had solved a related problem as an exercise in analytical geometry. All of a sudden, the solution of the present problem was obvious. It dealt with orthogonal transformations of frames of references. **h** is given as a vector sum of the bond vectors b_i, when all b_i are transformed to a common frame of reference. Only later I found out that Henry Eyring had published the desired solution already in the early 1930s [17].

The second task was to link the matrices of the Ising model, which represented the statistical weights of neighbour interactions along the chain, with the transformation matrices which represented the microscopic states of the chain. This involved some amusing mathematical acrobatics, forming matrices of matrices, but led to rather straightforward expressions.

That study was summarized in a series of papers [18–20]. When Irwin Oppenheim came to Rehovot in 1958, he became interested in the subject, and I had the pleasure to work with him in calculating the effect of the solvent on the chain statistics. We presented closed expressions for 'the potential of the average torque' acting on the chemical bonds along the polymer backbone through the combined effect of intra-chain and solvent-chain interactions [21].

My work on the statistical mechanics of polymers, particularly polyelectrolytes, was received favorably by my colleagues in the Polymer Department as well as abroad. Yet, I was constantly bothered by the question of its relevance. I felt that theories must stand the test of experiment: be capable of making non-trivial predictions, of correlating experimental facts in a non-trivial manner, or of putting to quantitative test, the validity of qualitative arguments related to experiment. Wall's experiments were an example of just such a confrontation. However, here experiments seemed to be an artifact a priori. I was eager to find valid but puzzling experimental facts that require a theoretical rationale.

Then one day I believed I had found such a puzzle. The ensuing theoretical analysis indeed led to a better understanding of titration curves of polyelectrolytes, but here again the puzzle was solved by exposing the observation as an artifact. This study is, however, particularly interesting because of the indirect and unexpected impact it had years later. In 1957, we [13] studied the effect of neighbour interactions of higher order (second neighbours, third neighbours, etc.) on the pH of polyelectrolyte solutions. A general theorem was obtained which stated that for any order of neighbour interactions, the titration curve of pH versus the degree of ionization α must be symmetric around the mid-point of ionization $\alpha = 1/2$, namely.

$$pH(\alpha) - pH(1/2) = -[pH(1 - \alpha) - pH(1/2)] \qquad (4)$$

This symmetry property was indeed observed for polyacids under all experimental conditions. However, some new titration experiments on polybases showed a remarkable asymmetry. At high pH's, where the degree of ionization α of the polybase was low, the 'buffering capacity' $\partial\alpha/\partial\delta$pH was much higher than at low pH's, where the degree of ionization was high. The only sensible explanation for such an 'anomaly' seemed to be that in polybases, neighbour interactions are stronger and of higher order than in polyacids, but are strongly screened by the counter-ions and their ionic atmospheres. Such qualitative considerations imply that ionized groups which are neighbours of high order, $m>2$, are supposed to interact appreciably only if the $m - 1$ groups between them are unionized. Otherwise, the interaction diminishes, or even vanishes.

The calculation of the partition function corresponding to this model seemed a difficult task. Luckily, Jack Warga, an excellent mathematician who came to Rehovot as a Weizmann Fellow, offered his cooperation. I still remember our lengthy arguments, and the many pages of calculations we wasted. To our great delight, however, the final solution was presented in three lines, it was so simple. It made use of a mathematical trick forming a 'quasi-grand partition function' which included all degrees of polymerization. (Quasi, since the degree of polymerization was not a dynamic variable).

Comparison with experiment was, however, disappointing. The theory predicted increase of the buffering capacity with the degree of ionization, contradicting the experimental results. The contradiction was ultimately resolved in favour of the theory. It was found that at high pH, the glass of the vessel inadvertently participated in the titration, thus creating the artifact of high buffering capacity. When the experiments were repeated in a polyethylene vessel, the titration curves were normal, namely symmetric around $\alpha = 1/2$. Consequently, our model had nothing to do with experimental facts, and seemed to be useless. At least for a while. A few years later, it experienced a blessed resurrection in a different context. For the time being, the short paper in which we published our theory for asymmetric titration [22] was hardly noticed.

My next scientific adventure involved me in research more closely related to molecular biology. During my sabbatical leave in 1959/1960, I stayed for a few months in Harvard with Paul Doty and his group, well known for its pioneering work in this field. Doty's group used to have coffee every afternoon in the seminar room, where daily events and problems were discussed. Soon after my arrival, as I came one day for the coffee meeting, Doty challenged me in front of the whole group: 'You are a theoretician', he said. 'Can you produce a theory to explain the following observation? When we plot the melting temperature T_m of DNA, as well as of synthetic polynucleotides versus the logarithm of salt concentration, c_s, we get always straight lines'. Now, the qualitative explanation was obvious. The DNA is a double stranded helix. Both strands carry negatively charged phosphate groups, which repel each other, thus destabilizing the double helix. Salt ions reduce this repulsion, acting as a screen. Thus the double helix will be more stable at high salt concentrations (c_s), namely T_m will increase with c_s, but why logarithmically? Drawing on my past experience, I improvised an answer. There is no reason to believe that every empirical linear relation can be derived by a theory. Let us go the other way, ask what linear relation can be predicted. At T_m, the free energy of the double helix and the random coils are equal, therefore their difference, ΔF, must vanish. Writing $\Delta F = \Delta H - T_m \Delta S$ we obtain

$$T_m = \Delta H / \Delta S \tag{5}$$

Assuming that ΔH has an electrostatic part ΔH_e, i.e., $\Delta H = \Delta H_0 + \Delta H_e$, and that entropy is independent of salt concentration, we obtain for the limit of high salt ($\Delta H_e = 0$) $T_{m,0} = \Delta H_0 / \Delta S$. The shift from high salt to low salt

$$\Delta T_m = T_m - T_{m,0} = \Delta H_e / \Delta S \tag{6}$$

Now, the simplest estimate of ΔH_e based on Debye–Hückel's theory is $\Delta H_e \sim \exp(-\kappa r)/Dr$, where κ is the Debye reciprocal radius (defined by $\kappa^2 = 4\pi \varepsilon^2 c_s / DkT$), and r is a measure of the distance between the strands. I recalled that Onsager used a similar estimate of ΔH_e in his study of phase separation in suspensions of tobacco mosaic virus. Therefore, the following linear relation may be predicted between log ΔT_m and κ

$$\log \Delta T_m = -\kappa r + \text{constant} \tag{7}$$

This linear relation, I noted, has an advantage over the empirical linear relation T_m versus log c_s, in that the slope of log ΔT_m versus κ, namely $-r$, must correspond to a physical quantity, the distance between the strands.

All the participants in this discussion, myself included, were surprised by the simplicity of this consideration. Carl Schildkraut volunteered to help me to check it on the available experimental results. The same afternoon we obtained the plot log ΔT_m versus κ and, lo and behold, the line was straight. However, the value of r was too small, only 5 Å, while the distance between the backbones of the strands was ~12 Å. Carl's enthusiasm was not abated. He argued that the experiments could be refined if divalent ions were carefully extracted. Indeed, after experimental procedures were refined, the new linear plot of log ΔT_m versus κ gave a straight line with a slope of ~12 Å! The result seemed very exciting, however there was one catch. The limiting melting temperature $T_{m,0}$, for which ΔH_e should vanish, could not be obtained experimentally, since high salt concentrations are known to introduce all kinds of disturbing effects. Thus we had to vary $T_{m,0}$ as an adaptable parameter. Doty and Schildkraut nevertheless thought that the results were most significant, and urged immediate publication. I argued that a lot of information was thrown out by taking the logarithm, and insisted on refining the examination of the theory by calculating T_m versus ΔF_e rather than log ΔT_m versus $-\kappa r$.

Later I realized what a fool I had been. The simple theory had a kernel of scientific truth, linked with beauty and simplicity, and would have received a wide response. The improved treatment could anyway come later. In fact, it came much later [23]. As we entered the improvement phase, all kinds of difficulties arose. The Debye–Hückel theory was good enough for the rough approximation but when we derived ΔS and ΔH_0 from the linear plot of T_m versus ΔH_e, the non-electrostatic enthalpy ΔH_0 came out too large, and we worked very hard to overcome this difficulty. Above all, I learned a lesson: the best is sometimes the enemy of the good; and another lesson: wise men learn from the experience of others, others learn from their own experience.

The theoretical and the experimental studies of the phenomenon of the helix-coil transition in poly-amino acids were initiated by John Schellman [24] and Paul Doty [25], respectively, in the mid-1950s, and raised great interest. A poly-amino acid is a chain of amino acids linked by peptide bonds, and differs from a protein 'only' by comprising one kind of amino acid along the chain, while a protein has a specific sequence of different amino acids, which endows the protein with its specific structure and biological function. When a poly-amino acid is dissolved in a mixture of solvents, appropriately chosen, whose molar ratio is varied continuously, it changes rather abruptly from an ordered helix to a random coil. Hence the term 'helix-coil transition'. By the late 1950s, the theory of such transitions had been perfected by Zimm and Bragg (ZB) [26] and by Gibbs and DiMarzio (GD) [27], applying statistical mechanical methods borrowed from the Ising model. Doty asked me to give a few lectures on these new theories. While preparing these lectures, I was intrigued by the subtle differences between the two theories, which were not immediately obvious, and which led to somewhat different results. ZB assumed the completely random coil to be the refence state. The nucleation of a helix was initiated by forming the first intramolecular hydrogen bond, and was linked with a large loss of entropy. GD assumed the completely ordered helix as the reference state.

The nucleation of a random coil required the simultaneous breaking of three consecutive hydrogen bonds. It occurred to me that the elementary conformational changes in the helix-coil transition are the internal rotations around the N–C and C–C′ bonds of the amino acid residues. I thought that by using the internal rotations as variables, rather than the hydrogen bond making or breaking, it would be easier to explain the theory to the students. Working out the theory I was surprised to obtain results which were not only conceptually more elementary, but quantitatively different from both theories. While ZB required a matrix of order 8, and obtained a secular equation of order 4, I obtained a matrix of order 4 and a secular equation of order 3. It seemed that a new insight was gained by this development which was worth pursuing further. Thus with Antonio Roig, then a student of Stockmayer at MIT, a third version of the helix-coil transition theory of polypeptides was worked out [28].

My interest in the theory of helix-coil transition made me eager to meet Bruno Zimm, whose work in the field of polymers I had followed and admired for years. We met at the home of a friend in San Diego, soon after Zimm's theory of helix-coil transitions of poly-nucleotide double strands was published [29]. His theory was an amazing tour-de-force, both conceptually and mathematically. He attached to the double strands, fictitious annexes; he constructed matrices of order 2^n (n being the chain length), then sent n to infinity; after pages of such manipulations he obtained a nice, simple secular equation. I asked Bruno how he had arrived at such an amazingly complicated model and yet an amazingly simple solution. His answer was perhaps as amazing: He explained that he had started with $n = 1$, then $n = 2$, etc. By the time he was at $n = 4$, the thing had become damned complicated, but nevertheless a pattern could be discerned which eventually led him on the right track. I expressed my admiration for his achievement. However, I added, since the result is so simple, there must exist a simple way to obtain it. 'If there is such a way, why didn't you find it?' he answered and from his tone I sensed what he had not said, namely 'and if you can't, why don't you shut up'. I realized how hasty and unfair my comment had been, and kept quiet. Driving home at midnight, I tried to understand what had made me believe that 'there must exist a simple way' to the solution of Zimm's problem. Suddenly, with a sense of 'déjà vu', I recalled that I must have encountered a very similar problem... and sure enough, I soon realized that Zimm's model of the helix-coil transition of double strands is isomorphous to the model of a polyelectrolyte molecule with high-order neighbour interactions screened by counter ions [22]. So the solution obtained by Warga and myself must apply to Zimm's model.

Next morning I called Bruno up and told him a simple way to obtain his result, one requiring only three lines, did indeed exist. We soon met to discuss the details, and developed the subject into a joint paper [30]. Thus the unsuccessful effort of finding asymmetries in titration curves of polyelectrolytes paid off in a rather unexpected manner.

The 'quasi-grand partition function' method was successful in handling the asymmetric titration curves of polyelectrolytes, as well as the order-disorder transitions in double stranded polynucleotides. However, my repeated efforts to apply the same method for the polypeptide helix-coil transition failed for a long time. Eventually, the difficulty was overcome by a natural, simple modification of the 'quasigrand' method, which yielded a powerful tool for the statistical mechanical analysis of polymers, namely the 'sequence generating function' (SGF) method [31]. Let me introduce briefly the main considerations which led from the 'quasigrand' to the SGF method, and point out how they differ from the corresponding considerations underlying other methods. A linear chain is commonly considered as a sequence of units, e.g., ionizable groups, peptide residues, nucleotide pairs. A microscopic state of the chain may be specified in terms of the states of these units, e.g., ionized or unionized, helical or non-helical etc. The SGF method specifies a microscopic state of the chain in terms of sequences. A sequence of length i is a segment of the chain, comprised of i units all in the same state. Consecutive sequences obviously represent alternating states, e.g., a helical sequence of length j following a non-helical sequence of length i. A microscopic state of the whole chain is specified by the number of alternating sequences and by their lengths, i.e., by $i_1, j_1, i_2, j_2, \ldots, i_s, j_s$, where all i_σs and j_σs are non-zero (except possibly at the ends i_1 and j_s). A major advantage of this way of enumerating the microscopic states of a chain is the possibility to attribute to a sequence of length i_σ, a statistical weight, u_{i_σ} (sometimes called equilibrium constant, or partition function of the sequence) without specifying a priori the particular dependence of u_i on the length i. Similarly, a sequence of length j_σ of the other type is supposed to have a statistical weight v_{j_σ}. Thus the partition function of the chain is given by

$$Z(N) = \sum_s \sum_{\{i_\sigma, j_\sigma\}} \prod_{\sigma=1}^{s} u_{i_\sigma} v_{j_\sigma} \tag{8}$$

with the restriction on the summation indices s, i_σ and j_σ given by

$$\sum_s (i_\sigma + j_\sigma) = N \tag{9}$$

The power of SGF is in the following theorem: let us define sequence generating functions

$$U(x) = \sum_{i=1}^{\infty} u_i x^{-i}, \quad V(x) = \sum_{j=1}^{\infty} v_j x^{-j} \tag{10}$$

Then the partition function $Z(N)$ is given in the limit of large N by

$$Z(N) = x_1^N \qquad (11)$$

where x_1 is the largest root of the equation

$$U(x)V(x) = 1 \qquad (12)$$

(Partition functions $Z(N)$ for any finite N were also derived, using all the roots of Eq. (12), see the application for poly-(isocynate) presented below). This theorem implies that this equation is identical to the secular equation derived for the same problem by the matrix method. The proof of this theorem is based on the same 'quasi-grand partition function', which was previously applied to the problems of the asymmetric titration and the DNA order-disorder. However, in these particular applications, only one of the two possible states was represented by the statistical weight v_j of a sequence of length j. The other state was still represented by the conventional method of attributing to each unit its own statistical weight. The SGF enumerates both states by the lengths of their sequences, thus introducing an element of symmetry in the basic concepts. Hence its simplicity and generality.

One further generalization incorporated in the SGF is linked with a little episode. Once I was invited to give a seminar lecture on SGF at the Rockefeller University. Mark Kac, the famous mathematician, could not attend, and asked me to tell him briefly the content of my lecture. He seemed amused but not particularly impressed, until I came to the last part, the formulation of SGF for the case where each unit can have more than two states. In this case, we may formulate a sequence generating function for each state, i.e., $U(x), V(x), W(x), \dots$, etc. Let us define a matrix which has zeroes along the diagonal and sequence generating functions as off-diagonal elements, one for each row, as follows:

$$\mathbf{M} = \begin{pmatrix} 0 & U(x) & U(x) & \cdots \\ V(x) & 0 & V(x) & \cdots \\ W(x) & W(x) & 0 & \cdots \\ \vdots & \vdots & \vdots & \vdots \end{pmatrix} \tag{13}$$

Then x_1, the largest root of the determinantal equation

$$|\mathbf{M} - \mathbf{1}| = 0 \tag{14}$$

satisfies Eq. (11), namely $Z(N) = x_1^N$. 'You see', said Professor Kac, 'up to this point, all that you told me was similar to the application of generating functions in the statistical theory of Markov chains, but this generalization is new to me'. The moral of the story: if you have a problem, better ask Mark Kac. If you do it yourself, however, you may gather enough momentum to extend knowledge one small step further. As Professor Racah said: 'it is more fun this way...'.

Let me conclude the SGF story by mentioning three applications of the method to problems that were too difficult to handle by other methods. One was concerned with binding of acridine orange to DNA, the second, with fine details of the helix-coil transition in polypeptides, and the third, with temperature dependence of optical rotation of stiff helical polymers.

Acridine orange is a large, flat heterocyclic molecule known to cause mutations during the replication of DNA, probably due to its intercalation between the base pairs of the DNA. Dan Bradley and co-workers analyzed in detail the large amount of spectroscopic and other data, and concluded that another type of binding existed. It competed with intercalation, becoming dominant at high degrees of binding. They suggested that it was an external binding of the positively charged acridine to the negatively charged phosphates of DNA [32]. Due to its large size, acridine could either bind along the DNA groove to more than one phosphate at a time, or, attach itself to a single phosphate, and the type of phosphate binding could change with acridine concentration. SGF appeared to be the only statistical mechanical method capable of calculating the partition function for such a complex model [33].

The second application involved a 'story behind the curtains' as to how SGF was instrumental in solving the problem, but was not even mentioned in the final version of the published results. I have referred to the fact that the theories of Zimm and Bragg [26], Gibbs and DiMarzio [27], and Lifson and Roig [28], considered different kinds of nucleation models in the study of the helix-coil transition of polyamino acids. Bixon and I were seeking a general comprehensive theory, such that the three theories would be derived as special cases. Using the SGF method, it was relatively easy to formulate the sequence generating functions for the various nucleations independently, then combine them using Eq. (14). The ensuing secular equation was indeed the master equation from which the three models could be obtained. It also provided us with the clue of constructing the appropriate matrix of neighbouring interactions, thus deriving the same result by SGF and by the matrix method. In the short paper we published on the subject, we presented only the matrix method, for the sake of simplicity and brevity [34].

The third application worth mentioning caught me by surprise many years later, after I shifted my interest far away from the statistical-mechanics of polymers. Mark Green of the Polytechnic University of Brooklyn NY, made an intriguing observation, and a friend of mine, Herbert Morawetz, told him that I was the only person on earth who could wrap it with a nice theory (what an exaggeration!).

Anyway, Mark traced me one late evening, in 1988, in a New Yorker hotel, and next morning we had an early breakfast in a nearby restaurant, where he told me his fascinating story [46].

Poly(isocyanate) is a polymer whose backbone is made of conjugated amide bonds. It can be presented briefly as $(-CO-N(R)-)_n$, where n is the length of the polymer chain, i.e., the number of repeat-units. When the side-chain R is aliphatic and non-chiral, e.g., $R = -CH_2-CH_2-\cdots-CH_3$, the polymer is devoid of any chiral center, and shows no optical rotation. Mark observed that if the first or the second CH_2 group in the side-chain is replaced by a chiral deuterated group $C*DH$, the polymer exhibits a strong, temperature-dependent optical rotation. In contrast, the corresponding deuterated monomer shows hardly any trace of optical activity. The explanation of this apparently amazing phenomenon is straightforward. The backbone of the polymer chain is very stiff because the amide bond has a partial double-bond character. It is distorted out of planarity in either a right-handed or a left-handed sense because of steric hindrances. When consecutive repeat-units are distorted in the same sense, they form a helix. Reversals of the helical sense are rare because a reversal requires a considerable strain free-energy ΔG_r. Thus, a very short polymer chain is expected to be made of a single helix, either right-handed or left-handed, while a very long helix is expected to contain sparsely distributed reversals and long alternating helical sequences between them. When the chiral deuterated group $C*DH$ is introduced, the right-handed and the left-handed helices become non-equivalent. The free-energy difference between them, ΔG_h per monomer, may be very small. Yet, the free-energy per helix of one sense may become significantly lower than that of the other sense when the helix is sufficiently long. The helix with the lower free-energy will be more abundant, according the Boltzmann distribution law, resulting in excess optical rotation. Mark wanted to know whether I could give him a theory which could present these considerations quantitatively. Of course I could. The SGF method yields easily the partition function from which the optical activity could be derived. The results of comparison between theory and experiment were truly astonishing [47]. The free-energy difference ΔG_h between right-handed and left-handed helical units appeared to be as small as about one calorie per mol, while the free-energy difference ΔG_r between helical and reversal units appeared to be as large as several kilocalories per mol. Furthermore, the theory predicted that the change of optical rotation with chain-length should be extremely slow. In short chains, where n is less than few hundred units, the optical rotation should depend on n and ΔG_h but not on ΔG_r. For very long chains, it should depend on ΔG_r and ΔG_h, but not on n. In the intermediate range, optical rotation is a rather sophisticated function of all three parameters, ΔG_r, ΔG_h and n. Careful fractionation of the chiral polymer by Professor Akio Teramoto and his colleagues at Osaka University, yielded excellent confirmation of the theory over the entire range of chain-lengths. Furthermore, it yielded valuable information about the temperature dependence of the free-energy parameters [48]. My fruitful cooperation with Mark led also to the application of the 'Consistent Force Field' (CFF) method discussed below. We calculated the conformations of the helical sequences and of the reversals. To our big surprise, the calculations predicted that the monomer units of the helical sequences are free to twist collectively within a wide range of torsion angles [49]. This prediction still awaits experimental proof.

As I returned to the Weizmann institute from my sabbatical leave in 1961, I was looking for a new theoretical approach to the study of biological molecules, which led me eventually to the CFF method to be discussed soon. However, before I found my way, my position at the Weizmann Institute took a sharp and totally unexpected turn.

Professor Amos deShalit, 14 years my younger and a close friend for many years, was already a world renowned nuclear physicist who founded the Nuclear Physics Department at the Weizmann Institute in the mid-1950s, and became the Scientific Director of the Institute in 1961. He was an extremely intelligent and charming person, a born leader, with an irresistible persuasive power. One day in 1963, as we met as if by chance, he said I must do him a great favor. 'You see, Shneior, I know that I am destined to lead the Institute for the rest of my life. However, I am thirty-five, still young to do good science which I love most. You are already close to fifty, and your chance to do good science is much smaller. And you owe much to the Institute. Where else in the scientific world could you start so late in life, and climb all the way from Ph.D. to full professor within such a short time?' His request was that I replace him as scientific director for, say, four years, after which he would take over. I consented after much soul searching. In retrospect, I know that it was a mistake, but at that time I thought, perhaps he was right? What could be better for an aging scientist who runs out of ideas than to move into scientific administration? In fact, in spite of the time and energy I spent in administration, I continued to be addicted to science and to do research in cooperation with colleagues and students during and after this period.

Statistical mechanics of chain polymers deals with models, which are simplifications of reality. Poly-amino acid chains are themselves oversimplified analogs of proteins. Therefore, the helix-coil transition theory tells us something about the general principles involved in the native folding and unfolding of proteins, but ignores the details. These details, however, are of great importance for the understanding of protein structure and its relation to function, one of the central problems of molecular biology. In order to calculate the equilibrium structure of a protein molecule, one must know the molecular forces. Intra-molecular forces are involved in bond stretching, bond angle bending, torsional rotations; inter-molecular interactions comprise van der Waals forces between atoms, repulsive at short contact range and attractive at longer range, as well as electrostatic interactions. A vast literature accumulated on the experimental and theoretical studies of these forces. Application of the results of these studies to biopolymers was, however, fraught with drastic simplifications, and was often severely criticized as being arbitrary and unreliable. It involved, in general, taking semi-empirical estimates of various molecular forces derived from different studies, without worrying about their mutual consistency. In order to remedy this situation, I proposed, in the mid-1960s, a different approach to the description of molecular forces, which was later named the 'Consistent Force Field' (CFF). The principal idea of CFF is as follows.

If one would represent faithfully, the energy of polyatomic molecules as a sum of functions of internal coordinates (bond lengths, bond angles and bond torsions), as well as contact distances between non-bonded atoms, then one could calculate, to a good approximation, all molecular properties that depend on this energy. Reversing the argument, by carefully selecting approximate functions and fitting their parameters to obtain optimal agreement with a large number of independent experimental data, it should be possible to obtain an optimal set of empirical functions.

The CFF is produced as follows: we choose a tentative set of empirical energy functions. The parameters belonging to these empirical functions are then subjected to a least squares optimization to obtain an optimal agreement between calculated and experimental results. If the optimal agreement is not sufficient for our purposes, we single out the energy functions responsible for the deficiency and substitute them by more appropriate ones. Functions which have a negligible effect on the least squares are deleted, and new functions may be added. Thus, one introduces into the discipline of empirical functions the Darwinian principle of the survival of the fittest.

The ultimate purpose of these studies was the application of empirical energy functions to molecular biology. However, the need for empirical data of high accuracy for the least squares optimization, diverted us to many problems of physical chemistry. These often possessed intrinsic interest and importance of their own. Since my emphasis here is on personal motivations, emotional and intellectual, rather then on reviewing the results in detail, let me single out, from the extensive work on CFF done in cooperation with my colleagues and students, a few illustrations.

One of the major sources of experimental information for our study of empirical functions was, at first, the extensive classic work of Jack Dunitz and his co-workers on the conformational analysis of cycloalkanes. In fact it was Jack who encouraged me to submit our first paper in this field to *Tetrahedron* [35]. Visiting Rehovot in 1966, he delivered a lecture on his recent work on cycloalkanes, where he pointed out an astonishing observation, derived from the X-ray diffraction of a substituted alkane ring, 1,1,5,5-tetramethylcyclodecane

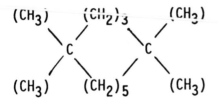

The upper part of the ring, including the 4 methylsubstituents, had peculiar, abnormal properties. A few bond lengths were much too short, some bond angles were much too wide; temperature factors were abnormally large. After the lecture, Jack and I agreed to examine the ring conformation using the energy functions for cycloalkanes obtained by

Bixon and myself. We could not find a single conformation which agreed with the X-ray diffraction, but we did find two conformations which, by superposition, yielded an 'average' conformation in excellent agreement with experiment. Jack immediately recognized the reason: the two conformations were randomly distributed in the crystal, thus producing the observed diffraction pattern with average atomic coordinates and large 'temperature factors'. These results were a great triumph for the method of empirical energy calculations. The triumph was further augmented when the calculated difference between the strain energies of the 1,1,4,4- and 1,1,5,5-isomers was examined experimentally. Theory predicted 1.25 kcal/mol difference between the two isomers. Subsequent calorimetric measurements confirmed the prediction, yielding 1.3 ± 0.3 kcal/mol! Incidentally, the preliminary note on the subject was an analog of Noah's arc, with 4 pairs of scientist-species: syntheticists from Prague, crystallographers from Zurich, calorimetrists from Amsterdam and theoreticians from Rehovot [36].

Cycloalkane ring molecules $(CH_2)_n$ of medium size ($n = 7,...,$ 11) comprise an interesting set. Contrary to cyclohexane $(CH_2)_6$, or to $(CH_2)_{12}$, they are highly strained, since their bond angles and torsional angles have to be distored in order to have the ring closed. Accurate measurements of the conformations and strain energies of these cycloalkanes were the main source of experimental data for optimizing our initial CFFs [35]. The success in calculating these observables culminated in our contribution to Dunitz's problem of the substituted cyclodecanes [36]. However, when we applied the same CFF to cyclopentane $(CH_2)_5$, we obtained an excessive strain energy of 12 kcal/mol, by 5 kcal/mol larger than the experimental value. Since cyclopentane is not an abnormally strained molecule, such a discrepancy, if not alleviated, would make the other successes of the CFF look like a fortuitous episode.

Actually, we knew what was wrong. In comparing calculated strain energies with observed strain enthalpies, we had only poor estimates for the contribution to enthalpy from molecular vibrations, which was neither experimentally, nor theoretically, available. A theoretical derivation of vibrational enthalpies from the same energy functions that yielded, hitherto, molecular conformations and strain-energies seemed, therefore, a natural extension of the consistent force field. However, I realized immediately that such an extension meant much more than improving our estimates of enthalpies. Conformational analysis and vibrational analysis were considered, at that time, as totally separate fields of research. Here was an opportunity to unite them, and to learn how conformations and vibrations affect each other. As I was considering the feasibility of such an enterprise, Arieh Warshel started his Ph.D. studies with me, and was willing, indeed eager, to meet the challenge.

In order to include the normal modes of vibrational frequencies of all the alkane molecules in our least squares optimization, we had to derive both the frequencies and the equilibrium coordinates, directly from the CFF functions. For this purpose, we replaced the classical normal mode analysis, based on symmetry coordinates, character tables of group representations, etc., by a straight-forward procedure, particularly suitable for fast computers with large memories [37a]. The energy of each molecule was expressed in Cartesian coordinates, the equilibrium coordinates were determined by the requirement that the gradient of the energy, i.e., all its first derivatives, vanish at equilibrium, and the normal modes of vibrations were derived from the Eigenvalues of the matrix of the second derivatives of the energy, properly weighted by the atomic masses. Our procedure had sacrificed the beauty of the classical theory but gained beauty of its own by its simplicity, by the fact that the particular symmetry of the various molecules was derived, not assumed and, more importantly, by the fact that the so-called 'force constants' of vibrational spectroscopy were explicitly dependent on the equilibrium coordinates of the molecules under consideration. The linkage of conformational and vibrational analysis to form a unified, consistent discipline had great potential, which the CFF method has started to exploit and extend in many directions. We applied the CFF through the years successively to a number of families of organic molecules, such as alkanes [37], alkenes [38], amides [39], lactams [40] and carboxylic acids [41]. The observable properties calculated from the empirical energy functions were strain energies of medium-size rings, molecular conformations, crystal structures, Raman and infrared vibrational spectra, heats of sublimation, dipole moments, thermal expansion of crystals, and so on.

As we chose amides for inclusion in the CFF [39], we encountered for the first time the intermolecular hydrogen bond $N-H\cdots O=C$, typical for crystals of molecules containing amide groups. Should we consider the hydrogen bond as a strong intermolecular interaction, or should we look upon it as a weak covalent bond? According to the then prevailing quantum mechanical calculations, the energy of the hydrogen bond was comprised of an 'exchange repulsion' positive component, and of electrostatic and 'charge transfer' negative components of about equal size. These components corresponded roughly to van der Waals, electrostatic and covalent interactions, respectively. After trying out a number of hydrogen bond potentials from the literature, we focused our effort on the model of a hydrogen bond potential which included a Lennard–Jones potential, electrostatic interactions between partial charges, and a Morse-like potential for the covalent contribution. In order to isolate the problem of intermolecular interactions, we confined our use of data to crystal properties alone, and after much effort, obtained good agreement with experiment. However, the energy parameters of the covalent Morse-like potential, as obtained by least squares, were erratic; standard deviations were larger than the values themselves. The whole situation was rather irritating: a lot of hard work and inconclusive results. I remember how, at one point, Arnie Hagler and Eduardo Huler got so fed up that we considered dropping the project and moving to a less unwieldy problem. In fact, however, the computer was sending us an important message which we failed to receive at first: there is no covalent bond contribution. Sure enough, when we dropped the Morse-like potential out of our calculation, we obtained as good agreement with experiment as before. Thus, we were forced to the conclusion that, essentially, the hydrogen bond is comprised of the same forces that comprise intermolecular interactions in general, with no significant 'charge transfer' or covalent contributions. We also had an answer to the question, why then is the 'hydrogen bond' significantly stronger than other intermolecular interactions. The strong electronegativity of N makes the NH bond highly polar, namely, it shifts the electron cloud from the hydrogen to the nitrogen. As a result, the van der Waals radius of the H becomes negligibly small. This allows the two dipoles, $N \overset{\leftarrow}{-} H$ and $O \overset{\leftarrow}{=} C$ to make a shorter contact distance. Coulson, with whom we discussed our results when he visited the Weizmann Institute, liked these ideas. However, the younger generation of quantum mechanicians either rejected or ignored them. After all, how seriously can one take empirical calculations? We too were hesitant. Maybe 'charge transfer' is negligible in the case of $NH \cdots O=C$, which is a particularly weak hydrogen bond, but will show up in the stronger $OH \cdots O=C$ hydrogen bond? Our subsequent extension of the CFF to crystals of carboxylic acids [41] supplied the unequivocal answer: the hydrogen bond in carboxylic acids had no covalent character either. The reason for its being stronger than that of amides was traced to the fact that the van der Waals radius of the oxygen is smaller than that of nitrogen, therefore the $OH \cdots O$ contact distance is smaller than the $NH \cdots O$ distance, hence

the shorter and stronger OH···O hydrogen bond. It is noteworthy that this time, the quantum-mechanicians had no
reason to object to our results. More elaborate ab initio calculations of the hydrogen bond of water reduced the charge
transfer contribution from ~8 kcal/mol obtained earlier, to
about 1 kcal/mol.

As I noted above, the empirical energy functions of the CFF
were first chosen tentatively. Most of them were carried over
from normal-mode analysis of vibrational spectra. In classical vibration analysis, it was always assumed that the forces
generated by molecular distortions are harmonic. Therefore,
the potentials were quadratic and bilinear functions of the
bond-lengths and the bond-angles. Such simple functions
could not reveal the full interdependence between conformational and vibrational properties of molecules. The reason is
very simple: The molecular vibrations are determined by the
second derivatives of the energy, and the second derivatives
of harmonic potentials are constants independent of the
molecular conformation. I was looking, therefore, for functions of bond-lengths and bond-angles which would give a
better insight into the real nature of the energy of molecules.
One such empirical function, known as the Morse potential,

$$V_M(b) = D\exp[-2\alpha(b - b_0)] - 2D\exp[-\alpha(b - b_0)]$$

represents, very well, the potential energy of a diatomic
molecule as a function of its bond-length b. Its adjustable
parameters D, α and b_0, have a well-defined physical meaning, and its second derivative is a decreasing function of b.
We applied it first to a single, highly overcrowded, branched
alkane molecule [42], then we optimized the whole force-
field with Morse functions replacing the harmonic functions
of bond-stretching [43, 44]. The new Morse-CFF was far
more physically meaningful than the harmonic CFF, and
brought a dramatic improvement of the power to predict and

explain properties of alkanes. The experimental bond-lengths, bond-angles and stretching frequencies are known to vary systematically along the series CH_4, $-CH_3$, $>CH_2$ and $>CH$. In the harmonic CFF, each member of this series required different empirical parameters for the stretching potentials, and the agreement between the calculated and observed frequencies was poor. In contrast, the Morse potential of the bond-length of the C–H bond was one and the same for all members of this series, and similarly for all C–C bonds. The agreement between the calculated and observed data was close to the corresponding experimental errors for all data, including bond-lengths, bond angles, stretching-frequencies and, most significantly, all other vibrational frequencies.

This study gave me considerable satisfaction. Obviously, the Morse potential represented, very well, the energy of polyatomic molecules over a wide range of bond-lengths. By using it, we obtained a broader and deeper understanding of the interrelation between conformations and vibrations. We also obtained a force-field which is as good or better than any other current force-field in predicting properties of the family of alkane molecules. The next challenge, I thought, should be to replace also the harmonic potentials of bond-angle bending by more physically meaningful functions. This would raise, significantly, the theoretical and practical value of the consistent empirical force-field. Regrettably, in spite of many trials, this goal has not been achieved as yet, and awaits perhaps the intervention of younger and wiser scientists.

The CFF studies were originally prompted by the wish 'to calculate the equilibrium structure of a protein molecule'. I recall a discussion at an international conferene in Napels, 1965, where Francis Crick criticized energy calculations of protein-like molecules as unreliable. He conceded my argument that such calculations may become reliable if the philosophy of the CFF is put into practice, but added: 'If you are going to do it, you better hurry-up'. Well, we are still in a hurry... When Michael Levitt came to the Weizmann Institute as a young 'predoc', he succeeded, for the first time, in obtaining, by energy minimization, an approximate equilibrium conformation of a protein, using our CFF [45]. Now, almost three decades later, many research-groups are still pursuing this goal, much progress is being made, but the goal is still far from being reached. I myself refrained from entering the race of computational protein folding. I think that empirical energy functions should be used mainly to obtain definite answers to good questions when other answers are not available, and to make meanigful predictions which can be tested by experiment. Consequently, I was looking persistently for experimental scientists who were confronted with problems which only the CFF could solve. Soon I found out that such scientists were a rare species. Most experimental chemists and biologists either did not have such problems or did not trust the CFF as an effective tool for their solution.

Then one day I got a lucky strike. I met, by chance, Avi Shanzer, a young organic chemist at the Weizmann Institute, and he told me about his work. He had an ingenious technique to synthesize macrocyclic lactones and was trying to use it for the synthesis of an ion-binder for lithium. He was proceeding by trial and error, but was, as yet, unsuccessful. I suggested that I could perhaps predict whether a molecule is a potential binder before it is synthesized. Avi would, of course, not believe it, so I suggested that the CFF may be challenged to yield the conformations of the failed trial-molecules, in order to understand why they failed. Avi agreed immediately, but only later he told me why: He had already, conformation of one of his compounds determined by X-ray crystallography. He gave me the chemical formula, and I calculated the molecular conformation. It was an easy task, because macrocyclic rings have relatively few degrees of freedom, but for Avi it was a big surprise to see how close were the calculated and observed conformations. Thus started a long, pleasant and fruitful cooperation on the design, synthesis, examination and selection of molecules which mimic biological ion-carriers.

Let me describe briefly the nature of our cooperation. In order to bind a given ion, a molecule should carry the right number of polar groups, and these should surround a cavity of the right size, with dipoles pointing toward the center of the cavity. Furthermore, in order to be functional, the molecule has to cross biological membranes or be recognized by specific receptors. The design of such a system is a formidable task, but it can be very much simplified and become feasible if one asks the right questions.

Avi and I, together with our co-workers, did it by joining experiment and theory in a mutually supporting strategy [50]. The feasibility of a molecule to function as an ion-carrier requires a finely tuned balance between the free energies of ion-binding and of ion-hydration, which are both very large. The free energy of binding depends, *inter alia*, on the strain energy associated with the conformational change imposed on the ion-carrier by the formation of the ion-complex. The strain energy difference between different carriers of the same ion is the main component of their free-energy difference because other components cancel out to a good approximation. Therefore, a molecule which is less strained, i.e., less distorted upon binding, has a better chance to be a good binder. We examined, with our empirical force field, whole families of potential carriers of the same selected ion. The molecules which were predicted to have the smallest conformational changes and strain energies were preferred for synthesis. Experiment determined the stoichiometry of binding as well as various conformational properties of the free and complexed ion-carrier. Calculations then helped to interpret and correlate the experimental results and put them into a coherent structural framework. Remaining discrepancies between theory and experiment guided us to reevaluate both. Thus, experiment and theory were linked in a mutually supporting, interactive and iterative manner.

For the sake of simplicity, we chose to imitate natural ion-carriers which possess structural symmetry, or synthesize symmetric analogs of non-symmetric ones, because symmetry contributes greatly to the economy of effort in calculations, synthesis and the interpretation of the observed properties of the product. Interestingly, structural symmetry does not always lead to conformational symmetry.

We constructed the families of potential ion-carriers by modular design. An ion-carrier is a composite structure, and may be conceived as a combination of several components, or modules. Each module may be modified independently, thus forming families of related molecules. All possible combinations of all modifications of the modules were considered *a priori* as candidates for synthesis, but most of them were eliminated by the theory, or modified after synthesis and evaluation, using the above mentioned guide-lines plus some chemical common-sense. Once an efficient ion-carrier was reached, systematic modification of the appropriate modules yielded families of ion-carriers of gradually varying characteristics, thus improving the performance systematically.

The results of this cooperative program were most gratifying. After a period of trial and error, we arrived at several basic structures which yielded biologically active ion-binders of efficiency comparable to, and in some cases even surpassing that of, natural ion-binders [51].

The question of the origin of life intrigued me ever since I was a teenager, when I heard for the first time about Darwin and his theory of evolution by natural selection. As a teacher in Mishmar-Ha'emek, I became more closely acquainted with Darwinism, since I was teaching biology, among other subjects. I realized that natural selection offered an unshakable and beautiful explanation for the evolution of higher species from lower ones. However, I noted also that it did not explain the biggest riddle of all: the evolution of the lowest species in the first place, namely, the origin of life from inanimate matter. This riddle, I felt, will remain for quite a time one of the great mysteries which only great minds can solve.

Even in my wildest dreams I did not see myself involved in this field of research. Yet circumstances pushed me gradually in this direction. It started with my accidental acquaintance with Manfred Eigen, which developed into genuine friendship. Around the mid-1960s, Eigen became fascinated with the problem of the origin of life and of natural selection. The time was, by then, ripe for a fresh view of the subject, because of the great innovations of molecular biology which unfolded the molecular basis of evolution by natural selection, and thus paved the way to understanding the molecular origin of evolution by natural selection. Eigen founded his ideas on these innovations, and did it with deep insight and creative imagination. I attended his lectures often, had many discussions with him, and had even the opportunity to offer a few comments related to his fundamental paper on self-organization of matter and the evolution of biological macromolecules [52].

I was enormously impressed by the idea that molecules which replicate themselves can be subjected to mutations, natural selection and evolution, just like biological species. It had the simplicity of Darwin's theory, and solved beautifully, although only in principle, what I thought to be 'the biggest riddle of all'. Consequently, I followed with interest, for many years, Eigen's studies, as well as other theories of the origin of life. However, the interest was mingled with some measure of frustration. There were plenty of theories on the scene beside Eigen's. They were all based partly on solid facts and ideas, and partly on models, scenarios, guesses, assumptions and hypotheses of personal choice. I felt that their statements should be examined according to three levels of credibility:

(a) Statements which are true beyond reasonable doubt, by objective criteria of science. These should be included in a general theory of the origin of life.

(b) Statements which might be, but are not necessarily, true. These should be recognized as conjectures.

(c) Statements which are not true. These should be proven wrong and be abandoned.

It seemed to me that only with the fulfillment of these requirements could the study of the origin of life become a sound scientific discipline. However, I realized how difficult a task it is to make such distinctions, and how unqualified I was to undertake it. Still, I continued to ponder over the subject and to follow the literature in my free time, but otherwise, I attended to my bread-and-butter research and refrained from active search for a general theory of the origin of life.

This passive interest went on for many years, until I was forced out of it. The editorial board of *Biophysical Chemistry* decided on a Manfred Eigen Festschrift, a special issue dedicated to Professor Manfred Eigen and scheduled for his 60th birthday, May 1987. In view of my personal friendship with Manfred, I was asked to participate. It was a request I could not turn down. I could either review my current research, which hardly fitted the Festschrift, or present some of my reflections on Eigen's theory as befitted a Festschrift in his honor, for whatever they were worth.

The outcome was a paper on 'chemical selection, diversity, teleonomy and the second law of thermodynamics' [53]. It was written hastily because I had to submit it in a short time, and I felt that it was a crude, unfinished product. It contained, nevertheless, a genuine effort to look upon Eigen's ideas from a different, more general point of view in the spirit of level (a). I asked myself: 'What are the most general statements one can make about the origin of natural selection?', and realized that spontaneous appearance of autocatalytic replication in an inanimate environment is a necessary first condition, irrespective of its chemical details. Without the spontaneous synthesis of at least a single auto-catalytic molecule, there could not be natural selection. Such an event may have been very rare. However, once having occurred, the probabilities of replication to macroscopic abundance and of occurrence of replicating variants were radically changed. Variants competed with each other for their common reactants, and were thus subjected, necessarily, to natural selection. The 'unfittest', which replicated at the slowest rate, were selected out and disappeared, while the 'fitter' variants 'survived'. This, I suggested, is the general character of the origin of natural selection, of which template-replication of information-carrying polymers is a special case. Thus, template-replication might have evolved by natural selection from simpler systems.

I further asked myself: 'What is the most general distinction between animate and inanimate matter which originated from natural selection?' According to Eigen, the origin of animate matter was directly linked to the generation of genetic information. I looked for a more general view and found it in Jaques Monod's monograph [54], were he elaborated on the concept of purposive behavior, or teleonomy, as the distinctive characteristic of animate matter. Monod defined as teleonomic 'all structures, all the performances, all the activities contributing to the success of the essential project'. I viewed the origin of animate matter and of teleonomy as the origin of two fundamental, closely linked properties. One is the specific structural complexity of biomolecules, which befits their function in the living organism. Such complexity could hardly ever appear by chance in inanimate matter, therefore it must have been the outcome of natural selection. The other is the purposive organization by which all these complex components are regulated to function in coordination for the sake of the survival of the organism and of the species to which it belongs. Purposive organization could, similarly, hardly ever appear by chance, so it must have been the outcome of natural selection. As the years went by, I continued to ponder over these problems. I felt that my Festschrift paper expressed, although hesitantly, some first steps toward a general theory of the origin life, and that such a theory is now within my reach. Indeed, my recent paper 'On the crucial stages in the origin of animate matter' [55], contains such a theory, or at least I believe it does. It shows that life and its origin can be viewed as a continuos physico-chemical process of replication, random variation and evolution by natural selection. The process began when some, as yet unknown, elementary autocatalyst occured spontaneously in a favourable environment, then replicated to macroscopic abundance and generated 'mutants' by random processes. Natural selection acted on such mutants because of their unequal rates of replication and decomposition. Gradual depletion of reactants which were common to all mutants slowed down replication. Those mutants whose replication-rate was reduced below their decompostition-rate were selected out by becoming extinct. As mutants reached macroscopic abundance, they enriched their environment with their decomposition-products and other direct or indirect sequel-products. Mutants which replaced their depleted reactants by sequel-products had a selective advantage, and these sequel-products became new reactants. Such mutants formed complex autocatalytic systems, and the sequels-made-reactants served as internal components. Subsequent evolution by the same mechanism generated gradually ever more complex autocatalytic systems with ever higher teleonomic properties. It led first to primitive metabolism and purposeful organization then, eventually, to cellular metabolism, cell division and encoded information.

At this point in time, as my scientific life-story has been told, my recent paper [55] is still in press. Will it be accepted by some of my peers? Will it be rejected by others? Or, worst of all, will it be just ignored? Whatever may be, I am still fascinated by 'the biggest riddle of all', and wish to continue to ponder over problems as yet unsolved with respect to the origin of animate matter.

References

1 W. Kuhn, O. Künzle and A. Katchalsky, Helv. Chim. Acta 31 (1948) 1994.
2 A. Katchalsky and S. Lifson, J. Polym. Sci. 11 (1953) 409.
3 R.M. Fuoss, A. Katchalsky and S. Lifson, Proc. Natl. Acad. Sci. USA 37 (1951) 579.
4 S. Lifson and A. Katchalsky, J. Polym. Sci. 13 (1954) 43.
5 A. Katchalsky and S. Lifson, Sci. Am. March (1954).
6 A. Katchalsky, S. Lifson, I. Michaeli and M. Zwick, in: Contractile Polymers, Chapt. I (A. Wassermann, Ed.), Pergamon Press, Oxford, 1960.
7 S. Lifson, J. Polym. Sci. 20 (1956) 1.
8 S. Lifson, Bull. Res. Coun. Israel 6A (1957) 119.
9 J.J. Hermans and J.T.G. Overbeek, Recl. Trav. Chim. Pays-Bas 67 (1948) 761.
10 G.E. Kimball, M. Cultmer and H. Samelson, J. Phys. Chem. 56 (1952) 57.
11 S. Lifson, J. Polym. Sci. 23 (1957) 431.
12 S. Lifson, J. Chem. Phys. 26 (1957) 727.
13 S. Lifson, B. Kaufman and H. Lifson, J. Chem. Phys. 27 (1957) 1356.
14 J.R. Huizenga, P.F. Grieger and F.T. Wall, J. Am. Chem. Soc. 72 (1950) 2636.
15 S. Lifson, J. Chem. Phys. 28 (1958) 989.
16 S. Lifson and J.L. Jackson, J. Chem. Phys. 36 (1962) 2410.
17 H. Eyring, Phys. Rev. 39 (1932) 746.
18 S. Lifson, J. Chem. Phys. 29 (1958) 80.
19 S. Lifson, J. Chem. Phys. 29 (1958) 89.
20 S. Lifson, J. Chem. Phys. 30 (1959) 964.
21 S. Lifson and I. Oppenheim, J. Chem. Phys. 33 (1960) 109.
22 J. Warga and S. Lifson, J. Chem. Phys. 29 (1958) 643.
23 C. Schildkraut and S. Lifson, Biopolymers 3 (1965) 195.

24 J.A. Schellman, C. R. Trav. Lab. Carlsberg, Sér. Chim. 29 (1955) No. 15.

25 P. Doty and J.T. Yang, J. Am. Chem. Soc. 78 (1956) 498.

26 B.H. Zimm and J.K. Bragg, J. Chem. Phys. 31 (1959) 526.

27 J.H. Gibbs and E.A. DiMarzio, J. Chem. Phys. 30 (1959) 271.

28 S. Lifson and A. Roig, J. Chem. Phys. 34 (1961) 1963.

29 B.H. Zimm, J. Chem. Phys. 33 (1960) 1349.

30 S. Lifson and B.H. Zimm, Biopolymers 1 (1963) 15.

31 S. Lifson, J. Chem. Phys. 40 (1964) 3705.

32 A.L. Stone and D.F. Bradley, J. Am. Chem. Soc. 83 (1961) 3627; D.F. Bradley, Trans. N. Y. Acad. Sci. 24 (1961) 64.

33 D.F. Bradley and S. Lifson, in: Molecular Association in Biology (B. Pullman, Ed.), Academic Press, New York, 1968, pp. 261.

34 M. Bixon and S. Lifson, Biopolymers 5 (1967) 509.

35 M. Bixon and S. Lifson, Tetrahedron 23 (1967) 769.

36 M. Bixon, H. Dekker, J.D. Dunitz, H. Eser, S. Lifson, C. Mosselman, J. Sicher and M. Svoboda, Chem. Commun. (1967) 360.

37a S. Lifson and A. Warshel, J. Chem. Phys. 49 (1968) 5116.

37b A. Warshel and S. Lifson, J. Chem. Phys. 53 (1970) 582.

38a O. Ermer and S. Lifson, J. Am. Chem. Soc. 95 (1973) 4121.

38b O. Ermer and S. Lifson, J. Mol. Spectrosc. 51 (1974) 261.

38c O. Ermer and S. Lifson, Tetrahedron 30 (1974) 2425.

39a A.T. Hagler, E. Huler and S. Lifson, J. Am. Chem. Soc. 96 (1974) 5319.

39b A.T. Hagler and S. Lifson, J. Am. Chem. Soc. 96 (1974) 5327.

40 A. Warshel, M. Levitt and S. Lifson, J. Mol. Spectrosc. 33 (1970) 84.

41a S. Lifson, A.T. Hagler and P. Dauber, J. Am. Chem. Soc. 101 (1979) 5111.

41b A.T. Hagler, S. Lifson and P. Dauber, J. Am. Chem. Soc. 101 (1979) 5122.

41c A.T. Hagler, P. Dauber and S. Lifson, J. Am. Chem. Soc. 101 (1979) 5131.

42 A.T. Hagler, P.S. Stern, S. Lifson and S. Ariel, J. Am. Chem. Soc. 101 (1979) 813.

43 S. Lifson and P.S. Stern, in: Intramolecular Dynamics (J. Jortner and B. Pullman, Eds.), D. Reidel, New York, 1982, pp. 341.

44 S. Lifson and P.S. Stern, J. Chem. Phys. 77 (1982) 4542.

45 M. Levitt and S. Lifson, J. Mol. Biol. 46 (1969) 269.

46 M.M. Green, N.C. Peterson, T. Sato, A. Teramoto, R. Cook and S. Lifson, Science 268 (1995) 1860.

47 S. Lifson, C. Andreola, N.C. Peterson and M.M. Green, J. Am. Chem. Soc. 111 (1989) 8850.

48   H. Gu, Y. Nakamura, T. Sato, A. Teramoto, M.M. Green, C. Andreola, N.C. Peterson and S. Lifson, Macromolecules 28 (1995) 1016.
49   S. Lifson, C.E. Felder and M.M. Green, Macromolecules 25 (1992) 4142.
50   S. Lifson, C.E. Felder, A. Shanzer and J. Libman, in: Synthesis of Macrocycles: The Design of Selective Complexing Agents (R.M. Izatt and J.J. Christensen, Eds.), Wiley, New York, 1987 pp. 241.
51   S. Lifson, C.E. Felder, J. Libmann and A. Shanzer, in: Computational Approaches to Supramolecular Chemistry (G. Wipff, Ed.), Kulver Acad. Pub., 1994, pp. 349.
52   M. Eigen, Naturwissenschaften 58 (1971) 465.
53   S. Lifson, Biophys. Chem. 26 (1987) 303.
54   J. Monod, Chance and Necessity, Alfred A. Knopff, New York, 1971.
55   S. Lifson, J. Mol. Evol. 44 (1997) 1.

G. Semenza and R. Jaenicke (Eds.)
Selected Topics in the History of Biochemistry: Personal Recollections, V
(Comprehensive Biochemistry Vol. 40) © 1997 Elsevier Science B.V.

Chapter 2

Keilin and the Molteno*

MAX F. PERUTZ

MRC Laboratory of Molecular Biology, Cambridge CB2 2QH, UK

Introduction

David Keilin, who was born on 21 March 1887, was Lecturer in Parasitology at Cambridge University from 1925 to 1931 and Quick Professor of Biology from then until his retirement in 1954. He first distinguished himself with the discovery of the life cycles of flies whose larvae develop parasitically in animals and plants, and are themselves parasitized by micro-organisms. Keilin then made his fame with the discovery and characterization of a system of coloured enzymes, the cytochromes, which turn the chemical energy gained by the combustion of foodstuffs into a form that organisms can use for growth, movement, reproduction and even for thought. Keilin did most of his life's work at the Molteno Institute, a pleasant small building on the Downing site next to the Anatomy School, built in 1920 from a donation by Mr. and Mrs. Percy A. Molteno, who lived in South Africa and took a keen interest in the parasitological research of Keilin's predecessor, Professor G.H.E. Nuttall.

We used to attend Keilin's lively lectures on haem proteins every year, I mean we, the permanents at the Molteno Institute, together with some undergraduates and the ever-

---

* Reproduced with permission from Cambridge Review, October 1987, pp. 152–156.

changing visitors who flocked there from abroad, attracted by Keilin's fame as a biochemist and parasitologist. He showed lecture demonstrations, a practice long since abandoned, and always took a special delight in projecting the porphyrin spectrum from a feather of the turaco bird. In fact, the spectroscope was his favourite instrument, not one of those hefty ones that chemists used, but a delicate little Zeiss hand spectroscope replacing the eyepiece of a microscope, with which he first discovered the absorption bands of the cytochrome system in the thoracic muscle of the fly *Gastrophilus intestinalis*. He used it to observe the absorption spectra of living matter. He recalled:

> One day, while I was examining a suspension of yeast freshly prepared from a few bits of yeast shaken vigorously with a little water in a test tube, I failed to find the four-banded spectrum, but before I had time to remove the suspension from the field of vision of the microscope, the four bands suddenly re-appeared. The experiment was repeated time after time and always with the same result. This first visual perception of an intra-cellular respiratory process was one of the most impressive spectacles I have witnessed in the course of my work.

Keilin used to tell us always to work with coloured proteins because their spectra were so revealing.

After his lectures Keilin sometimes asked me: 'Perutz, why can the same haem fulfil such different functions in haemoglobin, catalase and cytochrome c?' Implied in the question was: 'I am a parasitologist, but you as a chemist ought to understand this'. I was clueless at a time when the structures of all these proteins were still unknown, but I doubt that I could provide complete answers even today when they have been determined in detail.

Keilin's favourite

Cytochrome c was Keilin's favourite, but he never succeeded in crystallizing it. In 1955, Gerhard Bodo, a young Austrian biochemist working with John Kendrew, crystallized it from the muscle of the King Penguin and tried to determine its structure. Keilin's birthday was on the first day of spring, symbolic of his youthful exuberance and easy to remember, so that I always called on him to wish him many happy returns. That year I brought him the first X-ray diffraction pictures of cytochrome c. Keilin's delighted response was: 'Next year you will bring me model'.

He was born in Moscow, brought up in Warsaw, studied in Liège and Paris, arrived in Cambridge in 1915 and still spoke English with an endearing Polish accent and syntax. When doing his round of the laboratory, he would ask: 'How you are getting on Perutz?' That was not a perfunctory question, but sprang from a deep interest in my work and everyone else's at his laboratory.

The physiologist Gilbert Adair made the first haemoglobin crystals for me and later taught me how to make my own in his room in the Low Temperature Station, but his working habits were eccentric. He never washed any of his glassware nor let anyone else touch it until he actually needed it so that all his benches were cluttered as in an alchemist's shop. Sometime in 1938, Keilin offered me bench space in his laboratory which was always clean, tidy and warm. Keilin was no pedant, but he suffered from allergic asthma which was provoked by house dust and aggravated by cold. His asthma caused him much discomfort and he kept it in check with an adrenaline spray that may have brought on the heart attack which finally killed him.

Charles Darwin wrote in one of his letters from South America: 'In short, I am convinced it is a most ridiculous thing to go round the world when by staying quietly at home, the world will go round with you'. Some science professors are in their departments only when they are not attending committees in London or touring abroad, but Keilin was always at his post and accessible to his staff. Scientists from all over the world sought him, like an oracle, for his knowledge, wisdom and humanity. He would always *listen*, not only to the president of some foreign academy, but to anyone of us, his students, watching us benignly from behind his mahogany desk in his small office on the first floor, an office without a secretary, because she was downstairs and also answered the lab telephone which had only two or three extensions. So there was not that barrier that other professors erected between them and their staff, the secretary whom you had to ask to be admitted into their presence: you could just knock at the door and walk in if there was no-one else in already. Keilin was never too busy to see you when he was in his office, but we did feel shy interrupting him when he was doing experiments. In 1957, I called at the Molteno to give him my good wishes on his 70th birthday. He arrived a few minutes after me and found a pile of mail waiting for him, but put it aside saying proudly: 'I am not going to read it, I am going to do experiments.' He continued them right to the end of his life. He was averse to theorizing and disapproved of young Crick's flights of fancy, telling me to 'keep him to the bench'. I ignored that advice, because Crick was irrepressible, but had he followed it, DNA would not have been solved. There are many ways of doing science.

Outside Keilin's office and all along the passage between the laboratories hung photographs of the world's leading parasitologists, collected by Keilin's predecessor G.H.E. Nuttall, and himself: dignified elderly gentlemen, mostly, with magisterial beards and studious pincenez. As leader of the field and editor of the journal *Parasitology*, Keilin had known most of them personally. Besides his native Polish, he spoke fluent French, Russian and also German, so that he could converse and correspond in several languages at a time when English was not yet the scientific lingua franca. At one end of the passage was the tea room, where the monotony of dry biscuits was broken by an occasional cake to celebrate the publication of a paper. One day we had a real celebration.

Warburton and the Siberian tick

In 1931, when Keilin had succeeded Nuttall as Quick Professor of Biology, an elderly lecturer called Warburton com-
plained that he had been appointed before the University
Superannuation Scheme had been instituted, so that he had
no pension and would have to die in harness. When Keilin
told the University Treasurer that Warburton was in his
late seventies and had no pension, the Treasurer agreed that
in view of Warburton's advanced age the University could
afford to be generous. He failed to foresee that 24 years later
we would celebrate Warburton's 100th birthday! On that occasion he told us a wonderful story. In his prime, Warburton
had been the world's authority on ticks. One day in the
twenties, some of his students were eating their lunch of
bread and cheese when they found a tick in their butter.
They brought it to Warburton who identified it as a Siberian
tick. That discovery was to provoke a diplomatic crisis. The
students had bought their butter at Sainsbury's, not knowing where it came from. Impressed by entomology's detective
powers that could trace the butter's origin to Russia, they
told their story to a don who mentioned it to a visiting MP
and he in turn related it to a journalist. The outcome was a
headline in one of the London evening papers: 'Disease-
Carrying Tick Imported with Russian Butter'. Questions
were asked in Parliament, the horse-drawn milkcarts which
in those days also distributed butter in London bore placards
reassuring housewives that they carried no Russian butter,
the Soviet Ambassador called on the Foreign Secretary to
protest against the campaign of slander against his country's
agricultural exports and Pravda condemned Warburton's
deliberate lies. Years later, Russian parasitologists visiting
the Molteno Institute reproached Keilin for allowing it to
become a tool of anti-Soviet propaganda and refused to believe that Warburton was just an unworldly scholar who had
happened to come across a curiosity. Secure in his generous

pension, Warburton continued to live in good health in Grantchester to the ripe old age of 103.

Frail health

Keilin did not ride a bicycle, nor drive a car. In good weather he walked to the lab from the Keilins' comfortable home in Barton Road and on bad days Mrs. Keilin, also from Poland and a G.P., drove him to the lab before she started on her round of visits. His frail health prevented his engaging in any vigorous physical activity, and his asthma kept him from travelling. When he did not work in the lab, he edited the journal *Parasitology* or studied at home. To me, skiing and mountaineering holidays in the Alps and all kinds of physical activity at home were an essential part of well-being. I learnt to appreciate Keilin's cheerful endurance of his handicaps when an accident condemned me to lead a similar life of all work and no play for several years.

Keilin measured the respiration of tissues with Barcroft manometers manufactured in the Molteno's poky basement workshop by his general factotum, the rotund, easy-going laboratory steward and mechanic Charles, whose instruments had a habit of falling to pieces at awkward moments. There was no money to buy instruments except when Gerald Pomerat, the lean, dapper, meticulous representative of the Rockefeller Foundation, appeared like the rich uncle from New York and asked Keilin whether he needed anything for his research. I remember our joy when the Foundation bought Keilin the first Beckman spectrophotometer that measured optical densities directly, and we no longer had to work them out elaborately by comparison with known absorbents.

In 1944 or 1945, Keilin persuaded the Faculty Board of Biology B to let me give an annual course of eight lectures on what later came to be called Molecular Biology. They were held in the museum on the 2nd floor that housed disgusting

tropical parasites pickled in glass jars and which also served
as a lecture room. They were extracurricular, advertised on
notices to the various science departments and attended
mainly by research students and foreign visitors, rather
than undergraduates. In 1954, the University appointed me
a University Lecturer in Biophysics, and my lectures were
advertised in the University Reporter as part of the biochemistry course, which made me think the posting of notices unnecessary. I arrived in the museum full of expectation, soon joined by James, the scraggy, ginger-haired little
technician who regularly projected my slides, to await the
students. Five o'clock came but no students. James and I exchanged little jokes which became feebler as time passed,
but still not a single student appeared, to the joy, possibly, of
the plant virologists whose rooms opened onto the museum
and who could not run their roaring centrifuges while I lec-
tured. But here I was, a newly appointed University Lecturer without a single student to listen to his lecture. I
nearly wept with humiliation. After a while, I did find out
what happened. The Professor of Biochemistry who had
supported my appointment had also advised his students
that my lectures were too specialized for them.

Lack of tenure

Although Keilin was one of the most distinguished Cambridge scientists, he did not have tenure because the statutes of the Quick Professorship, the first research Chair to
be established in Britain, required his appointment to be renewed every 3 years. Apparently, the Quick Trustees had
made that provision because it had seemed inconceivable to
them that anyone could successfully concentrate on research
for a period longer than 3 years at a time. Keilin once as-
serted to me that this never worried him one little bit, but I
have wondered whether he really possessed such complete
self-confidence or said it only to console me for the insecurity

of my own position. Going backwards and forwards on my bicycle between the Cavendish Laboratory in Free School Lane and the Molteno Institute in Downing Street, I was the living link between Biology and Physics, but I was a chemist, incapable of teaching either subject, and neither the University nor any of the Colleges had a place for me. I was officially beyond the pale, but Keilin gave me confidence that my crazy attempt to solve the structure of haemoglobin would succeed, and he suggested W.L. Bragg's approach to the Medical Research Council which led to the foundation of the Research Unit for the Study of the Molecular Structure of Biological Systems. This was the forerunner of today's Laboratory of Molecular Biology. Fifteen years later, when I proudly showed Keilin our new laboratory, he said to me: 'Now you merely have to win Nobel Prize!' I protested that he deserved it first, but he waved me off with the resigned remark that all this was long past. Clearly, he should have shared either the 1931 Prize for Physiology or Medicine that was given to Otto Warburg 'for the discovery of the nature and mode of action of the respiratory enzyme' or that given to Hugo Theorell in 1955 'for his discoveries concerning the nature and mode of action of oxidation enzymes'. When Theorell's prize was announced, Keilin's long-term collaborator, compatriot and friend, Thaddeus Mann, went to see him to express his disappointment at his exclusion. Keilin took him affectionately by the elbow and said: 'When I stand at the Gates of Heaven, St. Peter is not likely to ask me whether I have brought with me Nobel Prize'. He was thrilled all the same when the Royal Society awarded him their highest honour, the Copley Medal, and he proudly showed it to all of us in the lab.

Peter Mitchell, who received the Nobel Prize for Chemistry in 1978 for his chemiosmotic theory, had this to say about Keilin in his Nobel lecture:

But let me first say that my immediate and deepest impulse is to celebrate the fruition of the late David Keilin, one of the greatest of biochemists and – to me, at least – the kindest of men whose marvellously simple studies of the cytochrome system, in animals, plants and microorganisms, led to the original fundamental idea of aerobic energy metabolism: the concept of the respiratory chain. Perhaps the most fruitful (and surprising) outcome of the development of the notion of chemiosmotic reactions is the experimental stimulus and guidance it has provided in work designed to answer the following three elementary questions about respiratory chain systems and analogous photoredox chain systems: what is it? what does it do? how does it do it? The genius of David Keilin led to the revelation of the importance of these questions.

Fig. 1 is a facsimile of the letter he wrote to me in his vigorous hand when Kendrew and I shared the Nobel Prize for Chemistry in 1962.

Keilin personified the saying of Montesquieu that knowledge makes men gentle, and also Bertrand Russell's remark that 'Einstein's is a kind of simplicity which comes of thinking only about the subject concerned, and forgetting its relation to one's own ego'. John Kendrew and I owe Keilin a tremendous debt, for he was one of the first to see the potentialities of our physical approach to biochemistry. Until the 1950s, we had no facilities for biochemical work in the Cavendish Laboratory; Keilin gave us bench space in his institute even though he was short of space himself, and he helped us to grow protein crystals. When Kendrew's and my research was in danger of closing down for lack of support by the University, Keilin suggested, and supported, Sir Lawrence Bragg's approach to the Medical Research Council which saved us. But to me his most important gifts were his confidence in me and the warmth of his friendship, which helped me to gain confidence in myself.

After his death, Francis Turner wrote of him in the Magdalene College Magazine and Record (I,13,1963):

David Keilin wore the mantle of his world-wide distinction as an invisible garment: what was visible was disarming modesty, true friendli-

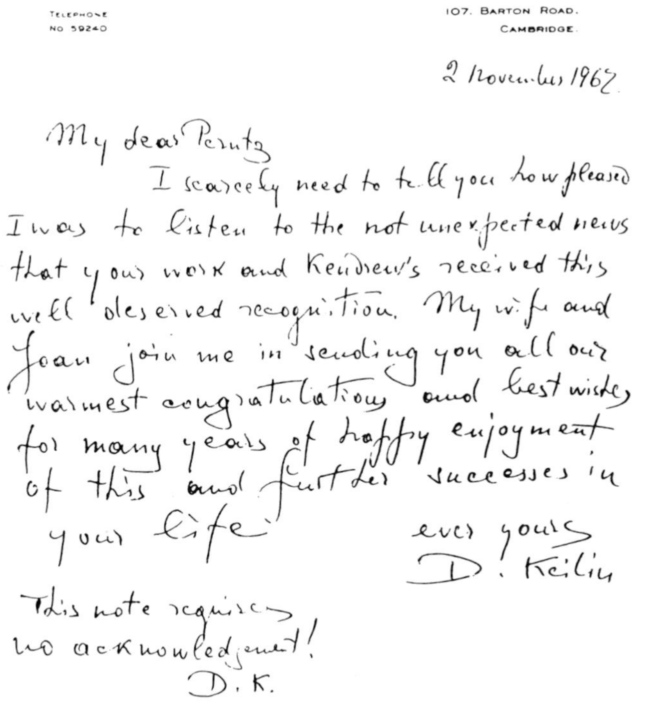

TELEPHONE
NO 59240

107. BARTON ROAD.
CAMBRIDGE.

2 November 1962

My dear Perutz

I scarcely need to tell you how pleased I was to listen to the not unexpected news that your work and Kendrew's received this well deserved recognition. My wife and Joan join me in sending you all our warmest congratulations and best wishes for many years of happy enjoyment of this and further successes in your life.

ever yours
D. Keilin

This note requires no acknowledgement!
D.K.

Fig. 1. Keilin's letter to the author offering congratulations for winning the Nobel Prize.

ness, and a mind so quick that he immediately understood all that his neighbour meant, however imperfectly expressed. But this understanding was not only quickness of wit: it arose from the deeply charitable view he had of human nature, which, together with his wide experience, gave him an unusual insight into those with whom he came in contact. He was a most remarkable and loveable person, patient and wise, with so great a spirit in so small a frame.

G. Semenza and R. Jaenicke (Eds.)
Selected Topics in the History of Biochemistry: Personal Recollections, V
(Comprehensive Biochemistry Vol. 40) © 1997 Elsevier Science B.V.

Chapter 3

An Australian Biochemist in Four Countries

E.C. SLATER

Department of Biochemistry, University of Southampton,
Southampton, UK

Introduction

An explanation of the title of these personal recollections is called for. Although it is 50 years since I left Australia, I have never changed my citizenship or nationality. More importantly, although Australia has, in many respects, changed out of recognition in the last half century and I have spent nearly 90% of my working life outside Australia, emotionally I still feel myself an Australian.

Early life in Australia 1917–1939

Family background

I was born in St. Kilda, a suburb of Melbourne, on 16 January 1917. On my father's side, I am a third-generation Australian, on my mother's side, a second-generation. I have been able to trace the Slater line for six generations to John Slater (died 1781). His son, my great-great-great grandfather, Joseph Slater (1775–1826), and his descendents lived in or nearby the village of Salwell, in the parish of Whick-

ham in County Durham in northeast England, close to New-castle, untill well into the 20th century. Joseph's eldest son, John Slater (1807–1877), was a miller; the other two sons were a grocer and butcher, respectively. One of John's sons, John Norvell Slater (1835–1870), my great grandfather, also a miller, emigrated to Australia in the 1850s and in 1860 he married an Irish lass, Anne Neleyn, born in County Clare in Ireland, in 1838 or 1839, at Inglewood, near Bendigo (then called Sandhurst), in the State of Victoria where the gold fields that had been discovered in 1851 heralded the great gold rush that was to transform Australia. He appears to have changed trade after emigration, since he is described on my grandfather's wedding certificate as a butcher.

My grandfather, Edward Slater, was born in Inglewood in 1863 (all four of my grandparents were born in this year), when the city of Melbourne was only 28 years old (but Melbourne University was already 11 years old). He was only 8 years old when his father died and knew nothing about his relatives in England, except his father's birthplace, until my father visited Swalwell in 1917. I know nothing about my great grandmother, Anne Neleyn Slater, except that she married again to a William Pollock.

At the age of 13, my grandfather was apprenticed to a printer in Bendigo to be trained as a compositor and moved to Melbourne when his apprenticeship was completed in 1880. Where he received his further education, I do not know, but I suspect that it was largely from his own reading. In a letter to my father, he refers to his father's Shakespeare as 'one of the few very things he left behind, and from which I obtained most of the knowledge of Shakespeare I have'. In any case, he was well read. When I knew him, he worked as a type-setter with the Melbourne daily newspaper *The Age*, and its sister weekly, *The Leader*.

In 1885, he married Sophia Brunton, who was born in 1862 in Aldershot, England, and emigrated to Australia shortly afterwards. Her father, Henry Brunton, was the pro-

prietor of the Brunton Temperance Hotel in the centre of
Melbourne. From all accounts, my grandmother was a very
active and intelligent woman, with strong views which she
did not hesitate to express. She was Treasurer in the first
Ministry chosen by A Woman's Parliament, formed by a
Woman's Political Association, 'with the double view of edu-
cating women into Parliamentary procedure, and exposing
the weakness of the present Parliaments run by men'.

At the turn of the century, my grandparents purchased a
block of land at Sandringham, a seaside suburb about ten
miles from the centre of Melbourne, overlooking Port Philip
Bay about a quarter mile away. In the course of time, four
houses were built on this block, two of which were still occupied by members of the family until the death of my sister in
1990.

My father, Edward Brunton Slater, was born in 1888, in
South Yarra, a suburb of Melbourne. He was originally
trained as an architect, but, by the time I was first conscious
of what he did, he was a Civil Engineer in the Victorian
Railways. Not surprisingly, with the sea nearby, sailing became his main hobby, as it later became mine, and many of
my early memories are connected with sailing.

My father and mother were married in 1915, in St. Kilda.
My mother, Violet Podmore, was the daughter of Henry
Podmore, son of Richard Podmore, manufacturer, and Mary
Smith of Stafford, England. Henry Podmore, my maternal
grandfather, who was born in Watford, England, in 1863,
was a surveyor. My maternal grandmother, Violet Alexandra
Taylor, whom he married in Kilmore, Victoria, in 1889, was
born in Melbourne in 1863, the daughter of Swanson, a tailor, and Elsie Taylor. I know little about my maternal
grandparents. My grandmother died before I was born and I
was very young when my grandfather died. My mother was
born on 22 May 1893, in South Yarra, also the birthplace of
my father, but spent much of her youth with her mother in
Sydney.

I had one sister, Patricia Violet Slater, 2 years younger than me, who, after serving in the Australian Army Nursing Corps in World War II, studied Nursing Education in Washington University, Seattle and had a distinguished career as Director of the Royal College of Nursing, Australia. She was largely responsible for introducing nursing education into Australian universities.

Schooling

My first recollections are of living in Elwood, a seaside suburb a little further down the coast from St. Kilda and about 3 miles from the centre of Melbourne. My first school (from 1923) was St. Bede's Preparatory School, run by the Church of England. When I was 9 years old, I moved to my father's old school, Wesley College, a Public[1] School. I do not remember much about the half-year I spent there, except that it entailed two quite long tram journeys on my own.

Towards the end of 1926, my father was appointed District Engineer for the Western District of Victorian Railways, with headquarters in Geelong, 45 miles south-west of Melbourne. For the next 8 years, we lived close to Geelong College, the smallest of the six Public Schools. Thus, shortly after my 10th birthday, I went to my fourth school, Geelong College, where I was to stay for 8 years and receive my basic education.

For its period, it was a good school, in some respects very good. I cannot say that it provided a strong atmosphere of learning and little emphasis was placed on preparation to enter the university, but there were some very good masters who provided a solid education. About two-thirds of a total of about 300 boys were day-boys, living in or around Geelong, a city at that time of about 40 000 inhabitants. They came each day to school, either like myself on foot, or by bicycle or

---

1 The illogical English name for a private school.

tram. Bruce Kennedy, who later became a good friend of mine, came on horseback from an outlying suburb. I cannot remember anyone being brought by car. The majority of the boarders came from the Western District of Victoria, the richest pastoral area of Victoria, many of them from families who had emigrated from Scotland. Among both masters and boys, names beginning with Mac as well as the Stewarts, Campbells and Camerons abounded.

Sporting activities were given great prominence. Indeed, after school hours, participation in football in winter and cricket or rowing in summer was compulsory. This compulsion did not worry me, since I was as keen on sport as any young Australian of that day. The only trouble was that I was not very good at it. Geelong College produced a number of very successful sportsmen. The most famous of my generation was Lindsay Hassett, who became captain of the Australian cricket eleven. Despite its lack of intellectual pretensions, it also produced some excellent scientists. The most famous of all was Sir MacFarlane Burnett, O.M. and Nobel Laureate. Sir Robert Honeycombe, who was in the Lower School when I was in the sixth (highest) form, became Professor of Metallurgy at Cambridge University and Treasurer and Vice-President of the Royal Society.

In 1927, Australia was prosperous, but, except for a short period during the Gold Rush, this prosperity had been, for a century, dependent on one product – wool. The fall in commodity prices, as a result of the great depression of the early 1930s, hit Australia particularly badly and very many were thrown out of work. Our family was less affected than many, but, as part of its deflationary policy, the State and Federal Governments cut the salaries of public servants – like that of my father – very heavily. It must have been very difficult for him, like many other parents, to pay the school fees for myself and my sister, who was at a girl's Public School. From 1930, I was able to lighten this somewhat by winning scholarships.

'Tam' Henderson, the science master, was of critical importance for my development as a scientist (and I think for
Honeycombe's also). In my last years at school, he steered
me towards chemistry, which I had not followed in the middle school. The son of a schoolmaster (whom I met in Dun-
fermline in Scotland in the 1940s), Tam had a broad Scottish
accent, which was, of course, not out of place in Geelong
College. He was an enthusiastic if unconventional teacher,
often impatient with the average boy, but able to arouse the
interest and curiosity of the brighter student. Unlike most of
the teachers, he played no part in coaching or organizing
sporting activities, and confined his extracurricular activities to running the Debating Society and supervising the
editors of the school magazine, *The Pegasus*. In this capacity,
he played an important role in developing whatever writing
style I might possess. In my last year at school, I was an editor of *Pegasus* and had to write an editorial. Mr Henderson
made so many textual corrections, in neat green ink, so as
completely to obliterate my draft written in blue-black ink.
At the same time, he gave me a reading list, including
popular-science essays by J.B.S. Haldane. Although I did not
know it at the time, Haldane was then Reader of Biochemistry at the University of Cambridge. He was known to Mr
Henderson as a fellow officer in the Black Watch regiment
during the campaign in the Middle East in the Great War
and was chiefly remembered for his ability to compose scur-
rilous limericks. I was, in fact, an avid reader and my
mother had seen to it that, from an early age, I read what
she considered 'good' books, such as Charles Dickens, Robert
Louis Stephenson and Walter Scott, as well as the more
usual boys' adventure stories. However, this reading had
had little effect on my style of writing and I do not remember any help in this respect from my teachers of English,
who had never torn my essays apart like Mr. Henderson's
green pencil.

Although I had already passed my matriculation requirements in 1932, I remained at school for two more years, in
the Honours Sixth Form, in order to increase my chances for
scholarships at Melbourne University, in which I was successful. During my last year at school, my father had been
transferred back to Head Office of the Victorian Railways in
Melbourne and our household moved to the third Slater
home in Sandringham. However, after the summer vacation,
I went into residence at Ormond College, which is attached
to Melbourne University.

University education

My 5 years in Ormond College played a most important part
in my development. The proximity of the lecture rooms and
practical classes meant that I was spared the long hours of
travelling that living in Sandringham would have entailed,
although I went home for most weekends. More importantly,
Ormond provided the intellectual atmosphere that was
missing at Geelong College.

My choice of Chemistry as main subject was dictated
chiefly by my success in this subject at school. My father had
been told by friends that the up and coming field was
Chemical Engineering, which was not a degree course at
Melbourne University, but a hybrid course of Chemistry and
Engineering had just been introduced, so I opted for that.
Since reading *Martin Arrowsmith*, by Sinclair Lewis, a
birthday present from my mother, I was really interested in
medical research, but nobody told me that it was possible do
this without being a practising doctor and I was scared stiff
of the responsibility of surgery, in which every General Prac-
titioner in Australia was trained. I had never heard of biochemistry.

Since the courses in Chemistry, Physics (still then called
Natural Philosophy) and Pure Mathematics were of about
the same standard as the Leaving Honours Course that I

had already followed twice at school, I rather coasted
through my first year, playing a lot of competition tennis
(and billiards in the evenings) and did not do as well in my
examinations as I should have done, but in the second year,
I surprised myself with a First Class in Natural Philosophy
and even more so in my final year for my Bachelor's degree
when I was top of my year in Chemistry. It was, in fact, a
pretty distinguished year. A class of twenty to thirty included Hugh Ennor (later Sir Arthur), who became first Professor of Biochemistry at the Australian National University
(ANC), Jack Legge (a friend from our schooldays at Geelong
College) and Bill Rawlinson, both of whom became Readers
at the Department of Biochemistry at Melbourne University,
and Dirk (later Sir David) Zeidler, who became a leader of
the chemical industry in Australia, President of the Australian Institute of Engineering and Fellow of the Australian
Academy of Science.

The engineering part of my course was, for me, rather a
waste of time. I found that the application of rather simple
mathematics was sufficient to pass the examinations with
reasonable marks, but I never had any feeling for engineering design; I was always glad, however, that I learned the
basics of draftmanship. In my third year, I was top of my
year in Metallography which I did find interesting – it was
really chemistry. The Melbourne University Chemistry Department was founded in 1886 and had gained a good reputation under the leadership of the first Professor, David
Orme Masson (later Sir), who was only 28 years old when
appointed and remained in function until 1923, and Professor Sir David Rivett, who left in 1926 to found the very successful Council for Scientific and Industrial Research
(CSIR). In 1935, the Department was housed in old buildings (the Organic Chemistry Laboratory in what had been a
gardener's cottage). The Professor was Hartung, who gave
spectacular demonstrations in his lectures to first-year students, and was very much involved in planning a new

building, which was completed shortly after I left Melbourne. Although its staff were not particularly distinguished, we received sound instruction in the basics of analytical chemistry from (Gus) Ampt, and in my third year, Associate Professor W. (Bill) Davies raised our enthusiasm for organic chemistry. Towards the end of my period, the school was greatly strengthened by the appointment of E. Heymann, a physical chemist and refugee from Hitler, and from Cambridge came J.S. Anderson, a leading inorganic chemist, who was later the Director of the National Physical Laboratories at Teddington in England, Fellow of the Royal Society and Professor of Inorganic Chemistry at the University of Oxford.

It was a lecture given by Bill Davies, towards the end of my third year, on the recent identification and structure determination by J.W. Cook of the carcinogenic hydrocarbons in coal tar that further stimulated my interest in matters related to medicine. This prompted me to write an article on cancer for the Science Review, a publication of the University Science Club, my first publication in a (semi)-scientific journal. I was not the only one so influenced by Davies' lecture. Geoffrey Badger (later Sir Geoffrey), who had been in the same class as me in the middle school at Geelong College but had moved to the Gordon Institute of Technology, and who had entered Melbourne University a year ahead of me and was then working for his Master's Degree under Bill Davies, was so enthralled that he went to Glasgow the following year to work under Cook. He returned to Australia after World War II as Professor of Organic Chemistry in Adelaide University and had a distinguished career, becoming President of the Australian Academy of Science in 1974–1978.

My Bachelor of Science Degree was conferred in March 1938 (the commencement of the university year in Australia). For the Master's Degree, we followed a practical course in physical chemistry run by Heymann (who detected my inadequate grounding in chemical thermodynamics which he attempted, unfortunately with only partial success, to remedy by prescribing a reading course). The degree was granted mainly on the basis of a thesis describing experimental research, without any written examination that I can remember.

First research

I was given two research projects by Davies – one purely organic and the second related to a project started by Davies a few years earlier, in conjunction with the CSIR, on the vitamin A content of some Australian fish livers. They had found that, in general, the livers of Australian fish are much richer in this vitamin than in fish found in the northern hemisphere. The school or snapper shark (*Galeorhinus australis*), which at that time formed a major part of the catch in the fishing industry, is a particularly rich source.

The purely organic chemical project was an unsuccessful attempt to extend a new synthesis of isoquinolines introduced by Davies to the synthesis of pyrroles, substituted indoles and methyl-iso-indole. However, this project gave me good experience of classical organic chemical techniques and how to search the literature. (A key paper, showing that the proposed protocol for the synthesis of pyrrole would not work, was discovered only during the course of the experiments).

Besides analysing fish livers from tuna, which had not been previously examined, I studied the seasonal variation of the vitamin A content of the snapper shark caught near Cliffy Island, off Wilson's Promontory, in the Bass Strait. The fish-liver project required the extraction of large fish livers with petroleum ether. In those days, little precaution was taken to protect oneself against the chemicals being used and I developed a nasty dermatitis on my hands, which troubled me for many years. After evaporation of the petroleum ether, the residual oil was saponified, the non-saponifiable fraction dissolved in chloroform and the vitamin A determined by measuring the intensity of the blue colour developed with antimony trichloride, in what was called a Lovibond Tintometer, which matched a colour with that given by a combination of standard glass slides. The vitamin A content was calculated from the known number of 'Lovibond units' given by pure vitamin A. Since the blue colour was only transitory, some practice was required.

The main point of scientific (and of some commercial) interest of this work was the finding of a clear maximum, in the winter, of the percentage of vitamin A in the liver oil. This is mainly due to a decrease in the percentage of fat in the liver. Davies thought that this was worth publishing and asked me to include a few vitamin D assays, which showed that the oil has little of this vitamin, that had been done for him by a Dr M.M. Cunningham from Wallaceville in New Zealand. So I wrote up the paper [1] and put her name first (because I thought that it was the custom to put the order of authors in alphabetical order). This is how my first scientific paper has, as first author, someone I have never met.

On re-reading my research report, it is clear that I obtained good training under Davies. I was also fortunate in being appointed part-time demonstrator to the practical course in organic chemistry, teaching students only 1 year behind me. A new arrival from England, Dr Hatt, who later became Head of the Organic Chemistry Division of the CSIR, had just taken over this practical course and I learned from him at least as much as the students under my care did.

At the end of 1938, I passed for my Master's Degree, with First Class Honours and again first place. Since I was awarded scholarships sufficient to support myself, I decided to stay on for a fifth year, in the hope of obtaining a scholarship to enable me to study for a doctorate in England. At that time, it was not possible to do a Ph.D. in Australia. There were, in fact, only two ways of obtaining funds to go to

England – either via a Rhodes Scholarship, or to win an 1851 Exhibition. I was not sufficiently prominent in sport, or in student affairs in general, to give me much chance in the first direction and, since there were only two 1851 Exhibitions for the whole of Australia, I would have been fortunate to receive one of these. I did not apply in 1938, hoping for a stronger candidature in 1939. Usually, Sydney and Melbourne Universities shared the two prizes, but in 1938 they were given to two students from the Organic Chemistry Department at Sydney University, who later became Sir John, Nobel Laureate and Lady Cornforth. Since war broke out before applications were due for the 1939 round, I did not make an application. I had made one other attempt to get to England. Tam Henderson suggested that I write to Hirst, a well known carbohydrate chemist at Bristol University, whom he knew from his student days. I received a charming letter from Hirst, accepting me into his laboratory, but regretting that he had no funds and hoping that I would be successful in obtaining a scholarship. I was immensely gratified at being accepted by such a distinguished chemist.

During my fifth year, apart from finishing off the fish-liver oil work, I took over from departing students a long-continuing and long-to-continue project. Davies had observed a crystalline deep blue coloured compound as a bye-product in his new method for the synthesis of 1-phenylisoquinoline. As a product of the Manchester school of Perkin, he was naturally interested in pigments and set his students to elucidate its structure. Osborne and Holmes, two M.Sc. students from my year, had already had a go when I was put on the project in 1939, and I had speculated on its structure and possible relationship to phthallocyanine in my M.Sc. thesis. From molecular weight determinations (by the old-fashioned Rast method) and elemental analysis, it was clear that some sort of dimerisation had taken place. I made little progress towards a solution of the structure and, after I left Melbourne, I thought no more about the pigment. Twenty eight years later, I received, in Amsterdam, a stencilled manuscript from a Dr A.V. Robertson, addressed to 'my co-authors'. Using the modern methods of high-resolution mass spectrometry, infrared and NMR spectroscopy, Dr Robertson had established the structure of the pigment to consist of two isoindole molecules linked by a methine bridge and a hydrogen bond between the nitrogen atoms. Thus, there is indeed an analogy to half a phthalocyanine molecule, except that the link between isoindole fragments is methine rather than azomethine. What we could never have suspected, since our analytical methods would have been too insensitive, is that a carbon atom is lost during the dimerisation. Dr Robertson generously put the names of all eight of the Bill Davies' students (in alphabetical order, so that his is the last-but-one, and mine the last) who had been engaged on the project, as well as that of Davies to the paper [2]. Nowadays, there is pressure from some quarters to require that all authors of a paper be responsible for all parts of it. The two examples of the publication of my first scientific work show the absurdity of a rigid application of this principle.

During my fifth year at Ormond (an unusually long period of residence, except for medical students), my college scholarship was continued, but Picken, the Master, asked me to give tutorials in Chemistry to first-year medical students. This was my first and one of my most difficult teaching tasks, since the brighter students were exempted from my tutorials and I was left to teach chemical calculations to those with little understanding of chemistry or ability in mathematics. The remembrance of my first struggles to come to grips with the language of chemistry were of great help in these first teaching tasks which I found interesting.

Australian Institute of Anatomy 1939–1946

The outbreak of war on 3 September 1939 closed the possibility of going to England. Since the relatively few chemists in Australia had been placed in a Reserved Occupation, there was no possibility of being called up into the armed services, so it became necessary to think about a job. Davies asked me if I would be interested in the position of biochemist at the Australian Institute of Anatomy, Canberra, which was attached to the Commonwealth Department of Health. W.J. Young, the Professor of Biochemistry, had been informed of the vacancy and since he had no one available to recommend, he had approached Davies. I was very interested in the prospect of moving towards biochemistry and regretted having chosen engineering subjects instead of biochemistry and physiology. Apart from my positive interest in problems related to medicine, I had become disenchanted with organic chemistry, as such. I remember, vividly, my reaction to seeing a paper by (Sir) Robert Robinson, entitled *The Structure of Strychnine and Brucine*, Paper 51 of the series. I felt that this was pretty slow going and that it was more interesting to find out how these alkaloids exerted their toxic effect. I felt, too, that it was time to leave Melbourne and to strike out on my own. I applied for the position and, in due course, found myself as Biochemist, Grade I, in the Commonwealth Department of Health, to work at the Australian Institute of Anatomy, starting on 4 December 1939. Professor Young (best known for his discovery of the Harden and Young ester, fructose 1,6-*bis*-phosphate, at the Lister Institute in London) gave me a crash course in the biochemistry practical course for medical students. So armed, at the beginning of December, I flew to Canberra in a De Haviland 86 aircraft of the Australian National Airways.

I immediately fell in love with Canberra. Situated on a tableland, about 2000 ft above sea-level, with mountains above 6000 ft to the west and low hills to the east, it was typically Australian landscape, with grand eucalyptus ('gum') trees fairly sparsely scattered over grassy fields (or 'paddocks' as we called them). The streets and avenues were lined with newly planted, mostly European, trees and flowering fruit trees, which were a picture in the spring, as were the gardens in the houses of the surrounding suburbs. A large area of what had originally been a sheep station had been set out for the construction of the capital city of Australia. When I went there in 1939, only 12 years after its inauguration as the capital, most of those years of depression, it was very incomplete. Instead of building outwards from the centre as most cities grow, Canberra was built from the periphery, with Government Offices, the Houses of Parliament and a single hotel near the middle and two shopping areas, several miles apart roughly equidistant north and south from the centre. There was a lot of empty space, some of which was destined to become a lake by damming the Molonglo River which ran through the centre. Although its height made the short winter fairly severe by Australian standards (but snow is seen only once every 5 years or so), it also tempered the heat in the summer, which is usually quite dry. The climate for most of the year is so good as to become almost boring. In 1939, the total population was less than 9000.

I lived nearby the Australian Institute of Anatomy in Beachamp House, a guest house that now houses the secretariat of the Australian Academy of Science. The Institute had been founded in the early 1930s, by Sir Colin McKenzie, who had made his name for work on the zoology of Australian marsupials. He offered his collection of anatomical specimens to the Australian Government with the proviso that the Government build a special museum to house them. The result was a fine building, the first permanent Government building to be completed in Canberra. It consisted of two large museums open to the public, with a number of laboratories and preparation rooms, offices and a fine library. For a later chapter in this story, the following quotation from a book published in 1938 is of interest: 'Many people regard the Institute as the real nucleus of a national research university'. There was also a reservation outside Canberra for some Australian marsupials.

After Sir Colin McKenzie's death, the Institute was taken over by the Commonwealth Health Department and Dr F.W. Clements was appointed Director. When I arrived, in addition to the Director, the staff consisted of a secretary, a zoologist (Bill Boardman), a dietician (Betty Willmott), a technician and artist for maintenance of the exhibits in the museum, two porters cum museum caretakers, and a lab boy used mainly for messages on his bicycle. I was given a large well equipped laboratory with ancillary rooms and a large office with a beautiful view over Canberra towards the Queanbeyan Hills which turned a vivid purple near sunset. The lab boy was available for washing up and other help in the laboratory.

In the first letter I received from Clements, he gave as my research project an investigation of the nutritional background for the, at that time, widely held view that the koala bear could only exist on certain types of eucalyptus leaves. (I understand that this has since been called into question). By the time I arrived, however, it was considered that a topic of more immediate relevance to the war effort was called for. The decision of the Ministry of Food in the UK, which was concerned with the nutritional well being of the British population in wartime, to authorise the addition of synthetic vitamin B_1 (and calcium) to flour for breadmaking had stimulated the Commonwealth Department of Health to investigate the extent of vitamin B_1 deficiency in Australia.

When I presented myself for work on the morning of Monday, 4 December 1939, Clements handed me a paper he had just seen, by Y.L. Wang[2] and L.J. Harris, in the *Biochemical Journal*, on the estimation of vitamin B_1 (also called thiamin) in urine by the thiochrome method and asked me to set up this method in the Institute. The thiochrome method had

---

2 Wang, Yin-lai became Director of the Institute of Biochemistry of the Chinese Academy of Sciences in Shanghai and a member of the first Council of the International Union of Biochemistry.

been introduced by a certain B.C.P. Jansen, whom I had heard of as the man who, in Batavia in the Dutch East Indies, had isolated the pure crystalline vitamin from rice polishings, the first pure vitamin to be isolated. The thiochrome method was based on the formation of thiochrome, which has a blue fluorescence when viewed under ultraviolet light, when thiamin is oxidized by alkaline ferricyanide. Jansen used a photoelectric fluorimeter,[3] made in his laboratory in Amsterdam, to measure the fluorescence, but since the construction of such an instrument was beyond the possibilities of many biochemical laboratories, Wang and Harris developed a simple procedure whereby the fluorescence of the thiochrome derived from the thiamin in the sample being analysed was compared with that from a standard thiamin solution viewed in two test tubes, side-by-side under an ultraviolet lamp.

It was a completely new experience for me, at the age of 22, to have at my disposal practically a suite of rooms, instead of the corner of a bench in a university laboratory. The disadvantage, of course, was that whereas I was used to asking advice from all and sundry in the Organic Chemistry Laboratory, here I had nobody to turn to. There was one other biochemist – Gordon Lennox, in the CSIR Division of Economic Entomology about a mile away, but he was soon to leave for Melbourne to set up a protein group in the Wool Research Laboratories. As a trained organic chemist, the first step I took was to purify, by fractional distillation, the isobutanol to be used to extract the thiochrome. Here, I made my first mistake (I mean in Canberra). Repeated fractionations failed to bring the boiling point up to the value reported in the literature – it stayed about 1.5° lower. Then the penny dropped. I had overlooked the fact that Canberra

---

3 This fluorimeter was still present in Jansen's Laboratory when I succeeded him in 1955.

is about 2000 ft above sea-level, which accounted precisely for the difference.

I returned to Melbourne for the Christmas vacation, during which I visited the public library to read Jansen's original paper [3], which had been published (fortunately in English) in a Dutch Journal unavailable in Canberra. Who could have foreseen that, about 15 years later, and on the other side of the world, the 22-year-old reader of this paper would succeed its author in his Chair?

When I returned to Canberra, I noticed that a young vivacious girl had joined the group of young people who were living in Beauchamp House. Marion Hutley was doing research on the middle ear of marsupials for her B.Sc. (Hon) course at Sydney University and had come to Canberra to study a specimen from a wombat at the Australian Institute of Anatomy. To make a short story short, we were engaged within 2 weeks. The most important decisions in my life were made almost instantaneously. Whether all were wise is arguable – certainly it is possible, in some cases, to speculate what would have been the course of my life if another had been taken. There has never been any question in my mind that the quick decision on 15 January 1940, the day before my 23rd birthday, to ask Marion to marry me was the best one that I ever took.

It took me some time to get the thiochrome method working to my satisfaction, perhaps because of the hot summer and perhaps partly because of the distraction of spending much of my time writing love letters, and every 2 weeks travelling to Sydney by the late afternoon train on Friday and returning sitting up in the overnight train, arriving in Canberra just in time to go to work on Monday morning. Clements was now anxious to have some measurements of the thiamin content of Australian wheat and flour and fortunately I was able, after a few months, to get things going sufficiently for these determinations. However, I found that the visual method of estimating the intensity of the fluorescence was so subjective that it was essential to have two independent determinations of every sample. Fortunately, there were now plenty of funds available from the Health Department for what was considered of national importance and Clements was a good advocate. By the middle of the year, a recent graduate in biochemistry from Sydney University, Joan Rial, who happened to be a good friend of Marion and was her bridesmaid at our wedding, joined me. Soon after, Judie Ritchie joined the team to analyse wheat products for calcium and phosphorus, and a little later Kathleen Lundie and Ian Robertson, to help with the increasing number of thiamin determinations. Because of visual fatigue, an individual could only cope with a limited number of visual determinations of the fluorescence in a single day.

We soon found that Australian wheat has a considerably higher thiamin content than English wheat, which fact was thankfully used by the Australian Government in persuading the Ministry of Food in Britain to buy Australian wheat. From the Australian public health point of view, the more important finding was that, partly because of the higher amount in the wheat used, but also because of different milling procedures, white flour used for bread in Australia contained much more of the vitamin than in England, some samples approaching that of fortified British flour. There was no question then of adding the synthetic vitamin to the flour. This work was published in the autumn of 1941. I also published a paper on the application of the Wang and Harris procedure to milk and cereal products [4].

The most important work, scientifically, that Joan Rial and I did in this period was the most thorough determination, up to then, of the thiamin (including thiamin pyrophosphate) content of human milk. We [5] found that the amount of the vitamin in the milk from mothers whose infants were growing at a normal rate is very low in the first few weeks and reaches a fairly steady level after about 11–12 weeks.

From about the middle of 1940, the staff of the Biochemical Department increased, so that, by the end of 1941, I had four or five more biochemists in my Department and two technicians. Important additions to the Institute were a medical graduate from Sydney University, Colin White, and Bill Lockwood from Lemberg's laboratory, whom I found very knowledgeable and stimulating colleagues.

Although we were living in the capital city of a country at war, and the alternative to the two cinemas for an evening's entertainment was to sit in the public gallery of the Houses of Parliament, World War II was far away, until one early summer morning in December 1941, the news came through of the attack on Pearl Harbour. Since anything seemed possible after the surprise attack of the Japanese, we spent the morning at the Institute removing the most valuable scientific apparatus to the basement. As usual, I went home for lunch and saw a column of smoke from a nearby house occupied by a member of the Japanese Embassy, and soon afterwards the occupant himself, walking unattended across the fields towards the Manuka Shopping Centre. The capital of a country now fighting for survival was still only a village!

Soon afterwards came the most devastating news for all Australians, the sinking of two British battleships off the coast of Malaya, an event that marked the end of the British Empire in the East. A journalist friend of ours kept us informed about events, such as Prime Minister John Curtin's historic conclusion that Australia must now look to America for its salvation. In this atmosphere, our work on nutrition seemed too far away from what was required for national survival and I asked for a transfer to the Munitions Supply Laboratories in Melbourne, mentioning a vacation job that I had had in the Chemical Defence Laboratories in 1937. Given the great scarcity of chemists, it was clear that I would not be permitted to serve in the armed forces. The transfer was soon effected and in mid-February, the day Darwin was destroyed by a single Japanese Air Raid, we took the train to Melbourne, leaving our new house in the hands of an estate agent for letting.

World War II work interlude (1942–1943)

The Chemical Defence Laboratories were now terribly over-crowded and lacked even the most basic chemical apparatus. In the first days, I was asked to do some routine chemical analyses, involving a titration. The only glass vessel I could find as titration flask was a used milk bottle. However, an extension to the laboratory was being built with wartime speed and when we moved into it I was appointed Deputy Head of one of the main groups, consisting of about twenty chemists and technicians, with half of these under my direct responsibility. My group was responsible for field detection of war gases, measurement of concentration of gases in ex-perimental gas chambers, and examination of captured en-emy equipment. We now had adequate equipment, but our experience for this type of work was very limited and the very rapid expansion of the laboratory was really beyond the capacity of the leadership, which seemed mainly concerned with questions of departmental competence. At least three bodies – the Australian Army, the American Army and our-selves – seemed to fight over who should examine a piece of captured equipment, with the result that parts of it were of-ten sent to all three.

In any case, we all worked extremely hard and had the feeling we were really doing something useful. Up to the outbreak of the Pacific War, the only real task of the labora-tory had been to supervise and treat the manufacture of gas masks and capes, for use by the Australian Army in the Middle East. Otherwise, it was dependent on the British, particularly on the results of investigations carried out at Porton Down on the Salisbury Plain in England and en-shrined in the secret Porton Reports. Now, however, we were engaged in a war closer to home in which the two main participants, the Japanese and the Americans, were not signatories of the Geneva convention forbidding the use of chemical weapons. It was not surprising that Australia was now being combed for chemists and the few who had slipped into the forces were transferred to a new Army unit for chemical warfare.

At times the work was exciting. I particularly remember a day in January 1943, when I was called from my birthday lunch at Sandringham to report to Maribyrnong where a captured Japanese 75 mm shell had been received. This episode is reported at the opening of a Horizon Programme of the BBC, shown just before hostilities broke out in the Gulf War in 1991.[4] After an explosive expert (Dick Gillis) carefully unscrewed the cap (there was no fuse) and removed a paper thimble (which chemists used in the laboratory for extraction by organic solvents in a Soxhlet apparatus), filled with what was obviously picric acid, a hole was bored into the side of the shell and the liquid contents decanted. We then retreated to the laboratory where we set up a fractional distillation which showed that the contents of the shell contained two liquids in equal proportions, quickly identified as mustard gas and lewisite.

Another exciting moment, was being shown a telegram from Churchill to Curtin stating that the Russians had captured evidence that the Germans had a new chemical agent. I had to memorize the systematic chemical name until I could scribble its formula on a scrap of paper, quickly destroyed when I realized that it was what is now known as a nitrogen mustard, closely related to mustard gas.

The secret Porton Reports from England, now replaced the scientific literature as my regular reading. As a biochemist, I was particularly interested in the development, by the team at the Biochemistry Department in Oxford under Professor

---

4 *Keen as Mustard*, Transcript of programme transmitted 14 January 1991, BBC, 1991, p. 4.

R.A. Peters, of 2,3-dimercaptopropanol, known to us under the code-name DTH (for dithiol), as an antidote to lewisite. Peters, Thompson and Stocken had the brilliantly simple idea that the poisonous effects of lewisite is due to its reaction with two thiol groups close to one another in an enzyme involved in the oxidation of pyruvate, and that lewisite could be removed from its site of action by a simple dithiol. Early in 1942, Jim Lincoln, in my group, synthesised DTH, following the procedure described by Stocken and Thompson in a Porton Report, for use in trials. This was the first sample of DTH available in Australia. The Americans changed the code name to BAL, for British anti-lewisite, and this name has stuck. As I have earlier reported in this series [6], BAL was to play an important part in my subsequent career.

Some time during 1942, I was told that a certain Major Gorrill had arrived from England to train the Australian Army in anti-gas measures and that he was getting together a group of physiologists for whom a small experimental gas chamber had been constructed in a room taken over from the Department of Physiology at the University of Melbourne. It was my job to fill this chamber with phosgene and to draw off samples for measurement of its concentration. On a Saturday morning, with the help of another chemist, Lilia Hunt, and a technician, all of us wearing gas masks, we were ready to let the gas into the chamber, but I could not get the valve of the phosgene cylinder to budge. In desperation, I detached the cylinder from the chamber and with a bubbler of alkali to trap any small amounts of gas which might escape, attacked the cylinder with a monkey wrench, with the result that the valve was suddenly freed and, without my noticing it until alerted by frantic signals from the technician, phosgene at high pressure was released in quantities far exceeding the capacity of the trap. By the time I managed to close the valve again, a large amount of phosgene had escaped, also into the surrounding rooms. Fortunately, it was Saturday and Professor Wright was not in his office. Thereby, I was saved from gassing a future Chancellor of the University of Melbourne and one of the founders of the Australian National University.

Gorrill's 'physiologists' turned out to comprise an anatomist, two biochemists and, indeed, one physiologist. The biochemists were Hugh Ennor and Jack Legge. I was only incidentally involved with this group, who had a much more interesting war than I did. When they moved to Townsville in Queensland, one of my team, Jim Lincoln, was detached to do the determinations of gas concentrations. After completion of a series of experiments with mustard gas in the gas chamber, Jim and some others were exposed to a small concentration of the vapour while cleaning out the chamber. They were badly blistered, temporarily blinded and suffered extreme nausea. This was the first indication that mustard gas was much more efficient in the tropics. This has been described in detail by Legge in the Horizon Programme of the BBC.[5]

As Legge has also described, Gorrill was a remarkable man, of the type who come to the fore in wartime, with a single-minded devotion to his task, which I always assumed to be to build up the defences against a possible Japanese attack by chemical weapons. The Brook Island Trials did not take place until early 1944 and it was only when I saw the BBC Horizon Programme in 1991, that I understood that General Douglas MacArthur, the Allied Supreme Commander in the South Pacific, was in fact considering the possibility of using of mustard gas to drive the Japanese troops from their foxholes, and that perhaps it was only the development of the atomic bomb that prevented this.

In ignorance of these developments, I had decided by the late winter of 1943 that I was wasting my time working on chemical warfare, since, in my view, chemical weapons would not be employed in the war. There may have been a

---

5 *Loc. cit.*, p. 5.

certain amount of rationalisation in my reasoning, since I was tired of my work at Maribyrnong. Although my relations with my immediate chief, Dick Neale, were excellent, I was not very popular with those above him and my vanity made me feel that my efforts were insufficiently appreciated. When, therefore, Clements informed me that he wanted me back in Canberra to take charge of a major project in analysing the Operation Ration for the Australian Army, I accepted.

Return to Canberra

We were very happy to return to our house and garden in Canberra, although having four lots of wartime tenants, it no longer possessed its pristine newness. Among the biochemists at the Institute, Kathleen Lundy had stayed on and the Department was soon extended by appointments of recent graduates of Sydney University – Cliff Kratzing, David Morell and, somewhat later, Keith Rienits.

Since the analysis of the Operation Ration did not supply much intellectual stimulation, I sought additional projects. I had been struck by Woods' hypothesis that sulphanilamide owes its bacteriostatic action to competition with *p*-aminobenzoic acid, a bacterial growth factor. There were some suggestions in the literature that the antimalarial Atebrin might act in a similar fashion by competing with the vitamin riboflavin. Since the allied troops serving in the Pacific were all receiving large quantities of Atebrin, it seemed advisable to include riboflavin among the vitamins to be analysed in the Operation Ration. The current analytical procedure, like that for thiamin, was also a fluorimetric one, based on the fluorescence of riboflavin itself. Since, however, foodstuffs and biological fluids contain numerous substances fluorescing in the same spectral region, most procedures were very non-specific and greatly over-estimated the riboflavin content. David Morell and I spent some time in adapting a procedure, introduced by Najjar, of extracting the riboflavin by pyridine-butanol. The main novelty in our method was to use controlled destruction of riboflavin by the bright Canberra sunlight on the roof of the Institute. We found that the kinetics of this destruction were so unusual that we could discriminate between destruction of riboflavin and fluorescent contaminants of the pyridine-butanol extract. This work was not written up until shortly before I left Australia and comprised my first publications in the *Biochemical Journal* [7, 8].

In my experience, the satisfaction that one gets from solving a scientific riddle is largely independent of the importance of the result, scientific or otherwise. A short paper by Morell and myself is a case in point. In connection with another investigation, it was necessary to store human urine for several weeks before analysis. Although riboflavin is quite stable, we found appreciable losses (up to 30%) of riboflavin in human urine, particularly if it was stored at low pH and at 4°. We found that this is due to crystals of uric acid, that were formed during storage, taking up large amounts of riboflavin. Much later, we learned that riboflavin forms a charge-transfer complex with compounds such as uric acid.

My most important research in this period also had its origin in the Woods hypothesis. Before the days of antibiotics, sulphonamides were widely used to combat bacterial infections. Indeed, the availability to the Australian troops in New Guinea of a small amount of sulphaguanidine, synthesised in the Organic Chemistry Laboratory at Sydney University to combat dysentery, was probably an important factor in stopping the Japanese advance to Port Moresby after they had successfully crossed the Owen Stanley Range to the north. Unlike the sulphonamides used in the treatment of non-intestinal infections, sulphaguanidine is poorly absorbed from the gut and has little side-effects. I was more interested in the possible causes of the side-effects, including nausea, of the easily absorbable sulpha drugs, such as sulphathiazole or sulphadiazine. These compounds have some chemical resemblance to thiamin, and it seemed to me possible that they interfere with thiamin metabolism. Since it was known that the urinary excretion of thiamin is a good measure of the nutritional status with respect to this vitamin, I decided to study the effect of sulphadiazine on this excretion during a period on a constant diet. For the latter, I chose to kill two birds with one stone, so to speak, by trying out the Operation Ration.

After 7 days on this ration, I undertook a standard course of sulphadiazine for the next 5 days, followed by a recovery period of 7 days. Even before taking the sulphadiazine, I found the ration, which included such exotics as carrot biscuits with peanut butter, a heavy fruit and nut block, and chocolate fortified with oatmeal and synthetic vitamin B_1 for breakfast and large amounts of barley sugar extremely heavy going, and I wondered how the troops would ever find time to consume it while fighting the Japanese, but I decided that at least the fruit and nut block was a useful missile. The most difficult part of the experiment for me was to consume the entire ration, while nauseated as a result of the sulphadiazine.

When I analysed the urine samples, I found a large increase in the excretion of thiamin both during, and subsequent to, the period of sulphadiazine supplement. Riboflavin excretion was unaffected. In view of this clear-cut effect, although its basis was obscure, I decided to send a letter to Nature, the usual manner then of communicating preliminary results, and was gratified to have it accepted [9].

Two possible explanations were apparent. Either the sulphadiazine prevents the uptake of thiamin into the tissues, perhaps by inhibiting its conversion to its coenzyme form, thiamin pyrophosphate, or the sulphadiazine has a 'sparing' action on the thiamin, such as had been described for high-fat diets. These explanations could be distinguished by analysing the tissues for the vitamin, since if the former explanation were correct, the amount of thiamin in the tissues would be lowered by the sulphadiazine treatment, whereas the reverse would be expected if the second explanation were correct.

Clearly, this experiment could not be carried out even on the enthusiastic experimenter and it was necessary to use animals, namely rats, a new experience for me. Cliff Kratzing joined me in these tedious experiments, which lasted months before we had a single result. The results of the first experiments clearly showed that the second of the two possible explanations was the correct one, but certain controls still needed to be done by the time I left Canberra and these I entrusted to Cliff Kratzing. Also we had, at that time, no idea why sulphadiazine has its sparing effect. While in England, I came across papers by Astwood and McKenzie showing that sulphadiazine, and the closely related sulphamerazine, but not other sulpha drugs, suppress thyroid action, so I asked Cliff to set up further experiments to compare the effect of different sulpha drugs. The results showed a clear correlation between the sparing action on thiamin and the effect on the thyroid described by Astwood and McKenzie. Given that it was known that the requirement for thiamin as a vitamin is related to the rate at which carbohydrate is consumed, it makes sense that the requirement would be lessened when metabolism is lowered, by depression of thyroid function. The work was finally published in 1950 [10].

Although the results of these extensive and time-consuming experiments were of little or no importance, I have never written a paper that gave me more satisfaction, since, within limits, it gave a completely rounded-off solution to the problem posed by the observation made with myself as experimental animal. 'Within limits'..... since no problem is completely solved. The solution of the question why specifically sulphadiazine has its action on the thyroid could be left to the endocrinologists. Why the requirement for thiamin should depend on the rate of biochemical reactions that it catalyses, as implied by the sparing effect of decreasing the rate of carbohydrate metabolism, although superficially it might seem reasonable, remained for me both mysterious and fundamental for nutritional science, but since I was then moving away from nutrition, I put this problem aside. I am not aware that it has as yet been solved.

There was not, at that time, a university in Canberra, apart from a small college affiliated with the University of Melbourne, mostly giving evening classes in economics or business administration. For a few years, until he took up an appointment as lecturer in the University of Melbourne, Bill Boardman ran a practical class in first-year zoology and Marion became demonstrator. We were enthusiastic members of the University Association, whose aim was to obtain a university for Canberra. In 1944, Florey visited Australia, at the invitation of the Prime Minister, to lecture on his work on penicillin and to report on the state of medical research in Australia. I remember vividly the excitement of his lecture in Canberra, which was given at our institute. Florey's recommendation, to set up in Canberra an Institute for Fundamental Medical Research was accepted by the Australian Government, but was later expanded into the creation of the Australian National University, with schools in physics and Pacific Studies, in addition to what became the John Curtin School of Medical Research. Visits by distinguished scientists from overseas were, of course, seldom. The only other one that I can remember, was Joseph Needham, Reader in Biochemistry at Cambridge, who was scientific attaché at the British Embassy in Chungking.

Shortly after the end of the war, Clements decided to run a course for nutritionists, in which I gave lectures on the biochemical basis of nutrition.

The Institute was built around three sides of a quadrangle, two of them occupied by the two museums, and the third by the laboratories, administration and library. Clements had plans for extensions to the Institute by completing the fourth side of the quadrangle and I spent some time planning these. Clements was very optimistic about the future of the Institute as a research laboratory for the Commonwealth Department of Health and had the support of the Director General of Health, Cumpston. Clements agreed that I should obtain further training abroad and that I could have unpaid leave of absence for this purpose. I first wrote to Elvehjem at Wisconsin (USA), the leading nutritional biochemist at that time, and was accepted in principle, but I would have to find my own support. The British Council then announced six scholarships for study in England, covering a wide field in science, humanities and the arts. The Imperial Chemical Industries had also made available some scholarships for overseas students at a number of British universities. I applied for some (two, I think) ICI Scholarships and a British Council Scholarship, and was successful in the latter. At the suggestion of Professor Priestley, Professor of Biochemistry at the University of Sydney, I wrote to Professor Keilin at the University of Cambridge, which was the Mecca for biochemists at that time, to ask for a place in his laboratory and was accepted. Although I had stated in my application my wish to work under Keilin, I was informed by the British Council that so many of their scholars wished to study in Oxford or Cambridge that it was their policy to send some to other universities and that Professor Krebs at the University of Sheffield had accepted me. I did not complain, since although Sheffield was not Cambridge, Krebs was also a very famous biochemist. However, I felt a little embarrassed and wrote to the British Council to the effect that, although I would be pleased to go to Krebs, I had already been accepted by Keilin. Since I had received no answer to this letter before we sailed to England, after leaving our house in the hands of a letting agent, we left Canberra in early August 1946 bound, as we thought, for Sheffield.

Cambridge 1946–1949

The trip to England, in the *S.S. Orbita*, was far from the comfortable sea-voyage that was the feature before World War II of 'going home', as Australians called a visit to England. It was run by the British Ministry of (War) Transport ('War' being crossed out) and was still fitted as a troop-ship without separate accommodation for married couples. Marion was allocated a cabin with thirty-five other women and I was in a six-berth makeshift cabin in the hold, which was so stuffy that, after the first week or so, I slept on deck curled up in two deck-chairs. However, it was an interesting trip. Highlights in my memory are: rounding Wilson's Promontory in a storm, when I could identify to myself (and everyone else in earshot) the islands along which I had sailed with my father; the night sky from my deck chairs on the boat deck; my first experience of the sun in the 'wrong' part of the sky, when I unconsciously incorrectly placed my deck-chair to stay in the shade; the cliffs of Aden at dawn (first sight of land for more than two weeks); and the trip through Suez Canal. Among the passengers were survivors of Japanese Internment camps, families who had been evacuated to Australia ahead of the Japanese advance, returning British servicemen, a contingent of war brides, and a large group of young academics, who like ourselves, were off to England for further study, among whom were some very bright people, including Reg Goldacre, Ron Bracewell and Joan Freeman (later Jelley). We formed a lifelong friendship with Joan, who has given a good description of life on the *Orbita* in her book [11]. She later became a recipient of the Rutherford Medal of the Physical Society. Ron Bracewell, who became Professor of Electrical Engineering at Stanford University, discovered the first practical algorithm now universally used in CAT scanners. We spent most of our time with this group, either in scientific discussions or playing chess.

In order to prepare myself for work in Sheffield, I read the long and authoritative review by Krebs in *Advances in Enzymology*, vol. 3, on what he called the tricarboxylic acid cycle, but which everyone else called the Krebs cycle. The day we passed Malta, with the ship's lifeboats swung out because the Mediterranean was not yet free of mines, the Purser gave me a letter, which must have come aboard in Australia 3 weeks previously. This was from the British Council stating that, since I had been accepted by Keilin, they had no objection to my going to Cambridge instead of to Sheffield! Although, if I had known, I would have spent my time reading Keilin's work, I never regretted my study of Krebs' review, since I would probably never otherwise have given it the same intensive attention as it received during the long journey to England. I was very embarrassed about this episode when I first met Krebs, but he was perfectly friendly and we became good friends. I have often wondered what my career would have been if I had joined Krebs instead of Keilin.

Molteno Institute

To my great surprise, I found that Jack Legge was also at the Molteno Institute. For the first 4 months, we shared a house with the Legge family. We were then able to move to a flat of our own at 5 Shaftesbury Road, which later housed the unit of the Medical Research Council, in which our daughter worked for a time, another of the circles in my life.

At my first meeting with Keilin, I told him that I hoped to obtain a Ph.D. under his supervision, after which I would return to Canberra in order to study the biochemical mechanism of diseases caused by vitamin deficiencies, and in particular the effects of drugs on nutritional requirements. It did not take Keilin long to realise how little I knew of modern biochemistry[6] and he gently suggested that perhaps it would be wise first to work on a topic closer to those at present pursued in the Molteno Institute. He suggested that I might be interested in taking up the study of the mechanism of cyanide resistance in yeast, but first, in order to gain a little more experience, I could undertake a small project, namely to repeat Hopkins and Morgan's experiment on the inhibitory effect of oxidized glutathione on succinate dehydrogenase and its reversal by glutathione.

A number of articles on the Molteno Institute, including one by myself on how I found it in 1946, have appeared in *The Biochemist*, the Bulletin of the Biochemical Society [12]. It owed its existence to the parasitologist, George Nuttall, who was Quick Professor of Biology in Cambridge, from 1906 to 1932. The Quick Chair in biology is a research Chair, unattached to a department and with no formal teaching duties. Until he persuaded Percy Alport Molteno, who had interests in South Africa, and his wife, Elizabeth Martin Molteno, to make a gift of £31 000 to build an Institute of Parasitology[7] (and an additional £10 000 to provide an income for its upkeep), Nuttall carried out his research in the Quick Laboratory in the Department of Zoology. David Keilin, an entomologist already well-known for his work in Paris on dipterous larvae, was appointed research assistant to Nuttall in 1915. He moved with Nuttall to the newly constructed

---

6 He asked me if I had studied David Green's monograph on respiratory enzymes. I answered truthfully that I had glanced at it in our library at Canberra, but had not studied it. I did not tell him why I had taken another book to read on the voyage to England!

7 A very full and (fulsome) account of the official opening of the Molteno Institute on 28 November 1921, including a long list of those present was published in *The African World* on 3 December 1921. Mr. C. Warburton, Demonstrator in Parasitology, deputised for Nuttall, who was ill. Warburton's name still appeared on the top left-hand corner of the attendance board at the Molteno Institute until his death, shortly before his 105th birthday, in 1951 or 1952. Soon after our arrival in Cambridge, there was a party in the Molteno Institute for his 100th birthday, which he celebrated by correctly solving the *Times* crossword.

Molteno Institute in 1921, and was appointed University Lecturer in Parasitology in 1925. When Nuttall retired in 1932, Keilin succeeded him, both as Quick Professor and Director of the Molteno Institute.

Paradoxically, it was in the same year that he was appointed Reader in Parasitology, that Keilin made the discovery – the role of cytochrome in intracellular respiration – that made him one of the leading figures in biochemistry of his generation. However, he retained his interest in parasitology and during my period at the Molteno Institute he published an extensive study of respiratory systems in dipterous larvae. Also, he was Editor of the journal *Parasitology* from 1934 to his death in 1963. When he was appointed Director of the Molteno Institute, 'and Biology' was added to the title of the Institute in order to take account of its wider field of interest. When Keilin reached the compulsory retiring age (65) in 1952, the link between the Institute and the Quick Chair was broken with the appointment of Wigglesworth, an entomologist working in the Department of Zoology as Director of an Agricultural Research Council Unit, to the Chair. After 1952, the current University Readers in Parasitology (Parr Tate, succeeded by Newton) were Directors of the Institute, which continued to house various research groups including that of Keilin himself, until the day of his death in 1963. The most important research to come out of the Institute in recent times was that of George Cross, now at the Rockefeller University in New York, on the molecular basis of antigenic variation in trypanosomes. After the retirement of Newton, the Molteno Institute ceased to exist, a Molteno Laboratory of Parasitology being incorporated into the Department of Pathology.

Our life in Cambridge at that time was not very comfortable, but it was very exciting to be living and working in such a famous and beautiful place. To most Australians, the first winter in England is quite a traumatic experience. It is not so much the cold (although the winter of 1946–1947 was exceptionally severe),[8] as the overcast skies and the early darkness, which we were quite unused to. Also, although we had expected that England would be very different from Australia, we were surprised to find that the English are very different from Australians and it took us some time to adjust to this fact. Food rationing in that first winter was still very severe, extending for a time even to bread, which had not been rationed even during the darkest days of World War II. Regular food parcels received from our families in Australia were a great help. The British Council had arranged for my entry into Trinity Hall and I dined one night a week there during term-time. I had one meeting with my tutor, but the college was not interested in older, married research students and I participated little in college life, except to use the tennis courts.

At Keilin's suggestion, I attended the lectures given to Part II (third year) students in biochemistry, as well as his own course on respiratory enzymes, which were officially for physiology students, but which were also attended by biochemistry students. These lectures, most of which were given by specialists, were of great help to me in filling my many gaps in my knowledge of biochemistry. The students for whom the lectures were intended were, in fact, greatly outnumbered by research students, visiting research workers and staff members, and the only criticism that I had was that there was a tendency for some of the younger lecturers to forget that the function of the lectures was to teach undergraduates and not just to impress their colleagues with their knowledge. I remembered this when I introduced similar lectures in Amsterdam. In any case, I found it a

---

8 Our flat was bitterly cold. The only heating that we had were from two electric radiators. Owing to a coal shortage during the Freeze, the use of electricity for heating and cooking was restricted during the day to 2 h around lunch-time. Joan Freeman [11] tells the story that she left her hot-water bottle in bed and when she tried to empty it the next evening nothing came out, because the contents were frozen.

great privilege to attend lectures by Malcolm Dixon and Edwin Webb on enzymes, Dorothy Needham on muscle, Kenneth Bailey on proteins, David Bell on carbohydrates, Marjory Stephenson and Ernest Gale on microbial metabolism, Philip George on enzyme kinetics, and Leslie Harris on nutrition. I still have the notes, typed out by Marion, that I made on these lectures. The weekly seminars (Tea-Clubs, they were called) in the Biochemistry Department, were also a delight. In my first term, both Fritz Lipmann and Carl Cori, legendary figures to me (I was later to get to know them quite well) gave talks.

For several weeks in early 1947, no heating was available for the university and it was impossible to do experiments. The contents of the reagent bottles froze on the shelves. I used the time in the library of the Biochemical Department, thoroughly wrapped up and with a heavy overcoat, reading all the literature on succinate oxidase and related topics. At that time, it was still possible to say that one was completely familiar with the literature in one's field. Over the next few months, I read every paper that Keilin had published, except those on parasitology, even when they did not immediately impinge on my research topic. I found this invaluable in future discussions with Keilin and followed the same principle when I was working in the laboratories of Ochoa and Chance.

One of our first purchases were bicycles, a must in Cambridge at that time, as indeed they had been in Canberra. Despite the cold and joining the grumbling of our Australian friends about what a terrible place England was, Marion and I loved snow and we had never seen it before in a town away from the mountains. Also I was buoyed up by the results I was getting. Measurements of succinate oxidase activity, by a simple manometric technique, could be done so quickly that every evening I had more results to work out and think about than after months of experiments with rats. Days (including all-day Saturday) were for making a Keilin and Hartree heart-muscle preparation or doing manometric experiments. Evenings were for calculations, writing up the results in a note-book and planning the next day's experiments. Often at the end of a long day, Keilin would waylay me, as I was on way home for the usual Australian six-o'clock dinner, to ask about my work. These talks with Keilin were, for me, the most wonderful memories of my stay in Cambridge. Apart from the fact that it was the first time anyone, let alone a great scientist, had shown such a knowledgeable interest in what I was doing, I learned so much from him about the way to do research. For me, trained as a chemist, it was a revelation to hear a biologist's approach to science. These discussions often meant that the tiny piece of meat that Marion had bought with saved-up coupons was over-cooked by the time I cycled home and 'Sorry, I was talking with the Professor' became a by-word in our family.[9]

Suddenly, there was the Cambridge spring with, first the crocuses and then the daffodils on the Backs, watching first-class cricket at Fenners, playing tennis, punting on the Cam on the Backs and to Grantchester, cycling to the neighbouring villages, drinking beer in the Orchard at Grantchester, listening to the Madrigals in May week, and going to the Arts Theatre where, often, new plays were tried out before being put on in the West End. I do not know how we managed to fit everything in, since I was, in fact, working long hours in the laboratory. Cambridge had so much to offer.

Since I have published, in another volume of this series [6] a detailed account of my first research in Cambridge, I shall be brief here. As already mentioned, my first project was to repeat with the Keilin and Hartree heart-muscle prepara-

---

9 Very good accounts of Keilin and the Molteno are given by Ted Hartree, Keilin's longest collaborator, in *Of Oxygen, Fuels and Living Matter, Part 1* (G. Semenza, ed., Wiley, 1981, pp 161–173) and by Max Perutz in the *Cambridge Review* (October, 1987, pp. 152–156).

tion, the experiments of Hopkins and Morgan on the inhibition of succinate dehydrogenase by oxidised glutathione (GSSG) and its reversal by glutathione (GSH). Sure enough, I found that GSSG slowly inhibited the succinate oxidase activity of the heart-muscle preparation. However, attempts to reverse the inhibition by incubation with GSH were only partially successful, since to my surprise I found that incubation with GSH of heart-muscle preparation, untreated with GSSG itself, caused a strong inhibition of the succinate oxidase activity. Since I observed that oxygen was consumed during the treatment with GSH, this inhibition could have been simply due to GSSG formed by the oxidation, but I was not satisfied with this explanation, since the inhibition by GSH was greater than that obtained with GSSG. In any case, it was clear that experiments on reversal should be carried out under anaerobic conditions and the GSH should be removed before measuring the enzyme activity.

I undertook a systematic study of the inhibition of succinate oxidation by different types of compounds, including an arsenical known to react with thiol groups in proteins, and the reversal of the inhibition by GSH and the thiol, 2,3-dimercaptopropanol. As previously mentioned, I was already acquainted with the latter compound, under the code names of DTH and BAL, as an antidote to the chemical-warfare agent lewisite. BAL was found to be even more inhibitory than GSH, but it was possible to overcome this by incubating concentrated arsenical-inhibited particles with BAL under anaerobic conditions, and then diluting about 50-fold before measuring the rate of succinate oxidation manometrically. This formed the basis of the fifth of a series of papers published in the *Biochemical Journal* in 1949 [13]. However, by the end of March 1947, I was much more interested in studying the inhibition by GSH and BAL. Since my stocks of GSH, which I had to isolate from yeast, were almost exhausted and BAL, a neutral compound, was much more convenient, all subsequent experiments were carried out with BAL.[10]

It was then known that biological oxidations require a specific enzyme (dehydrogenase) for the oxidizable substrate and a chain (called the respiratory chain) of cytochromes acting as electron carriers in the sequence $b \to c \to a \to a_3 \to O_2$. I was soon able to show that the site of action of BAL, unlike that of GSSG or arsenicals, is not on the dehydrogenase but elsewhere in the respiratory chain. The first thing that one did in the Molteno Institute with a new inhibitor of succinate oxidation was to look at its effect on the absorption bands of the cytochromes, making use of Keilin's microspectroscope. To my great excitement, I found that, after BAL treatment, the absorption band of reduced cytochrome b remained at full intensity after addition of succinate while those of cytochromes a and c had entirely disappeared. I showed this to Keilin and anyone else close enough to be called in to his microspectroscope. I concluded that the treatment with BAL had irreversibly inactivated an hitherto unknown component of the respiratory chain acting between cytochromes b and c.

It took a lot of further work to convince myself that the effect of BAL is specific and that the inactivation is brought about by the coupled oxidation by oxygen of BAL and a BAL-labile factor in the respiratory chain, and it was not until early 1948 that I was ready to publish a preliminary account in *Nature* [15]. The factor became known later as the 'Slater factor'.

During my first year in Cambridge, I became a member of both the Biochemical and Nutritional Societies and, in the next few years attended many of their meetings as well as some of those of the Physiological Society. I found these

___
10 This turned out to be a lucky choice. In later experiments, no inhibition was found with a purer preparation of glutathione. It is likely that the active agent in our preparations was an impurity of lower molecular weight, probably cysteinylglycine.

meetings most useful, not only in widening my knowledge of biochemistry but also in seeing in action well known names in biochemistry, as well as making friends with many contemporaries from outside Cambridge, for example, Bob Davies, Jo Stern and Henry McIlwain from Krebs' laboratory in Sheffield. These meetings also gave an opportunity of seeing other parts of the British Isles. My first glimpses of Wales, Scotland and Ireland were while attending such meetings.

In the summer of 1947, Marion and I travelled by ship to Stockholm to attend the International Congress of Experimental Cytology, organized by Rünnström, Director of the Wenner Grens Institute. This, the first of very many congresses that I have attended, was a wonderful experience, starting with the luxury of a double cabin in a ship of the Swedish Lloyd Line sailing from Tilbury to Göteborg, so different from the austerity of the *Orbita*, followed by the fascinating train journey to Stockholm. After the drab immediate post-World War II years in Britain, with its cities still showing great gaps, either still left as undeveloped bomb sites or covered in ugly pre-fab temporary shops, Sweden seemed to be out of this world, as was the hospitality of our Swedish hosts – there was a reception and/or a banquet every evening, including one in the magnificent Town Hall, famous for the Nobel Prize dinners. Since it was close to mid-summer, it was already broad daylight by the time we got back to the rooms that had been reserved for us in the centre of Stockholm.

This congress was the first international meeting in biology since the end of World War II and many colleagues from the victorious and neutral countries saw one another for the first time for many years. No-one from Germany or Japan were allowed to attend – I cannot remember whether there were any Italians. All the top figures in muscle physiology from before World War II, except Weber (in fact, an active anti-Nazi) from Germany, were there. There was a fasci-

nating lecture from Albert Szent-Györgyi, followed by a discussion that occupied the whole afternoon, so that the
Chairman, von Muralt, had to move the remainder of the
programme to the next day. Hugo Theorell reported some
experiments carried out with Britton Chance in his laboratory in Stockholm. For me, the most interesting paper was
that of Albert Claude, who reported that cytochrome oxidase
is located in the mitochondria. Among the many whom I met
for the first time at that meeting, I recall especially Britton
Chance, Arthur Kornberg, Roger Bonnichsen, René Dubos
and Sol Spiegelman.

On the boat trip back to England, we spent a lot of time
with two young Americans, Jack Buchanan and Dick
Abrams, then working in Theorell's laboratory. I travelled
directly to Oxford to attend the XVII International Congress
of Physiology. The most interesting paper from my point of
view was that by Ogston and Smithies on the thermodynamics of oxidative phosphorylation, in which they concluded that Ochoa's value of 3 for the P:O ratio for the oxidation of pyruvate could not be accepted. Although I was not
then sufficiently acquainted with the literature to have any
views on the matter, I talked with both Smithies, then an
undergraduate, and Sandy Ogston, so this meeting may be
considered to be my first introduction to the field of oxidative phosphorylation, which was to be the dominant theme of
my research career.

Britton Chance also attended the congress in Oxford and
afterwards spent several weeks setting up his self-made
photoelectric spectrophotometer in a room on the ground
floor of the Molteno Institute, in order to carry out some experiments with Keilin on catalase. In those days, such apparatus with amplifiers and voltage stabilizers occupied considerable space and looked most impressive in a laboratory
not acquainted with such sophistication. I can well remember Keilin standing in the doorway with a look of astonishment at what was replacing his micro-spectroscope. As re-

called elsewhere [14], a number of new faces also joined us at this time, including the first post-World War II American, Dan Arnon, already well-known as a plant physiologist and who was later to discover photosynthetic phosphorylation, Alfred Tissières and two Chinese, Chin and Chen-lu Tsou.

I returned to my study of the BAL-labile factor. As mentioned above and described in detail elsewhere [6], I was much concerned with the specificity of the effect of BAL and spent much time trying to define criteria for distinguishing between non-specific and specific inhibitors of the respiratory chain. These efforts and a description of other properties of the succinate oxidase system are described in two papers [16, 17], published together with the full paper on the BAL-labile factor [18]. An earlier paper [19] on the measurement of the cytochrome oxidase activity of the Keilin and Hartree heart-muscle preparation was an attempt, in the absence of any knowledge of the structure of the particles, to grapple with the problems of the accessibility of the large protein, cytochrome c, to the cytochrome oxidase firmly bound to the particles and the puzzling fact, already known from earlier literature, that some electron donors, such as p-phenylenediamine, could be oxidised by cytochrome c bound to the particles (the so-called endogenous cytochrome c), whereas others, such as ascorbate, could not. These five papers were first written as chapters for my Ph.D. thesis, which also included a long historical introduction, submitted in the spring of 1948. My examiners were Professor Sir Rudolph Peters and Dorothy Needham.

After finishing the five full papers for the *Biochemical Journal* on the succinate oxidase system, I turned my attention to the system responsible for the oxidation of what is now known as NADH (see Ref. [6]). The Molteno Institute had just received its first photoelectric ultraviolet spectrophotometer, which I found very suitable for studying the oxidation of NADH, catalysed by the Keilin and Hartree heart-muscle preparation, by following the disappearance of its absorption in the near ultraviolet at 340 nm. This was, in fact, an unusual use of a photoelectric photometer, since the heart-muscle preparation, being a suspension of particles, scattered light, thereby interfering with the measurement of the light absorption. However, by adding the same amount of diluted heart-muscle preparation to both the reference and measuring cuvettes, this interference could be overcome. Nowadays, this seems an obvious thing to do, but at the time it horrified spectroscopists, who were careful to remove all turbidity from the solution to avoid any light scattering. In fact, by the same procedure, I was even able to plot the absorption spectrum of the cytochromes in concentrated heart-muscle preparation without adding any dispersants. I believe that I might have been the first to do so, but did not publish it. Later, Britton Chance developed a differential spectrophotometer, which enabled much more accurate measurements.

I made a systematic study of the NADH oxidase system, repeating for this system all the experiments made with the succinate oxidase system by Keilin over the years and in my recent studies. Many of the findings confirmed scattered findings in the literature, but this was the first systematic study of the system. New was the finding that the BAL-labile factor is involved. This caused me quite a lot of trouble over the next few years, because it was widely believed that soluble reductases directly catalyse the reduction of cytochrome c by NADH and it was a period in which soluble enzymes held sway among biochemists. It is now known that the soluble enzymes then being studied catalyse other reactions not connected with the main pathway of intracellular respiration. An important difference between the succinate and NADH oxidase systems seemed to be the involvement of cytochrome b in succinate but not in NADH oxidation. I spent a lot of time staring down the microspectroscope. I was also able to observe, for the first time, the oxidation of

NADH by fumarate. Since this was a slow reaction, partially inactivated by treatment with BAL, I concluded that the BAL-labile factor links the succinate and NADH oxidase systems and proposed the following version of the respiratory chain:

Succinate dehydrogenase

$\downarrow$

Succinate → cyt. b → factor → cyt. c → cyt. a → cyt. a_3 → O_2

NADH → diaphorase

Diaphorase was the name then given to the enzyme responsible for the oxidation of NADH and it was believed that a soluble flavoprotein, isolated in the Molteno Institute by Bruno Straub just before World War II, that catalysed the oxidation of NADH by methylene blue, was identical with this enzyme.[11] More than 10 years later, it was found by Massey that the soluble enzyme was in fact part of the enzyme complex responsible for the reduction of NAD^+ by lipoic acid, a compound unknown in 1949–1950. It was still later before the flavoprotein responsible for the oxidation of NADH was characterised.

The above scheme was first reported at a meeting of the Biochemical Society in Oxford on 7 May 1949 and later to the 1st International Congress of Biochemistry in Cambridge. A longer summary was published in *Nature* [20] and the full accounts of this work in the *Biochemical Journal* [21]. This scheme was generally well received. Now it is clear that it was faulty in several respects. First, it is now

---

11 See F.B. Straub, *Of Oxygen, Fuels and Living Matter, Part 1* (G. Semenza, ed.), Wiley, 1981, pp 325–336.

known that it is not the factor that is the link between the succinate and NADH oxidase systems, but ubiquinone, which was not discovered as a component of the respiratory chain until 1957. The slow oxidation of NADH by fumarate is not due to the unfavourable energetics of reduction of cytochrome b by the factor, as I thought, but to the kinetics of mammalian succinate dehydrogenase. Cytochrome b is now known to be involved in the oxidation of NADH as well as of succinate. My difficulty in observing the reduction of cytochrome b by NADH is now explained by the Q cycle, proposed by Mitchell in 1977, according to which cytochrome b is on a shunt, being reduced by ubihydroquinone and oxidized by ubiquinone. Succinate dehydrogenase, which is shown as activating succinate in the above scheme, was soon shown by Tsou in the Molteno Institute to be itself a hydrogen carrier. Although its nature was not established until 1982, the existence and proposed function of the BAL-labile factor are correct, although the immediate electron acceptor of the factor is cytochrome c_1 rather than cytochrome c. I have referred elsewhere [22, 6] to my embarrassing mistake [23] in denying the existence of cytochrome c_1.

In the summer and winter of 1948, I presented two more papers to the Biochemical Society at Glasgow and London, respectively, on two of the chapters of my thesis. The Glasgow meeting was memorable for the announcement by Lester Smith that vitamin B_{12} contains cobalt, which he demonstrated by projecting a borax bead, which most of us had not seen since our first-year chemistry practicals. In the autumn, Marion and I travelled to Paris to attend the 8th Congress of Biological Chemistry, organised by the French Society of Biological Chemistry. This was a meeting of 'firsts': our first visit to the European mainland, not counting the Scandinavian Peninsular; our first visit to a country that had, until a few years previously, been occupied during World War II; my first paper given to an international meeting.[12] Although these meetings were planned for French-speaking biochemists and the previous one had been held in Liége in 1946, they were attended by many bio-chemists from other countries, including Britain and the USA. Indeed, they can be considered the fore-runners of the International Congresses of Biochemistry (and Molecular Biology). It was probably also the last of the relaxed meet-ings with the afternoon sessions beginning at 1530 h. Even so, we often arrived very late after an extended lunch hosted by our French friends.

The congress was opened by Gabriel Bertrand, a figure straight out of 19th century French science, looking just like Louis Pasteur. It was Bertrand who first introduced the word 'oxidase' into the biochemical literature. From the Mol-teno Institute, Joan Keilin, Professor Keilin's daughter, as well as Ted Hartree, Alfred Tissières and Leo Levenbook, also attended the meeting and, especially the first named, opened for us the doors to the French biochemical estab-lishment, including L. Rapkine, J. Roche, R. Wurmser, A. Lwoff and M. Polonowski. We visited the Pasteur Institute, where we met Jacques Monod, and the Institut de Biologie Physico-chimique, to which Jean Rosenberg had returned after a year in the Molteno. There we met Aubel, Marianne Grunberg-Manago and also 'Fitzi' Lynen, who happened to be visiting the Institute at the same time.

The next exciting happening was the 1st International Congress of Biochemistry in Cambridge on 19–25 August 1949, attended by most of the leading biochemists in the world. Although not as relaxed as the Paris meeting, it was still possible to follow the lectures that one was interested in

---

12 Jean Courtois, the Secretary-General of this meeting, often recalled that this was the first of many meetings that he organised and that my paper was the first that he received. My paper was given in English but translated into French in the published Proceedings of the Congress under the title *Un catalyseur respiratoire exigé pour la réduction du cytochrome c par le cytochrome b.* Hugo Theorell was Chairman of my session.

without having to rush from room to room. Ordinary communications were assigned 20 minutes, in place of 10 and
even 7 minutes that became customary, before uninvited
speakers were banished to poster sessions. Never having
seen Keilin at a scientific meeting, I was most impressed
with the fact that he attended every paper in the section devoted to haem compounds and biological oxidations and had
a question for every speaker. After Chen-lu Tsou's paper on
pepsin-digested cytochrome c, Keilin and Theorell had a vigorous discussion on the iron content of purified cytochrome c,
about which a controversy had existed in the literature for
some years. This was one of the few occasions on which
Keilin was wrong.

Australian National University

Marion and I were still hoping to return to Canberra. My
British Council Scholarship was renewed for a second year,
during which I was in regular correspondence with Clements
whom I had asked for the purchase of apparatus necessary
to continue the sort of work that I was doing in Cambridge.
Although Clements was supportive, it became clear to me
that Dr Metcalfe, who shortly before I left Canberra had
succeeded Cumpston as Director-General of Health, was not
as supportive of research as his predecessor. Moreover, the
Australian National University (ANU), which had just been
established in Canberra, was looking for staff. I wrote applying for a position with the John Curtin School of Medical
Research of the university and was appointed a Fellow.[13]
Professor Sir Howard Florey, still Professor of Pathology at
the University of Oxford, who was Director designate of the
John Curtin School, advised me that, as there not as yet any
facilities at the School in Australia, I should continue working at the Molteno Institute. The following year, a number of

13 The title was later changed to 'Research Fellow'.

research scholarships were offered by the School and both
David Morell and Cliff Kratzing, my co-workers from the
Australian Institute of Anatomy, were successful in obtain-
ing scholarships. David joined the Molteno Institute in 1949
and Cliff Kratzing went to Henry McIlwain at Maudsley
Hospital in London the following year. Keith Rienits, the
remaining member of the Biochemistry Department of the
Australian Institute of Anatomy, joined Colin White in
Zuckerman's Department of Anatomy in Birmingham.[14]
During the next winter, I made several visits to Oxford to
report to Florey. I was led to believe that I was being kept in
reserve for a senior position in the Biochemistry Department
at the Australian National University and Keilin told me
that he was sure, from his contacts with Florey, that I was
to be offered the Chair. This was the job I wanted more than
anything else. We had loved living in Canberra, where we
still had our house, and were very homesick for Australia
and our families. It was, therefore, a let-down to be told by
Florey that Hugh Ennor had been appointed to this post,[15]
although, in fact, Ennor's appointment was not unexpected.
He was well-known in Oxford where he had worked with
Len Stocken in the Biochemistry Department soon after the
end of World War II, and had the reputation of being a good
organiser. Pending a decision by Florey to return to Australia as Head of the John Curtin School, it was essential to appoint someone who was capable of planning and supervising
the new building in Canberra and equipping it and Ennor

---

14 David Morell, after obtaining his Ph.D. in Cambridge, joined Lemberg in Sydney. Cliff Kratzing, after obtaining his Ph.D. with McIlwain, returned to the Biochemistry Department of the John Curtin School of Medical Research in Canberra
and later became Reader in Physiology in the University of Queensland. Keith
Rienits, after obtaining his Ph.D. in Birmingham, joined the Biochemistry Department at the University of New South Wales. Both Cliff Kratzing and Keith Rienits
spent sabbaticals in my laboratory in Amsterdam.
15 The 'let-down' was literal. When I left Florey's room, I slipped on the stairs,
falling heavily with a great clatter, fortunately without hurting anything but my
already weakened self-esteem.

was considered to be ideal for this purpose. Indeed, in this he was very successful.

Since I had always got on well with Ennor while we were students together in Melbourne and during visits to Oxford in my first year in England, I still had plans to join him at Canberra, if he wanted me. However, in the absence of any communication from him, I became more and more uncertain about this.

Keilin was very keen that I should obtain some experience in an American laboratory before returning to Australia and arranged with Pomerat of the Rockefeller Foundation that I apply for a Rockefeller Fellowship for this purpose. Since I was now anxious to extend my experience to oxidative phosphorylation, I was fortunate to be accepted by Severo Ochoa in New York to work in his laboratory. Pomerat told me that the amount of the Fellowship was calculated as sufficient for a single person in New York, where living expenses were much higher than in England and suggested that I leave Marion in England. When I said that this was out of the question, he suggested that I ask the Australian National University to continue to pay part of my salary (in sterling) while in America. Since we had now pretty well exhausted our savings, I did so, arranging to meet the Vice-Chancellor of the university, Professor Copeland, to discuss my request during his visit to Oxford, which happened to coincide with the Meeting of the Biochemical Society at which I gave my first paper. He saw me at his hotel, 'The Mitre', in Oxford and rejected my request not only out of hand, but extremely rudely, shouting at me in a crowded hotel lounge for my effrontery.[16] When I tried to explain my situation (because I had to pay UK income tax, I was actually receiving less net salary than the ANU Scholars), he told me that I could not expect to live on my salary if I wanted to do pure research

---

16 I have refreshed my memory of this interview by consulting a letter to my mother written 2 weeks afterwards (22 May 1949).

and, anyhow, the ANU did not give a damn whether I could or not.[17] Although this episode dampened my enthusiasm for the ANU, I decided to reserve judgement until I saw Ennor, and went ahead with the application for a Rockefeller Fellowship and booked a passage on the *Ile de France*, leaving Southampton on November 10, which was the earliest booking we could get in the 'Steerage' Class.

Florey called a meeting for 6 August of all those connected with the John Curtin School, namely the three Professors: Ennor (biochemistry), Albert (medical chemistry) and Fenner (microbiology); the ANU Scholars and me, during which the scholars and I gave an account of our work. I spoke of my recent studies on the oxidation of NADH. During the meeting, I had a short and strained discussion with Ennor.

Jumping forward a little chronologically, while I was in New York, I wrote to Ennor asking for clarification of my position in the ANU, pointing out that, in view of my experience and age, I could not reasonably be expected to return to Australia on a short-term Fellowship, but that I would want a position in his Department. Shortly afterwards, I heard that the ANU was advertising for a Reader in Biochemistry. Since I thought that such a position would be relatively independent, I decided to make one last attempt to come to terms with the ANU and applied for the position, with the support of both Keilin and Ochoa. Late in April, I received two replies to my letter. One from Copeland stated that, after 12 months in Cambridge succeeding my leave of absence in America, the ANU Council wished me to return to Aus-

---

17 Many years later, I read in a review of an autobiography of the historian Hancock, who expected to be appointed as Head of the School of Pacific Studies, that on the day before my meeting, he had had a similar experience with Copeland during a discussion on a park-bench in London, during which, as Hancock put it, 'all my dreams of returning to Australia disappeared'. In fact, some years later, when Copeland was no longer Vice-Chancellor, Hancock was appointed to that position.

tralia to work for 1 year in Ennor's laboratory. A similar letter from Ennor concluded with the paragraph:

> Some doubt has arisen in the minds of the senior members of this university as to whether you will fit into a new Department as a permanent member.

This I took as an unmistakable sign that I was not wanted.

More than 40 years later, I was fascinated to read, in a chapter entitled 'The Australian Connection' in a biography of Florey [24] by T.I. Williams, an account of a meeting in 1957 between Florey and Ennor. Now that the building of the John Curtin School was completed, Florey was offered the Directorship. He was prepared to accept this appointment, but, in the first instance, for only 1 year, and had already been granted leave of absence from Oxford University for this purpose. This conditional acceptance gave some difficulties to ANU, and Ennor, who was then Dean of the Medical School, was sent to Oxford by the Council of the university to discuss the matter. Williams refers in detail to a 'stormy and unhappy' meeting in Oxford on 19 March 1957 attended also by three members of Florey's staff, whom he planned to take with him to Canberra.

As Williams puts it:

> Thus, on an unhappy note, ended a dream that had lasted 13 years; there was no further prospect of Florey returning permanently to Australia.[18] ...the university was a major factor in his life for almost the whole of the period with which this biography is concerned. It made great demands on his time and attention – as well as on his affection and loyalty as an Australian.

---

18 He remained associated with the university, opening the John Curtin School in March 1958, and being non-resident Chancellor of the university from 1965 until his death in 1968.

So had my dream of returning to Australia ended 7 years previously. At the time, the events described by Williams were taking place, I was in Amsterdam and the Australian National University figures only briefly in what follows in this account.

USA 1949–1950

At the time of my stay in New York, Severo Ochoa was Chairman of the Department of Pharmacology of the New York University College of Medicine, housed on the 6th floor of Osborne House at 26th Street on First Avenue.[19] However, all his research was in biochemistry. Two floors below the Pharmacology Department, Pappenheimer and Ef Racker were also doing biochemical research in the Microbiology Department. Pappenheimer was at that time very interested in cytochrome b, having recently proposed that diphtheria toxin is apocytochrome b, that is cytochrome b without its haem group. Racker was working on glyceraldehyde phosphate dehydrogenase.

Ochoa had moved away from oxidative phosphorylation and was now working together with Jo Stern on what was then called the condensing enzyme, responsible for the synthesis of citrate in the Krebs cycle, and, together with Seymour Korkes and Seymour Kaufman, on the oxidative decarboxylation of malic acid. Sarah Ratner was carrying out her classical work on the enzymes involved in the urea cycle and Giulio Cantoni was just starting his work on the mechanism of methionine synthesis. However, all were interested in biological oxidations and oxidative phosphorylation and there were lively discussions at lunch in the Medical School canteen, in which Ef Racker also participated. The intensity of these discussions, and what seemed to me

---

19 Later he became Professor of Biochemistry housed in the new Bellevue Hospital.

the tacit assumption in New York that he who shouted the loudest won the argument, took me aback at the beginning, but they were on the whole very stimulating and taught me a lot about the way science was conducted in America. Even in the days before fax and e-mail, results obtained in one laboratory became rapidly known elsewhere through a very efficient grape-vine. During my stay in New York, I visited a number of laboratories all over the country and I was always asked about what the laboratory that I had just visited was doing. After such visits, I was closely quizzed by the other members of the lunch group at Osborne Hall. I have a photograph taken by Seymour Korkes of one such occasion after a visit to D.E. Green's laboratory in Madison with the caption on which he wrote, 'You were telling us about D.E. and all the little Greens'. I found that people worked much harder in America than was general in England.

In the report [25] of a symposium held on the occasion of my 60th birthday in 1977, I have written:

Although ...30 years ago quite a lot was known about the electron-transfer chain, practically nothing was known about oxidative phosphorylation, and I do not recall that it was ever a topic for discussion in the Molteno Institute. The mechanism of the esterification of phosphate in glycolysis was, of course, well-known, and I can recall with pleasure Dorothy Needham's lectures on this. It was vaguely recognised – mainly because of Meyerhof's calculations – that another mechanism must be operative in respiration, but the direct demonstration of oxidative phosphorylation by Engelhardt and Kalckar, and the strong but indirect evidence by Ochoa and Belitzer and Tsibakowa of the existence of what later became known as respiratory-chain phosphorylation, made very little impact on us in the Molteno Institute. Since Ochoa had reported that no phosphorylation could be obtained with NADH as donor, and since I knew from Keilin and Hartree's, and my own work, that phosphate is not necessary for the oxidation of succinate or NADH by the Keilin and Hartree heart-muscle preparation, it was difficult for me to relate my own work on the NADH oxidase system to oxidative phosphorylation. It was Al Lehninger's demonstration of oxidative phosphorylation with NADH as substrate, using what must have been pretty bust-up mitochondria[20] that changed all that and led me to join Severo Ochoa's laboratory to learn about such strange substances as ATP.

Although Lehninger's demonstration of oxidative phosphorylation coupled with the oxidation of NADH by oxygen was generally accepted, it was puzzling that he was unable to find any phosphorylation when NADH was oxidized by ferricytochrome c, since this reaction covered a large traject of the respiratory chain. I decided to reinvestigate this apparent inability of cytochrome c to act as an electron acceptor for oxidative phosphorylation. Although Lehninger had used a preparation of rat-liver mitochondria as the source of the oxidative phosphorylation system, I was more familiar with heart preparations and Ochoa had used a rat-heart homogenate, so I used washed granules obtained by differential centrifugation of an extract obtained by grinding cat-heart mince with sand and phosphate buffer. I followed the conventional procedure of using glucose in the presence of hexokinase as an ATP trap, but, against current practice, used ADP instead of AMP as the primary phosphate acceptor, since I found that this was much the more effective. Since the amount of ferricytochrome c that could be used as electron acceptor was limited by its high molecular weight, the usual method of measuring oxidative phosphorylation by determining the disappearance of inorganic phosphate

20 The reference to 'bust-up' mitochondria was Lehninger's demonstration, subsequent to his original studies, that NADH does not penetrate mitochondria unless they are swollen and cytochrome c is added. In fact, it was shown some years after my talk in 1977 that Lehninger had not succeeded in demonstrating phosphorylation coupled to the oxidation of NADH by the mitochondria. NADH does not penetrate the inner mitochondrial membrane, the site of oxidative phosphorylation. The phosphorylation that he measured was mainly due to the oxidation of ferrocytochrome c formed by reduction of ferricytochrome c by the NADH, catalysed by a NADH–ferricytochrome c reductase in the outer mitochondrial membrane and partly due to endogenous substrate in the mitochondrial preparation [26].

could not be used. Accordingly, I devised a method, based on Racker's recently published spectrophotometric method of determining phoshohexokinase activity, directly to determine hexose monophosphate.

Following Ochoa's most recent paper on oxidative phosphorylation, I used α-ketoglutarate as substrate. I used two methods to stop the re-oxidation of ferrocytochrome – anaerobiosis and addition of cyanide. With both methods, I was able to show phosphorylation with ferricytochrome c as acceptor. Indeed, since the P:2e ratio approached the values obtained with oxygen as acceptor, and I thought that there was reason to expect that the values with ferricytochrome c might be under-estimated, I concluded in a preliminary report that was published in *Nature* [27]:

> These experiments suggest that all the phosphorylation which occurs when α-ketoglutarate is oxidized by oxygen occurs in those steps of the hydrogen transfer reaction between substrate and cytochrome c.

As will be discussed below, subsequent work showed that, although the main conclusion, namely that the complete respiratory chain is not necessary for oxidative phosphorylation, is correct, three quarters, rather than all, of the phosphorylation occurs between the substrate and cytochrome c.

During my stay in New York, I visited other biochemical centres in Boston, Cambridge, Philadelphia and Bethesda, all near enough to be reached by the Greyhound bus. During these visits I met many who were already, or were to become, leading figures in biochemistry. In Boston, I met John Edsall and Fritz Lipmann and his group, including Earl Stadtman; in Cambridge, George Wald who was just beginning the work on the biochemical basis of vision that was to bring him the Nobel Prize; in Philadelphia, Britton Chance, Lucile Smith and Otto Meyerhof; in Bethesda, Arthur Kornberg and Bernie Horecker.

On 17 January, I gave a seminar to Britton Chance's Department at the Johnson Foundation in Philadelphia, on the reaction between cytochrome c and cytochrome c oxidase, attended also by Lucile Smith, who had just completed her work on the preparation of purified cytochrome c oxidase solubilized with cholate. In March, I gave two seminars in New York, one to the Polytechnic Institute in Brooklyn and one to the Enzyme Club. The latter was quite a nerveracking experience, since this club contained all the leading biochemists in New York and previous lecturers were distinguished visitors. There were several hazards to overcome. First, the strongest cocktails that I have ever come across that were served before the lecture. Secondly, it was customary to interrupt the lecturer with questions put with full New York vigour. However, they were kind to me and the lecture was well received. Phil Siekevitz refers to it in an historical account of the Enzyme Club.

In April, I attended, together with Ochoa and a number of members of his laboratory, the meeting of the Federation of American Biological Sciences in Atlantic City. These meetings were at that time the most important biochemical gatherings in the world and gave a very complete picture of the latest developments. I gave a 10-minute paper on my Cambridge work.

I finished my experiments in New York with what I called a 'Sunday-best' or publication experiment, suitable for publication in a single table. Immediately afterwards, I set off for the biochemist's 'Cook's tour' visiting laboratories in Rochester, Cleveland, London (Ontario, Canada), Chicago, Madison, St. Louis, Pasadena, Pacific Grove, Palo Alto (Stanford University) and Berkeley.

I travelled mostly by train, except for a side visit by bus to Niagara Falls, where I crossed the Canadian border on a Sunday. Here I was held up some time by the Immigration authorities on the US side, which caused much interest among the other bus occupants. Much later, I read that, on that day, the hunt was on for one of the atomic spies, who was described as a biochemist from New York, which perhaps accounted for the interest of Immigration, since I was described as a biochemist in my passport. Between St. Louis and Pasadena, I travelled by the Santa Fe railway with a side visit for one day to the Grand Canyon.

I gave a seminar, entitled the *Enzymic reduction of cytochrome* c, in Rochester, Chicago, London, Madison, Pacific Grove and Berkeley. This talk combined both the work that I had done in Cambridge and the just completed work in New York. The talk at Madison was a more formal evening lecture. In Chicago, I met Al Lehninger, Konrad Bloch and Gaffron. I spent 2 weeks in Madison, doing some experiments at the Enzyme Institute, at that time completely occupied by David Green and his large group, including Osamu Hayaishi and Henry Mahler. I also met Van Potter and Henry Lardy. Although I was very impressed by the equipment in the Enzyme Institute, especially a room full of refrigerated centrifuges (we did not have one in the Molteno), I was not impressed by the work there. This was the unproductive cyclophorase period of Green's work.

Claude Liébecq, with whom I had shared laboratory space in the Molteno Institute (see Ref. [13]) was now working in St. Louis in Carl Cori's laboratory and I met both Carl and Gerty Cori. I also met Mildred Cohn (who told me about her experiments with ^{18}O-labelled phosphate to study the mechanism of action of phosphatases), Robert Crane, Ed Hunter and Lowry. I was there for the week-end and I spent a terribly hot Sunday with the Liébecqs and the Hunters picnicking and swimming in a river outside St. Louis. We returned to the city to hear that the Korean War had broken out.

California reminded me of Australia. James Bonner, the plant physiologist, and Walter's[21] elder brother, was my host

---

21 Walter Bonner joined the Molteno Institute shortly before we left for New York.

at Pasadena. The visit to van Niel's marine biological laboratory at Pacific Grove, then the Mecca for microbial biochemists, was memorable. At a moment's notice, I was invited to
give a seminar 'on the rocks'. A blackboard was propped up
against a rock at the sea's edge just outside the laboratory.
The students and staff members sat on the rocks and I had
to talk loudly enough to drown the noise of the Pacific Ocean
breaking on the rocks below. I gave a short account of the
New York work. Dan Arnon was my host at Berkeley. There
I met the Nobel Prize winner, Stanley, and a future Prize
winner, Calvin who had recently established the carbon
pathway in photosynthesis, now known as the Calvin cycle.

About the middle of July, I returned to New York, a 3-day
journey. I had a Pullman car, air-conditioned with sleeper,
so it was quite comfortable. In New York, I picked up Marion to travel to Philadelphia for an intensive 2 weeks of experiments with Britton Chance. With the help of a student,
Tom Devlin, to grind the heart-muscle mince with sand in
the absence of the mechanical pestle and mortar used in the
Molteno Institute, Lucile Smith and Helen Conrad (later
Davies), I made a heart-muscle preparation for Brit and I to
follow the reduction of the cytochromes in his photoelectric
spectrophotometer (only single beam at that time) during
anaerobiosis induced by adding succinate, and to study the
effect of cyanide. Here I met once again what appeared to be
the anomalous behaviour of cytochrome b, a problem that
was to plague me for so long.[22]

When they heard that I was without job prospects, Ochoa
and others asked me if they could put my name forward for
possible positions in the new Departments that were being
created in America. Britton Chance offered me an Assistant
Professorship at the Johnson Foundation. However, Marion
and I did not find the American way of life at that time con-

---

22 These experiments were, in fact, the first that Chance carried out with preparations of the respiratory chain. My collaboration is acknowledged in his paper [28].

genial and did not think that we could possibly make our home in America.

The summer crossing of the Atlantic in the *Ile de France* was much more pleasant than the trip to America. We stopped in Cambridge only long enough to deposit our luggage and then crossed the North Sea in luxury in a Danish boat to Denmark, in order to attend the XVIII International Congress of Physiology in Copenhagen, where we had a wonderful time, reminiscent of the congress in Stockholm 3 years earlier. I gave a paper on my work in New York, at the conclusion of which, I had the great pleasure of meeting Vladimir Engelhardt from Moscow, whom I consider to be the discoverer of oxidative phosphorylation.

Return to Cambridge 1950–1955

After returning to Cambridge, I was able to take up, once again, my ANU Fellowship, but in view of my decision not to return to Ennor's Department, an early concern was to find alternative support. Through the good offices of Keilin and the Chairman of the Council, Lord Rothschild, the alternative support came from the Agricultural Research Council (ARC), who offered me a 3-year contract to undertake research on intracellular respiratory enzymes in the Molteno Institute. The letter from Rothschild, dated 23 December 1950, in which the offer was made further stated that, during the period of this contract, 'we would consider and discuss the possibility of you taking over a Biochemistry Department at one of the ARC Institutes'. The appointment commenced on 1 May 1951. In addition to my salary, there were funds for bench fees (required by the University of Cambridge), travel and for purchase of chemicals.

During the discussions with the ARC, I received a friendly letter from Hans Krebs stating that he had heard from Keilin that I was looking for possible openings and suggesting that I visit Sheffield with a view to discussing a possible position with him. When I told him of the offer that I had received from the ARC, he replied that he quite understood and would probably have made the same decision himself under the same circumstances. Thus, for the second time, I preferred Keilin's laboratory to that of Krebs! Especially in view of my unhappy experiences with the ANU, I was immensely encouraged and flattered by the support I received from such distinguished figures.

In any event, although I did not join their permanent staff, I was supported by the ARC for exactly 4 years and I owe much to all involved – Keilin, Rothschild and the ARC – for the opportunity to continue undisturbed research for this period, during which I built up my 'scientific capital', as Keilin put it. My only teaching duties were to give in each of the 5 years, 1951–1955, four of Keilin's lectures to senior students in Physiology and Biochemistry at noon on Saturdays in the Lent term.[23] This made me a member of the Regent House and gave me dining rights at my college, Trinity Hall, for one night per week.

It took a long time before I could get going again on my experiments on oxidative phosphorylation. My first task was to devise a procedure for preparing rat-heart mitochondria with the equipment available in the Molteno. While I was away, a deep-freeze cabinet and what was then considered a high-speed centrifuge (a table-top Sorvall) had been acquired. After several trials, I found that by pre-cooling the centrifuge head in the deep-freeze for a specified period (I have forgotten for how long) it was possible to carry out the differential centrifugation of the rat-heart homogenate with the temperature remaining between 0° and 5°, which was satisfactory.[24] Before I could continue experiments on oxidative phosphorylation, it was also necessary to prepare ATP

---

23 I had also given these lectures in 1949.
24 Early in 1953, Keilin obtained a grant for the purchase of a refrigerated centrifuge, which with great difficulty and damage to the staircase was finally installed in my old room at the top of the Institute.

from rabbit muscle and hexokinase from yeast, both time-consuming operations. In the absence of a cold room in the Molteno, I used one of the storage rooms in the Low-Temperature Research Laboratory, nearby. It was also necessary to prepare the ADP from the ATP by reaction with glucose and hexokinase.

Fortunately, I was now given assistance. Francis Holton,[25] who had joined Keilin while I was in New York, was assigned to me as research (graduate) student and, in, May 1951, an Australian, Ken Cleland[26] was sent to me from the Zoology Department by Rothschild, with whom he had been working. Cleland returned to Australia early in 1953 and was replaced by a technical assistant, Ann Searle, a 16-year-old local girl. Thus, after 4 years as a one-man team, I now had three working with me. As always, Marion was my unpaid secretary, typing my letters, the many drafts of my papers and the neat copies of the results of all experiments, as well as keeping my card index of the literature. Also, I took a keen interest in the work of others in the Molteno, specially that of Chen-lu Tsou,[27] and, in the course of the next few years, I published together with Walter Bonner, Alfred Tissières, Hans Laser and Stan Lewis.

Enzyme kinetics

The first of these collaborations was with Walter Bonner on the inhibition of succinate dehydrogenase by fluoride. My main interest in this was Walter's finding that phosphate greatly increases the inhibitory action of fluoride, and we joined forces in the hope that a further investigation of this

---

25 After obtaining his Ph.D., Francis joined Popjak in London and later was appointed Reader in Biochemistry at the Veterinary School, London.
26 Ken Cleland became Professor of Histology and Embryology at the University of Sydney.
27 See Ref. [29] for Tsou's recollections of this stay in the Molteno Institute and his subsequent life.

requirement might provide a clue towards the mechanism of oxidative phosphorylation.

It did not. However, it did resolve conflicting conclusions in the literature concerning the locus of action on respiratory systems of fluoride, one of the longest studied enzyme inhibitors. We were able to show that either fluoride or phosphate alone acts as a classical competitive inhibitor of succinate dehydrogenase. The two together are, however, much more effective, suggesting the formation of fluorophosphate in the active site of the enzyme. Using the classical procedures of enzyme kinetics, we measured the various inhibitory constants. However, there remained a problem. We confirmed a previous finding by Massart that the degree of inhibition is greater when the complete succinate oxidase system is measured than when its component, succinate dehydrogenase, is determined by the classical procedure with methylene blue as acceptor. We resolved this by showing that the kinetics of the enzyme follows Briggs-Haldane kinetics, rather than the, at that time, widely accepted Michaelis-Menten. This was demonstrated by a linear relationship between V (the velocity at infinite substrate concentration)/v (the observed velocity) at different values of V, obtained by varying the nature of the acceptor and its concentration. This enabled us to calculate the various rate constants.

Shortly after our paper [30] appeared in print, I received by messenger a hand-written note from the Professor of Botany of Cambridge University, requesting clarification of certain points. I knew that the Professor of Botany was named Briggs, but not that this was the Briggs of Briggs and Haldane. I replied and delivered the letter to the porter at the Botany Department, about 100 yards away. The same day, so far as I recall, I received a second letter again delivered by messenger expressing satisfaction with the clarification, but raising additional points to which I quickly replied, whereupon a third letter arrived, again agreeing with our

explanations but querying the actual value that we had calculated for the rate constants. By this time, I thought that a
personal discussion would be more productive, so I requested
an audience with Professor Briggs. During a pleasant conversation, I was able also successfully to defend our calculated values.

The paper and a more general exposition during a symposium [31] on enzymes organized by the Faraday Society in
1955, aroused considerably more interest than we had expected and we had the satisfaction of seeing it included in
Dixon and Webb's book [32]. Nearly 40 years after the publication of our paper, I was very surprised but pleased to receive an invitation from the Institute of Scientific Information to write a commentary on this paper, since it had been
cited so often as to be ranked a 'citation classic.' As we wrote
in the commentary [33], this surprised us because its main
message has long become incorporated into conventional biochemical wisdom and this message was scarcely sufficiently
novel to warrant citation 40 years later, nor did the interest
in fluoride toxicology and pharmacology seem a sufficient
explanation for the high citation. However, we had completely forgotten that, in this paper, we had introduced
rather trivial, although convenient, modifications to spectrophotometric methods for measuring succinate dehydrogenase and succinate oxidase. It appears that our procedure is
still widely used for measuring the activity of this enzyme as
a mitochondrial marker for the preparation of cell membranes and this probably accounts for the high current citation score, which exaggerates the importance of the paper
and points out one of the dangers of relying too much on citation scores.

Stability of isolated mitochondria

The continuation of my work on oxidative phosphorylation
started in New York developed along two main lines. Ken

Cleland and I studied the stability of the morphology and biochemical properties of isolated mitochondria.[28] It was well known that their ability to carry out oxidative phosphorylation was quickly lost at room temperature. Mainly by observations with the phase-contrast microscope, Cleland showed that this biochemical damage was correlated with swelling of the mitochondria even in isotonic medium and, based on an effect of calcium on the respiration of sea-urchin eggs that he and Rothschild had observed, got the idea that this might be due to calcium in the medium. He found that it could be prevented by the addition of citrate, which would be expected to bind calcium. Asking around for a good calcium binder that was not at the same time a substrate, we were told by Al Lehninger, who was on a sabbatical year in Cambridge, about the recently introduced ethylenediamine tetraacetate, then known by its trade name, Versene, but nowadays generally by the acronym EDTA. This compound was indeed found to be very effective, as we published in a Letter to *Nature* [34]. We found that mitochondria isolated in the absence of EDTA contained large amounts of calcium, apparently largely derived from blood during the isolation procedure. A surprising finding which has not been adequately explained was that, although the isolation procedure yielded little more than half of the mitochondria present in the heart, all the calcium in the heart was present in the isolated mitochondria. This work was the subject of a communication to a joint meeting at Brussels of the Biochemical Society with the corresponding Society in Belgium and a major paper in the *Biochemical Journal* [35]. At the request of the Editor, 2–3 pages were removed from the discussion dealing with the possible physiological significance of uptake of calcium by sarcosomes. The section that was removed con-

---

28 At this time, we referred to the respiratory granules of heart muscle as 'sarcosomes', following Retzius (1890). It was only after we convinced ourselves that Retzius' sarcosomes correspond in every respect with mitochondria in other tissues that we followed other biochemists in referring to them as mitochondria.

tained the suggestion that the recently discovered Marsh relaxing factor might be particulate, bound to the sarcosomes. It is now known that it *is* particulate, but bound to the sarcoplasmic reticulum.

Although we can claim to be the first to show the avid uptake of calcium by mitochondria, we missed an important point, namely that the uptake is energy-linked. Since isolated mitochondria took up calcium in the absence of substrate or ATP, I assumed that it was a simple binding and not energy-linked. I was not aware of the fact that even thoroughly washed mitochondria contain considerable amounts of endogenous substrate. Later work by others showing that either oxidizable substrate or ATP is necessary for the uptake of calcium opened up a whole chapter in mitochondrial research.

In New York, I had isolated the mitochondria in phosphate buffer. The first preparations of liver mitochondria described by Hogeboom and co-workers were prepared in 0.88 M sucrose, which was thought to maintain the *in vivo* morphology of the mitochondria. Ken Cleland undertook a systematic study of various isolation media and in particular of their tonicity on the morphology and biochemical properties of the mitochondria. These studies were reported to the 2nd International Congress of Biochemistry in Paris, and later as full papers dealing with the morphology and biochemistry, respectively [36, 37]. It was concluded that the best isolation media is isotonic sucrose containing EDTA.

These studies were particularly relevant to the question whether mitochondria are essentially a gel, as David Green believed at that time, or behaved as an osmotic system bounded by a semi-permeable membrane. Cleland's observations, which confirmed those of the 19th century microscopists and of Claude in liver mitochondria, clearly demonstrated the presence of a membrane in mitochondria, as well as the ability of the mitochondria to swell in hypotonic media or on ageing, in the absence of a calcium chelator. Unknown to us at that time were the recently published observations by thin-section electron microscopy by Palade and Sjöstrand that clearly showed two membrane systems – a boundary membrane and an internal convoluted membrane system. The boundary membrane that Cleland observed in swollen mitochondria was derived from unfolding of the inner membrane. Although the studies with the electron microscope soon made Cleland's observations with the light microscope redundant, our studies made a contribution towards the development of the concept that the mitochondrion is an osmotic system, as evidenced by the introductory paragraph of a paper that Peter Mitchell gave to a symposium of the Society of Experimental Biology at Bangor in 1953 [38]. In a paper that never received the recognition that it deserved, Cleland showed how the permeability of the mitochondrial membrane could be measured by following the swelling of the mitochondria by measuring the light scattering in a spectrophotometer [39].

Another aspect of these studies was to establish the origin of the Keilin and Hartree heart-muscle preparation. Cleland observed that, although in hypotonic medium the mitochondria swelled just like erythrocytes, they did not burst if they were not disturbed, but remained as large vesicular structures. Shaking the suspension, however, led to fragmentation of the vesicles. We concluded, therefore, that the Keilin and Hartree heart-muscle preparation consists of fragments of the mitochondrial membrane [40]. Nowadays, these fragments are referred to as sub-mitochondrial particles derived from the mitochondrial inner membrane.

Oxidative phosphorylation

Francis Holton joined me in continuing the work started in New York. First I wrote up the experimental procedures developed in New York in two papers, one dealing with the spectrophotometric method of measuring hexose phosphates,

ADP and ATP [41], and the other with the procedure for measuring the P:O ratios [42]. We also studied the kinetics of oxidative phosphorylation and established that the substrate is ADP, not the widely used AMP [43].

The method that I described for determining the yield of oxidative phosphorylation (the P:O ratio) was more accurate than those used at that time or even later, depending as it did on the formation of the product of oxidative phosphorylation, rather than the disappearance of inorganic phosphate. It was this that made possible a study of the kinetics of oxidative phosphorylation, as well as the use of low concentrations of electron acceptor, such as cytochrome c. Other workers in the field later developed the more convenient procedure based on the incorporation of inorganic ^{32}P into ATP or, when hexokinase was added, into hexose monophosphate.[29]

Holton and I undertook an exhaustive examination of the P:O ratio coupled with the oxidation of α-ketoglutarate to succinate. I was strongly against the current assumption that, since the experimental procedure tended to underestimate the true value, the highest value obtained was likely to be the correct one. I considered this unscientific, since aberrantly high values could also be obtained by experimental errors [44]. We reported all our results in a histogram and, since they were mostly clustered just under 3, we [45] concluded that the correct value is 3, not 4, as generally believed, and we criticised the methods used in the papers reporting the higher values. Our contemporaries believed that our lower values were due to our using damaged mitochondria and the value of 4 became generally accepted, and indeed, at the time of writing, is to be found in the textbooks. However, from the most recent data [46], the value

---

29 The measurement of the incorporation of inorganic ^{32}P into ATP, as used by Krebs, overestimates the yield, since it does not take account of an isotope exchange which takes place in the absence of electron transfer. The incorporation into hexose monophosphate is free from this error, provided sufficient hexokinase is used, and we used this method in later work in Amsterdam.

may be calculated to be 3.25, so that we may now claim to have been nearer the truth.

At the same time, I re-examined the conclusion that I had drawn in New York that all the phosphorylation is between substrate and cytochrome c and that there is none between cytochrome c and oxygen. This seemed to be confirmed by preliminary experiments showing little phosphorylation between ferrocytochrome c and oxygen. However, after Lehninger successfully demonstrated phosphorylation in this reaction using swollen liver mitochondria, I confirmed this with swollen heart mitochondria, finding a P:O ratio equal to 0.34 [47].[30] In any case, since Lehninger had showed unequivocally that the reaction between cytochrome c and oxygen is coupled with phosphorylation, it was clear that my earlier conclusion must be wrong. This was confirmed when I found [48] a previously undetected leak past cytochrome c when cyanide was used to block cytochrome oxidase, and that the P:2e ratio with ferricytochrome c as acceptor is definitely lower than with oxygen as acceptor, although well above 2. This was confirmed with succinate as substrate.

This was my first bad mistake. The positive result of my work was the first direct demonstration of what became known as Site-II phosphorylation. To Lehninger belongs the priority of the discovery of Site-III phosphorylation.

---

30 It is now clear that for the reaction with both ferricytochrome c and ferrocytochrome c, the outer membrane of the mitochondria must be ruptured to allow the cytochrome to come into contact with the electron donor and acceptor, respectively, in the inner membrane. In my experiments with the ferricytochrome, only those mitochondria with a ruptured outer membrane but still intact inner membrane would have participated in the reaction. Ferrocytochrome c, on the other hand, would be rapidly oxidised by any fragments of the inner membrane that would have been insufficiently intact to carry out oxidative phosphorylation, thereby lowering the P:O ratio. Only by swelling the mitochondria to rupture the outer membrane of otherwise intact mitochondria and make the intact inner membrane accessible to ferrocytochrome c, was it possible clearly to demonstrate phosphorylation.

Stan Lewis, from the Pest Infestation Laboratory at Slough, spent a year at the Molteno Institute to study the respiratory systems in insects. Since it had been reported in the literature that fly-muscle mitochondria do not carry out oxidative phosphorylation, Stan and I joined forces to check this and not surprisingly we were able to show that they did carry out oxidative phosphorylation with α-ketoglutarate as substrate [49], and I thought that we should clinch this by showing its sensitivity to the uncoupler, 2,4-dinitrophenol. Since I usually planned experiments with eight manometric flasks (the full complement of my manometer bath) to measure the oxygen uptake, we decided to use a range of concentrations of the uncoupler. I can well remember my astonishment, while carrying out the determinations with the manually operated spectrophotometer on the series of deproteinated extracts, when I found that, although as expected the amount of hexose phosphate was much lower in the experiment with the lowest amount of dinitrophenol, the amounts then increased with increasing amounts of dinitrophenol [50]. I thought intensely while doing the routine manual operations involved and it was this thinking that led me to develop the mechanism of oxidative phosphorylation that I had had in the back of my mind for some time. The explanation of the increasing amounts of phosphorylation was simply that, with the particular preparations of insect mitochondria that we were using, the rate of oxidation of α-ketoglutarate was limited by the rate at which the postulated high-energy intermediate of oxidative phosphorylation reacted with phosphate and ADP. Dintrophenol removed this block (at that time I thought by promoting the hydrolysis of the high-energy intermediate) and, although it also stopped the synthesis of ATP by respiratory-chain phosphorylation, this was more than compensated by the increased rate of oxidation of α-ketoglutarate, since, as Judah had shown, the substrate-linked phosphorylation is not affected by dinitrophenol. The point is that the phenomenon that I found so stimulating to my thinking and development of what became known as the chemical mechanism of oxidative phosphorylation was accidental – the use of a preparation of somewhat damaged mitochondria that exhibited good respiratory control as measured by the stimulation of respiration by uncoupler, but which gave low P:O ratios.

In 1953, I published the paper [51] in *Nature* on the mechanism of respiratory-chain phosphorylation that became probably my best known paper.[31] The mechanism was formulated by the reactions:

$$AH_2 + B + C \Leftrightarrow A{\sim}C + BH_2 \qquad (1)$$

$$A{\sim}C + P_i + ADP \Leftrightarrow A + C + ATP \qquad (2)$$

where AH_2 and B are adjacent members of the respiratory chain and C is a component required for the redox reaction and which remains liganded to the oxidised product in a high-energy compound which, in the second equation, reacts with ADP and phosphate to make ATP. It was further proposed that dinitrophenol uncouples oxidative phosphorylation by promoting the hydrolysis of A~C and may therefore stimulate respiration even in the presence of ADP and phosphate if the rate of this hydrolysis exceeds that of reaction 2. The three 'substrate-linked' oxidative phosphorylations of glyceraldehyde 3-phosphate, pyruvate and α-ketoglutarate, respectively, could be described by these equations, C being in each case a thiol compound, either a specific cysteine residue in glyceraldehydephosphate dehydrogenase or Coenzyme A. Thus, there was nothing very novel in extending the formulation to respiratory-chain phosphorylation. If this had been the only reason for postulating the above reaction sequence, I would never have bothered publishing it. Indeed,

---

31 B.H. Weber [52] has stated that between 1954 and 1984, there were 384 citations (less self-citations).

even though there were additional reasons that I shall now describe, after I wrote the article I hesitated before sending it to *Nature*, wondering if it were not all too trivial.

In fact, I wrote the article as the result of hearing a paper presented at a meeting of the Society of Experimental Biology in Bangor, Wales, in the spring of 1953, in which it was reported that 2,4-dinitrophenol inhibits the uptake of glutamate by the strict anaerobe *Streptococcus faecalis*. During the discussion, I asked how was it possible that 2,4-dinitrophenol, whose action is known to be as an uncoupler of oxidative phosphorylation, could be effective in a strict anaerobe in which oxidative phosphorylation does not exist. The answer that I received, namely that I could not expect to explain what happens in bacteria by studying isolated mitochondria, made me anxious to think up a possible explanation. I got the idea that ATP made by glycolysis could, by driving reaction 2 in reverse, make the same intermediate, A~C, as is made by respiration in reaction 1, and that this intermediate could be used by the cell for energy-requiring processes. It was this idea, that the energy of respiration can be conserved *and utilized* without the intervention of inorganic phosphate, that I considered sufficiently novel to warrant publication.

I collaborated with Alfred Tissières, who was working on the respiratory chain in *Azotobacter vinelandii*, the most active respiratory system known, to demonstrate oxidative phosphorylation in particles prepared by differential centrifugation of a sonic extract [53].

In May 1952, I was invited as one of a group of about 20 'Englishmen' (which included 5 Australians) to a meeting with Danish biologists in Copenhagen. I gave a paper on my work with Cleland, but my main memory is of a visit to Copenhagen by Otto Warburg just after the meeting. I attended a formal lecture that he gave on photosynthesis, followed by a discussion meeting at the Carlsberg Institute, and Marion and I attended a luncheon party given by

Linderström-Lang for him at the Carlsberg Institute. It was fascinating to see and talk with Keilin's great protagonist from the 1920s (see Ref. [54]).

In the summer of 1952, I attended the 2nd International Congress of Biochemistry in Paris. It was not a memorable meeting, except for the announcement by Gunsalus of the role of lipoic acid in pyruvate and α-ketoglutarate oxidation, which had not yet been published in the general literature. Since I had been invited to write the Chapter on Biological Oxidations for the *Annual Review of Biochemistry* [55], I included this in my review, with permission of Gunsalus, thereby obtaining a scoop on the normal literature! I gave two papers to the congress, in a session that was marked by the Chairman reading a telegram from Britton Chance apologising for not being present to give his paper, because he was leading in the competition for a sailing gold medal at the Olympic Games, which he duly won.[32]

In the spring of 1953, I was invited to give a paper at the meeting of the German Biochemical Society in Mosbach, and lectures in Lund and Stockholm, five lectures in all. On my way to Germany, I attended a joint meeting of the Biochemical Society and the Belgian Biochemical Society in Brussels. This was a memorable trip. I travelled by ferry and train to Brussels, train to Heidelberg, Mosbach and back to Frankfurt, air to Stockholm, train to Lund and Copenhagen, and back to England by train and ferry. It was most interesting to see Germany for the first time. Eight years after the end of the World War II, the area around the station of Hamburg was still devastated, and the whole centre of Frankfurt was still full of single-story temporary shops. However, Professor Felix, whom I visited in Frankfurt after the Mosbach meet-

---

32 Many years later it was realised that his short abstract described the so-called oxidant-induced reduction of cytochrome *b* which was not fully explained for 25 years.

ing, was living in a newly built house of the size and quality above that permitted by building regulations in the UK.

I called upon Richard Kuhn at the Max Planck Institute in Heidelberg. Owing to his importance in German science during the World War II, Kuhn was still a controversial figure. I found him friendly, but a tour through the Institute was a quick march with all doors left open behind him and his young associates clicking heels when he spoke to them.

Mosbach was a lovely mediaeval town, untouched by World War II, as was Heidelberg. The invitation to this meeting was the most important one that I had received to date and I prepared the paper for publication [56] with care. However, I did not take enough time with the preparation of my talk, sorting my slides in the train on the way to Heidelberg and getting some out of order, and I went over my time. I was unhappy with my performance, but it was a good lesson and I took much more care in future. I met many well-known European biochemists for the first time, including the Christian of the classical Warburg and Christian papers, Th. Bücher, Ernst Helmreich, Otto Hoffmann-Ostenhof and Pierre Desnuelle.

My hosts in Stockholm were Olov Lindberg and Lars Ernster, at the Wenner Grens Institute. I gave a talk at the Karolinska Institute in Stockholm and made a memorable visit to Lundergärdh, one of the fore-runners of the chemiosmotic theory, in his private laboratory in Uppsala.

Offer of Amsterdam Chair

Already in 1951, there was talk of my being appointed to head the Biochemistry Department at the Institute of Animal Research being set up by the ARC at Babraham Hall, outside Cambridge. Indeed, in July 1951, I had had some discussion with Sir Alan Drury, then Director of the Institute, about joining the Biochemistry Department at the beginning of 1952, but Keilin had urged me not to do so unless

I was to be Head of the Department from the outset. Nothing further was heard about this at the time. Since, in any case, it was clear that we should not be returning to Australia, we decided to sell our house and bought a vacant block at 3 Bulstrode Gardens, not far from where we living in Madingley Road. Building started after the excessively cold winter of 1953–1954, but, because of post-World War II shortages and restrictions, by the time the house was completed we were no longer in Cambridge. However we got a lot of pleasure, as we had previously in Canberra, in collaborating with the architect in designing the house and supervising its building.

Early in 1954, I was very pleased to be invited to join the Hardy Club, a rather exclusive club in Cambridge consisting of biologists mainly in the middle rank of seniority, but including some FRSs and younger Professors.

During 1954, my future once again became uncertain. The ARC was now undecided whether or not to go ahead with the development of Babraham as a major research institute. To give themselves breathing space, they made short-term appointments of senior scientists who had retired from their previous posts, including Sir Rudolph Peters, who had just retired from the Biochemistry Chair at Oxford, as Head of the Biochemistry Department. Rothschild told me that Peters would be at Babraham for 3–4 years, after which he would be replaced by a senior biochemist and that I would have a good chance to be appointed. However, Peters did not want me there at the moment.

Although neither Sir William Slater (no relation), the Secretary (Administrative Head) of the ARC, nor Rothschild put me under any pressure, I felt that it was important to consider positions outside Cambridge, even though this would once again mean abandoning our house, which was still not completed. In early September 1954, I was surprised to receive a letter from Professor Bruno Mendel in Amsterdam stating that my name had been mentioned (by Keilin, it transpired) in connection with a vacancy for the Chair of Physiological Chemistry at the University of Amsterdam. Shortly afterwards, Mendel wrote that my name had been placed on a list of candidates and, after a visit by Professor Formijne, member of the selection committee, to Cambridge, I heard in mid-October that the faculty had nominated me as No. 1 and in mid-January I received a telegram from the Burgomaster of Amsterdam congratulating me on my appointment. It was a great thrill that this was addressed to Professor Slater. My colleagues in the Molteno Institute were generous in their congratulations, but expressed their sorrow that I was leaving them. Keilin, I think, had mixed feelings. He was pleased at the offer, but I think hoped that it would stimulate the ARC to make a counter offer. Indeed, Rothschild had suggested that I might be interested in taking over the Biochemistry Department at the ARC Field station at Compton. I received a number of friendly letters from my future colleagues in Amsterdam, including my predecessor, Professor B.C.P. Jansen.

As usual in the Netherlands, the appointment was published in the newspapers and I was called by Reuters in London for some biographical details.[33] All this had happened very quickly, but, of course, we could not possibly accept the appointment without first visiting Amsterdam to inspect the laboratory and get some idea of what was expected of me and of the possibilities offered by the position. I was completely unfamiliar with how universities were run on the continent. In the meantime, we started to learn Dutch by gramophone.

Marion and I travelled by the night boat, Harwich – Hook of Holland, our first visit to The Netherlands. We were met at the Amsterdam Central Station and told that we had an

---

33 I was also interviewed by a reporter for the *Sydney Morning Herald*, who wanted to know whether I spoke German or Dutch and whether I was doing secret work.

appointment before lunch with the Burgomaster of Amsterdam, Mr D'Ailly, and the Alderman for Education, Mr Roos. This started an extremely busy 2 weeks of visits, consultations and dinners with the members of the Medical Faculty who, at that time, consisted only of the Professors, all of whom were Department Heads.

The first surprise was the unusual status of the university. It was, at that time, entirely funded by the City Council. A Professor was appointed by the City Council, meeting in public, but nobody told me, probably because it was considered a formality, that the appointment had to be confirmed by the Queen. Salaries and other payments were dealt with by the City Education Department and the Professor had direct access to its Director. However, shortly before my appointment, the office of Kansalier-Directeur of the university had been created preparatory to the university taking charge of its own administration. The first holder of this office, which corresponded roughly to that of Registrar in English universities, was Frans Bender, who became a good friend of ours and godfather to our daughter, who was born in Amsterdam in 1959.

Our first object was to decide whether we would accept the appointment. We approached the idea very positively, Marion even more so than I. The title of Professor, the novelty of an Australian being appointed in continental Europe, the idea of living in Europe, which had fascinated us on our visits (although these had not included Holland), the challenge of learning a new language, but above all, the chance it gave to build up my own school of research, which I had hoped to do in Canberra, were all in favour of acceptance.

There were, however, formidable difficulties. The Laboratory of Physiological Chemistry was housed in three buildings grouped around the Jonas Daniel Meijerplein. The main laboratory shared with the Laboratory of Physiology and a Police Station an old building that had once been a leper

house.[34] On the other side of a busy road was a small hut,
housing two research laboratories and, a little further back,
an old typical Amsterdam warehouse (a listed building),
which contained the animal house and the laboratories for
the practical classes for the medical students. The layout in
the main building was bizarre. There were virtually no corridors, rooms opening on to one another, or to steep stair-
ways to another floor. The furnishings of the laboratories
reminded me of the old Organic Chemistry Laboratory at
Melbourne University in the former gardener's cottage. My
future colleagues in Amsterdam were rather apprehensive
about my reaction to the way the laboratory was housed, and
had emphasised that there were plans for a new building,
but I surprised them when I told the Burgomaster, during
my first meeting with him, that the last thing I wanted to do
soon after taking up my appointment was to have to plan a
new laboratory. In fact, I found the existing building to have
a certain charm, a feeling that was to grow with the years.
Also, the total area was quite impressive. Thus, this was not
a main worry.

My reception by the senior staff gave me more cause for
concern. Professor Jansen received me courteously, but it
was clear that he was disappointed that his nominee from
his own staff had not been appointed to the Chair. One of
the senior staff members had already accepted an appointment in America, another was negotiating a position at the
Shell laboratories and a third had accepted a position as
clinical chemist. The fourth, Professor Radsma, who had
been repatriated from Batavia, where he had been Professor
of Physiological Chemistry and had been interned by the
Japanese, planned to stay on, which was a great relief, since
he did a large part of the teaching to medical students.

---

34 There is a famous painting in the Rijksmuseum in Amsterdam of a group of
distinguished looking ladies and gentlemen from about the 16th century, called the
Regents and Lady Regents of the Leper House in Amsterdam, whom I consider my
predecessors.

Although the laboratory was well equipped for research in nutrition, including an excellently equipped and run animal house, it lacked much of the apparatus necessary for my work. It was clear that I must make the adequate provision for more equipment a condition for my acceptance.

The biggest headache was, however, the revelation that the laboratory housed an Institute of Nutrition, funded by the Netherlands Foundation for Nutritional Research, which had been set up by Jansen as a vehicle to supplement the quite inadequate funding by the Municipality. I was at first surprised and elated to find that no less than 17 highly trained (or undergoing training) laboratory technicians were working in the laboratory. The bubble was pricked when I discovered that only one of them was on the staff of the laboratory. There were two secretaries, one, working in the room next to the Professor and whom appeared to me to be the *de facto* Department Head, was a member of the staff of the Institute. The secretary who was on the staff of the laboratory was, in addition, in charge of the joint physiology and physiological chemistry library, where she had her desk and which could be reached only through the joint lecture theatre when there were no lectures. The problem was aggravated by the fact that much of the ongoing research was supported by grants to the Institute and that most members of the technical staff were receiving from the Institute supplements to their university salary. Nearly three quarters of the money required for supplies and chemicals came from the budget of the Institute.

After 2 weeks in Amsterdam, I presented to the Presidium of the university a memorandum on the future development of the laboratory and the Institute, with a list of requirements. I stated that I would be pleased to accept the position, starting 1 May, provided that the memorandum was accepted. I also added the condition that suitable living accommodation be found.

Since it was clear that I could not immediately put a stop to the support funnelled through the Netherlands Foundation for Nutritional Research, I proposed that the Institute for Nutrition should remain associated with the laboratory and that Professor Jansen, who had been Director of the Institute should be appointed Advisor.[35]

My list of requirements comprised creation of posts for an additional member of the academic staff[36] and a secretary, and for the transfer of 8 of the 16 positions for laboratory technicians from the Institute to the laboratory, the construction of a cold room, a trebling of the budget for chemicals and supplies to f30 000 per year (then about £9500), an increase of the library budget to f5000 per year with an extra grant of f1000 for the purchase of books, an immediate equipment credit of f80 000 to be granted for the purchase of specified large pieces of equipment and the promise of an additional f125 000 for equipment to be purchased over the next 5 years. In drawing up this list of requirements, I had a lot of help from the staff of the laboratory and also from my colleagues in the medical faculty, especially the Secretary, Professor Hagedoorn. All urged me not to be modest.

We returned to Cambridge awaiting the response of the university authorities to our list of requirements. On 30 March, I received notification that my appointment had received royal approval and soon afterwards that my list of requirements, even the allocation by Housing Department of the Municipality of an apartment, had been approved, and we decided to accept the appointment. In order to comply with UK currency controls, we decided to travel to The Netherlands as tourists with the restricted tourist's currency

---

35 I left for further discussion with the Institute the question whether I should succeed Professor Jansen as Director. In fact, soon after I took up my appointment, it became clear that it was not possible to continue the association with the Institute, which subsequently moved to Wageningen. How I solved the problem of supplementary salaries to the technical staff will be discussed later.
36 I had in mind a candidate for this post, but he later declined.

allowance, to return to the UK in August and officially emigrate in September, when we should be allowed to transfer more funds to the Netherlands as emigrants. While waiting for an answer from Amsterdam, I was asked by a colleague, who had turned down the offer of the Chair of Biochemistry in Zurich, whether I would like to have my name put forward. I replied that things had gone too far in Amsterdam. I think that it is unlikely that I would have been successful ahead of the distinguished German Professor who was appointed, but one wonders what would have happened!

Amsterdam 1955–1985

Starting in Amsterdam

My first task in Amsterdam was to learn about the work in the laboratory and to understand the university system in the Netherlands. Marion's first task was to learn enough Dutch to be able to shop, since, at that time, few shopkeepers spoke English. From my first day in the laboratory, I had to deal with official correspondence in Dutch, which I managed with a dictionary, together with the help of my secretary and the senior academic staff. However, I had too many other worries to give much attention to speaking the language, except for one weekly evening lesson, together with Marion, given by Hagedoorn's secretary. The secretaries, academic staff and students spoke English, but talking to most of the technical staff had to be via an interpreter. In fact, because the Dutch spoke English with little accent, in some cases I over-estimated their ability to understand what I was saying and everyone was too polite (or, in those faraway days, too over-awed by a Professor!) to admit that they had not understood me. I suppose that my Australian accent did not help! In any case, when I decided at the beginning of 1956 to ban English, and ordered everyone to try to under-

stand my Dutch and to speak to me in Dutch that I understood, mutual comprehension increased and research by my
senior technician suddenly progressed. Within 2 weeks, I
was reasonably fluent in what I, at least, called Dutch. Although, unknown to me, the Law on Higher Education in
The Netherlands specified that all lectures should be in
Dutch, my appointment was not conditional on my learning
Dutch, contrary to transatlantic rumours, and no-one even
mentioned this informally. It was just taken for granted by
everyone, including myself, that I should give lectures in
Dutch as soon as possible. For the first year and a half, my
lectures to medical students were in English. Switching to
Dutch was sudden and un-premeditated when, having prepared my lecture in English and dealt with my morning
Dutch correspondence, I realized on my way to the lecture
room that I was rehearsing the lecture in my mind in Dutch.
Despite numerous grammatical mistakes, the lecture went
reasonably well, except for some American students who had
found my lectures the only ones they could understand!
However, having burnt my boats, I prepared my next lecture
in toto in grammatical Dutch. This was a disaster, since,
afraid of not sticking to my text, I read from it and got
through the material for the next two or three lectures in
one go. Lacking any spontaneity or fervour, it was terrible.
Having learned from this mistake, I changed my technique
and enjoyed lecturing in Dutch to medical students.

The educational system for medical students was quite
similar to that with which I was familiar in Australia. The
first year in the basic science subjects was looked after by
the Faculty of Science. Biochemistry (then called physiological chemistry), together with the other basic pre-clinical
subjects, was given in the second and third years. There
were, however, two important differences. First, owing to
limited teaching laboratory space, the 120 students were divided into 9 groups and the same practical class was given 9
times over a two-week period, the tenth day being required

for preparation for the next exercise. The advantage of this system, from the point of view of the laboratory, was that it justified the appointment of a large number of part- and full-time staff, mostly older undergraduates or graduate students. The other difference was that examinations were oral, which were quite new to me. This involved for us about 300 half-hour oral examinations each year. At first, Professor Radsma took all these examinations, with me often sitting in, to learn the ropes. Afterwards, we divided them equally, as we did the lectures. From my point of view, the oral examinations provided an excellent feed-back of how the lectures were getting across to the students, but I found their subjective nature a serious fault of the system. For example, shear boredom after 3 weeks of oral examinations made it difficult not to put more difficult questions in the fourth week. Later, I introduced a written examination as the primary test, with an oral as a second chance. Still later, multiple-choice examinations, which I abhor, were introduced for all subjects, but by then I had left teaching medical students to others.

The education system for science students was completely foreign to my experience. Although the first 3 years study was similar to that in England or Australia, leading to a B.Sc., which is a professional qualification, in The Netherlands it led to a candidate's diploma, which allowed the candidate to study for a doctoraal diploma, but was not itself a professional qualification. The doctoraal diploma corresponded roughly to the M.Sc. Degree in Melbourne and most English universities (Cambridge being the exception), but, whereas study for a M.Sc. requires a research project in one sub-discipline, for example organic chemistry, a doctoraal diploma required one in two sub-disciplines of chemistry and in a non-chemical subject (usually physics or biology), as well as stiff oral examinations. Another novelty for me was the fact that the oral examinations were taken not at fixed times, but whenever the student felt sufficiently prepared.

As a result, it was the norm to take 3–4 years to progress from the candidate's diploma to the doctoraal. Although a doctoraal diploma is recognised as a professional qualification, it is not a degree, but gives the right of the holder to present a thesis for a doctorate, which is the only degree proper in the Netherlands.

Even more alarming was to find that biochemistry was not considered a discipline in its own right, but, under the name physiological chemistry, as a sub-discipline of chemistry, to be taken only as part of the doctoraal examination, either as main or subsidiary subject.[37] Moreover, there was no formal teaching of biochemistry. Students were expected to pick it up by reading text-books during their research project. The hour in the time-table allocated to physiological chemistry was used for lectures on world nutritional problems give by Professor Dols, a 'Special' Professor, appointed by The Netherlands Institute of Nutrition. I replaced this course by *capita selecta* lectures on respiratory enzymes. Even after I started giving my lectures to medical students in Dutch, I continued using English for the *capita selecta* lectures for 10 years, in order to give the students experience in this language. After the first Bari Meeting in 1965, I decided that, in order to give the graduate students more experience in taking part in scientific discussions, all seminars and research discussions would henceforth be in English. Since this removed the reason behind my giving the *capita selecta* lectures in English, I switched to Dutch for these lectures.

The way in which the doctor's title was awarded in The Netherlands was also completely strange to me. At that time, the functions of supervisor, examiner and the conferer of the degree, which were clearly separated in England, were combined in a single Professor, the Promotor. When I took up my appointment, four members of the laboratory were

37 It was not until the mid 1960s that biochemistry became a subject for the doctoraal examination. It is still not a discipline independent of chemistry or biology.

preparing a thesis, with Professor Jansen designated as Promotor. However, according to the law as it then stood, he lost his right to be Promotor when he became emeritus[38] and I had to replace him, despite the fact that two of the theses were ready to be printed and the dates for their defence were already fixed. My problem was not so much in judging their merits, since I could rely on Jansen and his senior staff who had supervised them and the first thesis by B.N. Bachra on plasma phosphate was in English, but in speaking the words in which I, according to my legal right, conferred the doctor's title. My Dutch lessons consisted largely of reading the pronouncement, which included one of the most difficult sentences for an English speaker to pronounce. I got through this, but I think that my teacher taught me the wrong emphasis on one phrase, since, shortly before I left Amsterdam, I was told by a colleague that everyone waited for the way I spoke this phrase. Everyone enjoyed hearing this mistake so much that they never told me about it!

My first big worry was what to do with The Netherlands Institute of Nutrition. It soon became clear to me that, even if the grants for specific research projects were continued, there was an insufficient financial basis to maintain salaries unrelated to specific projects. When I indicated that I was not prepared to accept the Directorship of the Institute, its Board decided to move the Institute to Wageningen, which was a great relief to me. I was able to transfer to the laboratory staff those laboratory technicians (most of them) who wished to be transferred, but this left the problem of the supplements to their (very low) salaries that the other technical staff of the laboratory were receiving from the Institute. This was solved by the university agreeing to supplement their university salaries for a limited period, during

---

38 The law was later changed giving this right for 3 years after retirement. When the retiring age was reduced from 70 to 65 years, the Government generously extended this period of service without-pay to 5 years!

which I could make proposals for accelerated annual incre-
ments or promotions. In the event, nobody suffered a loss of
salary.

In this and in many other problems I had to contend with
in my first year, I was greatly helped by the fact that I was a
foreigner. I got away with things that would have been impossible for a Dutchman, because the authorities felt under
an obligation to someone they had brought from abroad and
who was confronted with difficulties not of his own making.
The support of Frans Bender was of particular importance.

Marion and I were immediately struck by the old-
fashioned formality, almost Victorian, of Dutch society, at
least in university circles.[39] Inaugural lectures were especially formal occasions and the new Professor was expected
to be 'at home' to receive colleagues and their wives for the
two Sunday afternoons after the lecture. The monthly evening meetings of the Medical Faculty, that at that time consisted only of about 20–25 Professors, opened my eyes to the
still existing trauma of the German occupation that had
ended only 10 years previously. Many of the Professors had
played a prominent role in resistance. One was, at the time
of the liberation in May 1945, in prison in The Hague,
awaiting possible execution. One could feel the coolness between these and some who had been less heroic. After one of
our 'at home' days, a colleague told me that that was the
first time he had spoken to Professor X from another faculty,
since the day he had to make him more frightened of the resistance than of the Germans and persuade him to hand
over a list of students who had been deported to Germany.

I was committed to three major contributions to scientific
meetings in the summer of 1955, all on different aspects of
my recent work The first was the main paper on the mecha-

---

39 Only later did I come across Heine's saying that if the world was coming to its
end, he would go to Holland, because there everything happened 50 years later.
While this was apposite in 1955, it was certainly not so in the late 1960s.

nism of oxidative phosphorylation at the 3rd International Congress of Biochemistry in Brussels. During this meeting, I met for the first time, Wang, yin-lai, who reported his discovery with my old friend, Tsou, Chen-lu, that succinate dehydrogenase is a flavoprotein. Another friend, Tom Singer, had independently and simultaneously made the same discovery and it was my pleasure to chair the session in which both reported their work.

The rest of the summer we spent in our newly completed house in Cambridge, travelling to Oxford for the other two meetings. The first was a Faraday Society discussion on the physical chemistry of enzymes to which I have referred above. My only other memory of this meeting is the impression made by a confident young man, R.J.P. Williams, who presented a discussion comment on the factors determining the redox potentials of haemoproteins.

The second Oxford meeting was a symposium of the Society of Experimental Biology on mitochondria and other cytoplasmic inclusions, in which I gave a paper entitled Sarcosomes (Muscle Mitochondria), in which the importance of the integrity of the mitochondrial membrane for oxidative phosphorylation is stressed. I quite enjoyed reading this paper recently!

First five years in Amsterdam 1955–1960

Until the newly ordered equipment arrived, it was not possible to continue many of the lines that I had been pursuing in Cambridge. Also, in the first few months the only staff member available for help in my own research was a senior technician, J. Bouman. Paul Greengard, who had joined me in Cambridge shortly before my appointment in Amsterdam, spent part of the summer in Amsterdam, and after the summer, I was able to appoint my technician, Ann Searle, from Cambridge. From Professor Mendel, I received Dave Myers, a Canadian, who had been working with him on esterases and whom I was able to appoint temporarily to the extra staff position that I had obtained. Before the end of 1955, I had received 7 applications from within the Netherlands from those wishing to obtain a doctorate, of which I was able to accept 3, and 5 from outside the Netherlands for a post-doctorate stage or a sabbatical year, 3 of which I accepted. In addition, more and more students were choosing biochemistry as a subject for their doctoraal examination, which involved a research project. Thus, by the beginning of 1956, manpower was not a limiting factor in developing the various lines I had been following in Cambridge.

In my original paper on the mechanism of oxidative phosphorylation, I had speculated that, since a lot of lipid is present in the Keilin and Hartree heart-muscle preparation, one might look for a lipid-soluble component as the hypothetical coupling factor C. A lipid-soluble vitamin was an obvious candidate and I had already asked Dr K.L. Blaxter at the Hannah Research Institute in Scotland to analyse the preparation for vitamin E, Professor K. Dam of Copenhagen for vitamin K and Dr E. Kodicek in Cambridge for vitamin A. Less than $0.03\,\mu$mol vitamin K and $0.01\,\mu$mol vitamin A per mg protein were found, too small in relation to the concentration of the electron-transfer carriers to make an involvement in oxidative phosphorylation likely. Preliminary analyses by Dr Blaxter, however, although inconclusive, were sufficiently encouraging to warrant further investigation and, immediately after my arrival in Amsterdam, I asked J. Bouman to take this up. He was able to show the presence of appreciable amounts of α-tocopherol ($0.4\,\mu$mol/ mg protein). Since α-tocopherol can be oxidised to α-tocopherol quinone, we also treated the non-saponifiable matter with ascorbic acid in HCl to reduce it to α-tocopherol, in order to obtain what we called the total α-tocopherol content, which amounted to about four times the amount of α-tocopherol itself [57].

This opened up what became a major line of the laboratory in the next few years, with, as it turned out, disappointing results. When F.L. Crane published, in 1957, that heart particles contain large amounts of a quinone, later called ubiquinone or coenzyme Q, it became apparent that the extra α-tocopherol that we measured after reduction by ascorbic acid-HCl might originate from the new quinone, whose structure was then unknown, rather than from α-tocopherol quinone. Harry Rudney, who spent a sabbatical year with us in 1957–1958, Jan Links, recently appointed to the laboratory, Bouman and I set about determining the structure of the new quinone, in glorious ignorance of the fact that the big guns of Merck in the USA, working with Green's group, and Hoffmann La Roche in Zurich, working with R.A. Morton, were competing in the same race. Indeed, in a paper submitted in April 1958, we could show that a compound indistinguishable from α-tocopherol by our analytical procedure (one-dimensional paper chromatography) was obtained by treating the new quinone with ascorbic acid-HCl and we suggested that the latter is structurally related to α-tocopherol quinone, but that the side chain was more like that in vitamin K [58]. A few months later, at the 4th International Congress of Biochemistry in Vienna, I gave an invited paper on the possible role of vitamin E in respiratory-enzyme systems reporting these and later studies [59]. During the discussion of this paper, the Merck group and Morton announced the structure of the quinone, determined mainly by NMR spectroscopy. We were gratified to learn that our proposed structure, even though incomplete, was right so far as it went. A full account of our work was published some years later[60]. In this paper, we showed how to distinguish between the acid-reduction product of ubiquinone (which we named ubichromanol), ubiquinol and α-tocopherol and more rigorously established the presence of α-tocopherol. We also showed that all the α-tocopherol of heart muscle is in the mitochondria. Dilution experiments with labelled α-tocopherol showed the absence of detectable amounts of α-tocopherol quinone.

This lack of any evidence for the presence of an oxidized form of α-tocopherol in mitochondria made it an unlikely candidate as a redox component and, although I continued to keep open the possibility of it having a coupling role in oxidative phosphorylation, by 1962 I had accepted Tappel's view that it functions as a lipid antioxidant.

About the same time as our interest in α-tocopherol, Martius proposed a specific role for vitamin K in oxidative phosphorylation and gave a Plenary lecture on this at the 3rd International Congress of Biochemistry in 1955. Since, as mentioned above, analyses carried out for us failed to reveal any vitamin K in heart particles, we were never very keen on this idea, but I asked my graduate student, Jo Colpa-Boonstra, further to investigate the possibility. Although she found that reduced vitamin K_3 (which lacks the long side-chain of the natural vitamins K) is rapidly oxidised by heart particles, we could not support the idea that vitamin K plays any role *in vivo* [61].

In retrospect, it can be said that, although both Martius and we were right in our idea that a lipid-soluble quinone is an electron carrier in the respiratory chain, we bet on the wrong quinone. The credit for the discovery of the quinone that *is* involved – ubiquinone – belongs to F.L. Crane and D.E. Green and their colleagues. All that remains of our work is the first identification of α-tocopherol in mitochondria.

Alfred Tissières joined us for a few months in order to make use of a very large Sharples centrifuge present in the laboratory to collect sufficient cells for the isolation of large quantities of cytochromes c_4 and c_5, which was his main interest at that time, and to complete the work on oxidative phosphorylation catalysed by the respiratory granules obtained from this organism that we had started in Cambridge. I asked Gerda Hovenkamp to assist him in this work, and she returned with him to Cambridge for a few weeks. After Tissières left to join Jim Watson in Harvard, she continued this research in Amsterdam for her Ph.D. thesis. My special interest in this work lay in the fact that, because of the very high rate of respiration, the rate of phosphorylation is the highest known, even though, like all preparations from bacteria, the P:O ratio is relatively low. Alfred's special concern was the question of the origin of the particles. Electron micrographs of our preparation by J.R.G. Bradfield showed the presence of large amounts of small granules, also seen in the intact cell. Our conclusion, citing the summary of our paper [62] was:

> The respiratory chain in extracts of *Azotobacter vinelandii* is localized in small particles. It is possible that these small particles are identical with granules of about the same size which can be seen in the cytoplasm in electron micrographs of the whole cell. It cannot be excluded, however, that they might be derived by the disintegration of the cell membrane. In any case, these granules differ in several respects from mitochondria, which do not appear to be present in bacterial cells.

Subsequent work by Tissières in Watson's laboratory, showed that the granules seen in the electron microscope are ribosomes and made it probable that, as first suggested by Mitchell and Moyle in 1956 the respiratory activity of our preparations resided in fragments of the cytoplasmic membrane (vesicles), which may not have been visible in the electron microscope pictures. Gerda Hovenkamp made a detailed study of the properties of the respiratory particles and obtained her degree in September 1959, after which we stopped this line of research.

The first grant that I obtained from The Netherlands Organization for Pure Scientific research (ZWO) was to try to identify the BAL-labile factor and in May 1956, I was able to appoint Dirk Deul for this purpose. Since his work is described in a previous contribution to this series [6], I can be brief here. He obtained his doctorate in July 1959, my first student to obtain the doctorate of science. His most important finding was made a little later, together with M.B. Thorn, who was on a visit to the laboratory from London. This was the discovery of what I later called the 'double kill'. It was known that either treatment with BAL, or addition of antimycin, blocks the *reduction* of cytochrome *b*. Deul and Thorn [63] now found that addition of antimycin after BAL treatment prevented the *oxidation* of cytochrome *b*. The significance of the 'double kill' did not become apparent until about 20 years later. No further attempt was made to identify the BAL-labile factor until 1979.

Naturally, I was anxious to get back to the study of the mechanism of oxidative phosphorylation, but, in lacking a refrigerated centrifuge, it was not possible to make mitochondria. When I paid a courtesy call on Professor H.G.K. Westenbrink at Utrecht, he kindly offered me a loan of a centrifuge, until the one that I had my ordered arrived. I was still not set up to measure P:O ratios, but Lardy's discovery of the 2,4-dintrophenol (DNP)-induced ATPase opened up another approach, requiring only a simple colorimeter to measure inorganic phosphate. I asked Dave Myers with his experience of esterases to undertake this research.

According to my hypothesis, the ATPase activity is explained by the reactions (see above)

$$ATP + A + C \Leftrightarrow ADP + P_i + A \sim C$$

$$A \sim C + H_2O \xrightarrow{\text{(DNP)}} A + C$$

Since one might expect a different ligand C for each of the three phosphorylation sites, we hoped that we might be able to detect three different ATPases. A detailed investigation by Myers of the effect of pH and various activators and inhibitors seemed, indeed, to show the presence of three enzymes with different pH optima [64]. It is now clear that this conclusion, although it seemed plausible at the time (and even now reads plausibly if one closes one's mind to what we now know) was an over-interpretation of a system, the complexity of which we greatly under-estimated. In addition, we were misled by one of those coincidences that plague the scientist. We considered the possibility that the effect of pH was on the concentration of undissociated dinitrophenol, but rejected it, since p-nitrophenol is about as effective as 2,4-dinitrophenol at the same concentration, despite the fact that, because of differing pK's, the concentration of undissociated p-nitrophenol near neutral pH is about 1000 times that of 2,4-dinitrophenol. Later work by Hemker and Hülsmann showed that undissociated 2,4-dinitrophenol is about 1000 times more lipid soluble than undissociated p-nitrophenol and, since it is the amount of undissociated phenol in the lipid membrane that determines its activity, the effects of lipid solubility and pK cancel out [65]. They found a single optimum at pH 6.9.

In order to be able to answer a question put after I presented these results at a conference in Japan at the end of 1957, I asked a young medical student, who worked as a volunteer for 4 months in 1957–1958, to determine whether the effect of pH on the concentration of the active species of ATP complicated the interpretation of the results. The table summarizing his measurements at 3 pH values, 3 concentrations of dinitrophenol and 6 concentrations of ATP, which I included in the printed report of the discussion of my paper and which showed that this was not a complicating factor, is typical of the care with which this student (Piet Borst) carried out his experiments.

Of course, it is now known that there is only one mitochondrial ATPase and that its activity depends on a number of factors, including an endogenous inhibitor, many of which are affected by pH. A lasting impact of Myers work was his procedure, based on that of others, for the preparation of rat-liver mitochondria, for which his paper was frequently cited in subsequent papers from our and other laboratories, and his rather overlooked finding that the requirement for Mg^{2+} could also be met by Mn^{2+}, Co^{2+}, Zn^{2+}, Fe^{2+}, Sn^{2+}, Cd^{2+}, Ni^{2+} and Be^{2+}. Also important was his finding that the ATPase activity is not affected by respiratory inhibitors or by the redox state of respiratory-chain components. This showed that it did not involve the 'A' in my formulation and made it necessary to introduce an additional reaction. Following a proposal by Chance and Williams, the symbol 'C' was replaced by 'I' (for inhibitor) and the original scheme was now expanded to

$$AH_2 + B + I \Leftrightarrow A{\sim}I + BH_2$$

$$A{\sim}I + X \rightarrow A + X{-}I$$

$$X{\sim}I + ADP + P_i \Leftrightarrow X + I + ATP$$

and X~I became the uncoupler-sensitive intermediate.

Myers' work was taken up by Wim Hülsmann, a medical graduate waiting for a vacancy to specialize in surgery. His work seemed to support our conclusion of three ATPases by the observation that, when the P:O ratio with heart mitochondria was plotted against the pH, the same three peaks as found for the ATPases could be observed with glutamate as substrate, two with succinate and one with ascorbate. The number of peaks was the same as the expected number of phosphorylation sites with these substrates [66]. The different pH optima are especially clear when the P:O ratios with one substrate are subtracted from those with a second substrate in difference curves. Even now the conclusion looks valid to me, whatever it might mean.

In 1957, a report appeared describing the preparation, from disintegrated mitochondria, of a naturally occurring uncoupler of oxidative phosphorylation which was named mitochrome, since its spectrum resembled that of a haem compound. Wim Hülsmann, together with Harry Rudney and Bill Elliott, who, like Harry, was spending the sabbatical year 1957–1958 in Amsterdam, undertook to examine what appeared to be an exciting finding. They confirmed the uncoupling activity, but found that the substance or substances responsible could be extracted from the haem compound into iso-octane [67]. Further studies showed that the compound showing the haem-like spectrum was a degradation product of the cytochromes and was of little further interest. The uncoupling lipid, which we nicknamed 'mitolipid', looked more promising as a new line of investigation. This proved to be the case, but only indirectly.

At this time, Warburg's hypothesis, that uncoupling of oxidative phosphorylation has something to do with cancer, still had its supporters. Mitochondria isolated from tumours had low P:O ratios. For this reason, it seemed not unreasonable to ask those funding cancer research to support an investigation into a new naturally occurring uncoupler. I was successful not only in this application but in persuading Piet Borst to interrupt the clinical stages of his medical studies to work for his doctorate on this project. However, by the time he started his research in September 1959, Hülsmann, in collaboration with chemists from Unilever, had shown by gas chromatography that 'mitolipid' consists of long-chain fatty acids, 60% of which are unsaturated. Together with a student, J.A. Loos, Borst showed that unsaturated fatty acids are much more effective uncouplers than saturated [68].

Since the original research project was now effectively finished, it was changed to an examination of the claim that mitochondria isolated from tumours have a low P:O ratio. Borst was also able to put that idea to rest by isolating structurally intact mitochondria from ascites tumour cells with a normal ability to carry out oxidative phosphorylation. This preparation was found to be especially suitable for studying the mechanism by which NADH formed outside the mitochondria by glycolysis delivers its reducing equivalents to the mitochondrion. Borst discovered in isolated mitochondria what is now recognised as a major pathway, the aspartate shuttle, although he rightly pointed out a thermodynamic difficulty in accepting this shuttle as the mechanism in the intact cell [69]. This difficulty was only resolved many years later with the discovery that the aspartate:glutamate carrier is electrogenic.

As recalled recently [70], a mistaken conclusion led Borst to the discovery of an unexpected cyclic mechanism of the oxidation of glutamate [71], which opened up a new line of research that was to occupy the laboratory for many years.

I was still much worried with the fact that we were finding lower P:O ratios than those reported by others. I was particularly concerned with the reports in the literature claiming a value for 2.0 for the oxidation of succinate to fumarate, since our values were much lower. Together with Paul Greengard and a newly joined graduate student, Koen Minnaert, we submitted a paper to the *Biochemical Journal* detailing our findings in Cambridge and Amsterdam. We explained the higher ratios obtained by others to the further oxidation of the products of the oxidation of succinate. The paper was rejected, a new experience for me. Accordingly, I asked a doctoraal student, I. Betel, to determine the (fumarate + malate) concentrations at the end of our measurements of the P:O ratios, at that time a tedious undertaking, which was nevertheless completed and included in the paper which was re-submitted and now accepted [72]. The new measurements supported the criticisms that we had made in our rejected paper. Nowadays, it is widely [46], although not universally [73], believed that the true ratio for succinate is 1.5, close to the average value (1.57) that we reported for rat-liver mitochondria.

An American post-doc, Jack Purvis, who spent 2 years in the laboratory from August 1956, undertook the search for the postulated high-energy intermediate of site-1 of oxidative phosphorylation, which, according to the chemical hypothesis, was NAD+ bound to the hypothetical C (or I). To our great excitement, he found that the NAD^+ + NADH content of mitochondria oxidising substrate in the absence of ADP appeared to be much less than the total NAD content, measured in the presence of uncoupler [74]. I reported this at a Ciba symposium, where it aroused great interest. Others, notably Martin Klingenberg, threw doubt on this finding, ascribing Purvis' results to a fault in measurement of NADH. Although I considered Purvis' analytical procedure to be well controlled, I had gradually to accept that the evidence for a NAD~ compound was unsatisfactory. The final proof that no such compound exists in detectable quantities was provided in 1964, by Karel van Dam [75], when he showed that the amount of NAD^+ formed immediately after addition of ADP to 'State 4' (resting) mitochondria is completely accounted for by the NADH present before addition of ADP. I still have no satisfactory explanation for Purvis' findings, but suspect that an artefactual formation of a C-4 adduct of NAD^+ was the reason.

One of my first graduate students, Jo Colpa-Boonstra, who impressed me with the care and neatness with which she carried out her experiments, was given the difficult project of studying the role of cytochrome b. In particular, I asked her to re-examine my finding in Cambridge that NADH appeared to be a poor reductant of cytochrome b. Her clearcut experimental findings remained without a satisfactory explanation until Mitchell's proposal of the Q cycle nearly 20 years after she graduated. Of remaining significance is the first determination of the effect of pH on the redox potential of cytochrome b [76].

With the help of a grant from the newly formed Foundation for Chemical Research in The Netherlands, I was able to appoint Cees Veeger to undertake research on flavoproteins, in particular succinate dehydrogenase. In early experiments, it was found that extracts of heart muscle containing succinate dehydrogenase also contained an enzyme catalysing the oxidation of NADH by ferricyanide, but not by dyestuffs, which was puzzling, since, at that time, it was accepted that the first step in the oxidation of NADH by the respiratory chain is catalysed by 'diaphorase', characterised by its ability to catalyse the oxidation of its substrate by dyestuffs. As already mentioned, a flavoprotein with this property had been isolated in the Molteno Institute by Bruno Straub and I had studied this preparation briefly in Cambridge. At this early stage in Cees' research, Vince Massey showed that Straub's enzyme catalyses the reduction of NAD^+ by lipoic acid, and proposed that this is the physiological function of the enzyme, which is now known as lipoamide dehydrogenase. As a consequence, Veeger's research now switched from succinate dehydrogenase to lipoamide dehydrogenase. He showed that the classical 'diaphorase' activity with dyestuffs is largely an artefact caused by an irreversible modification of the protein by heavy metals. Our practice of using EDTA to guard against the effects of traces of heavy metals explained the low activity with dyestuffs found by Veeger. When I told Vince about these results, he suggested that Cees should join him in Sheffield and I was able to arrange grants for this purpose. Their collaboration led to a considerable advance in establishing the mechanism of action of the enzyme [77]. Cees obtained his doctorate *cum laude* in December 1960.

The first applicant from outside Amsterdam for a place as graduate student was Koen Minnaert, from Utrecht University. He worked from 1956 to 1960 on cytochrome oxidase. He made a very thorough study of the kinetics of the enzyme and his name has a lasting place in the literature ('Minnaert Mechanism IV') for his theoretical examination of possible mechanisms to explain what seemed the paradoxical discrepancy between first-order kinetics when the oxidation of ferrocytochrome c is followed spectrophotometrically, and Michaelis and Menten kinetics with respect to cytochrome c in the system reducing agent – cytochrome c – cytochrome oxidase – oxygen [78]. He also measured the redox potential of cytochrome a.[40]

In the summer of 1957, I received an application from another medical graduate, Harry Hulsmans, to be allowed to work in the laboratory as an unpaid graduate student preparatory to specialising in internal medicine. I asked him to investigate the distribution of enzyme activity in fractions of a heart-muscle homogenate obtained by differential centrifugation, following the procedures developed by Albert Claude and Christian de Duve for liver. In particular, I wanted to know whether heart muscle contains what was then known as 'microsomes', which may seem a strange question in 1995, but it underlines how little was known about protein synthesis in the mid 1950s. Indeed, Hulsmans found, in a small fraction sedimenting at high speeds, a concentration of enzymes characteristic of microsomes in liver, as well as of RNA [79]. Most important, the largest incorporation of amino acids into proteins was found in the mitochondrial fraction. This confirmed Mel Simpson's discovery in 1955, of protein synthesis in mitochondria. Hulsmans' observations became important to us when the existence of mitochondrial protein synthesis was later called into question when it was found that subsequent work by one of Simpson's collaborators (not involved in the initial work) was fraudulent.[41]

Another new field for the laboratory was opened up in an unexpected way, when in the summer of 1959 I was asked by an Amsterdam paediatrician to measure the glucose-6-phosphate dehydrogenase activity of blood from a patient

---

40 After defending his thesis in 1960, Koen obtained a fellowship to work with Lucile Smith in Dartmouth, USA. He returned to a position in the biological section of the Philips Laboratory in Eindhoven, where he continued his work on the redox potential of cytochrome a until his early tragic death.

41 After obtaining his doctorate in May 1960, Hulsmans specialised in internal medicine and was later appointed to the Foundation Chair in this subject at the newly established University of Maastricht.

with suspected favism. It so happened that I had just received an application from a medical graduate, T.L. Oei, to work in the laboratory and I handed over the project to him. This led to an examination [80] of 100 members of a single family. This was later extended to 547 members of the family, including many who had emigrated, possibly one of the largest studies of this type ever undertaken [81]. The mutation was traced back as far as a lady born in 1796. The final chart is a tribute both to the accuracy of personal records in The Netherlands and the family morals in the village of Sloten, now swallowed up in the western suburbs of Amsterdam.

Oei also studied glcyogen-storage diseases, an abnormality first discovered by Professor S. van Creveld, who was still Professor of Paediatrics at the university. In collaboration with van Creveld, Hülsmann found that leukocytes can be used to test the deficiency of phosphorylase characteristic of a type of glycogen-storage disease [82]. This research was later extended by Frans Huijing under the leadership of Hülsmann. It was a source of great satisfaction that they were able to establish that this was the specific enzyme deficiency in two of the first patients that van Creveld described in 1934.

Up to the summer of 1960, 13 foreign visitors had worked, or were still working, in the laboratory, as post-docs or on sabbatical leave – Jack Purvis, Harry Rudney, Bill Elliott, Walton Geiger, Bob Berne and Frans Jöbsis from the USA; Bill Chefurka from Canada; Leslie Wheeldon and Meg Bailie from Australia; Jan Michejda and Lech Wojtczak from Poland; and Bodil Kruse Jacobsen from Denmark.

After the departure of Wöstmann and van den Linden already in 1955, the only senior staff members were Professor Radsma and, after his doctorate, R. Nunninkhoven, who stayed on for a few years. Radsma, who did a lot of the teaching, carried on his own research on some effects of ascorbic acid on respiring preparations. Nunninkhoven helped with the management of the laboratory and deter-mined the amino acid composition of horse-heart and two yeast cytochromes c [83]. Since my attempts to attract suit-able staff from outside were unsuccessful, I had no senior staff, apart from Dave Myers and Alfred Tissières for short periods to help with the supervision of the research of the doctoraal or graduate students or of the post-doc and other visitors. This is the reason that most of the projects in the early years derived from my work in Cambridge.

Fortunately, there were not many doctoraal students in the first year or two. By the time the number started to in-crease, the graduate students had progressed sufficiently far with their research to be able to supervise a student re-search project related to, or more often forming a part of, their doctorate project. By the end of 1958, when my first graduate students were graduating, I was looking to them to take up appointments in the laboratory. Wim Hülsmann, who, after graduating *cum laude* in December 1958, decided to remain in biochemistry instead of specialising in surgery, his original intention, received a Stipendium from ZWO to spend a year in New York with Fritz Lipmann. After his re-turn, I was able to appoint him to a position on the staff and in 1962 obtained from the medical faculty a new position of 'Lector' (corresponding to Associate Professor or Reader) in Medical Enzymology, to which position he was appointed.

It was more difficult to keep my other star pupil, Piet Borst, in biochemistry. After completing his thesis, also *cum laude,* in May 1960, he had to finish his clinical stages to complete his medical degree. Not surprisingly, given his in-telligence and family background,[42] he became interested in clinical medicine and less inclined to return to biochemistry. In the meantime, I was able to arrange a Rockefeller Fellow-ship for him to work with Severo Ochoa, whom I urged to make him enthusiastic again for a career in biochemistry.

---

42 His father was Professor of Internal Medicine at the University of Amsterdam.

This he did by setting him to work with Charles Weissmann on RNA viruses. While he was in New York, a new Lectorship in the laboratory was created by the Medical Faculty to teach biochemistry to dental students. Borst was appointed to this position in 1964.

Before this, I was able to appoint Cees Veeger to a senior position, immediately after receiving his doctorate (once again *cum laude*) in December 1960. Cees was given the task of starting a course in biochemical techniques, to be taken by students before they undertook their research project. Until then, following my own experience, they had been thrown in at the deep end, so to speak. Cees quickly built up an active group of graduate students studying flavoproteins, and took three of them, as well as this research topic, with him when he was appointed to the first Chair of Biochemistry at the University of Wageningen, in 1963.

Second five years 1961–1965

The period, 1961–1965, was critical in the development of the laboratory. Recently, I took as my theme for an after-dinner speech, 'the story of a year'. The year in question was 1961, a year of great significance in the history of the development of bioenergetics. It was also a critical year in the history of the laboratory in Amsterdam. Hülsmann and Veeger were now established with their own research groups, studying fatty acid oxidation and flavoproteins, respectively. Post-docs and those spending a sabbatical year in the laboratory in the period 1961–1965 included Joseph Tager from South Africa, about whom much more later; George van Rossum, Cherry Tamblyn and Bob Reid from England; Alan Snoswell, Robin Currie and Cliff Kratzing from Australia; Zbigniew Kaniuga from Poland; Dagmar Siliprandi and Sergio Papa from Italy; Marti Koivusalo from Finland; Harvey Penefsky, Richard Guillory, John Howland, A.M. Pappenheimer, Darrell Haas, Richard Tobin, Isaac Harary and Jack

Davis from the USA; and Dorothy Dow from Canada. The second generation of able students who started their graduate studies during this period included: Monne van den Bergh, Ab Kroon, Bob van Gelder, Bertus Kemp, Ed de Haan, Rob Charles and Jan de Vijlder, who all became Professors in The Netherlands universities, Danny Der Vartanian and Frans Huijing, who both went on to distinguished academic careers in the United States, and Emil Christ, who joined Unilever. I supervised van den Bergh, van Gelder and, together with Hülsmann, Kroon. Veeger supervised Der Vartanian, and Hülsmann supervised Christ and Huijing, as well as Coen Hemker, who had started his graduate studies at the end of 1959. Like Hülsmann and Borst, Coen was a medical graduate who planned a clinical career, in Coen's case in paediatrics, after completing his thesis in April 1962, but who was also seduced into a career in biochemistry. Coen became the founding Professor of Biochemistry and later Rector in the University of Maastricht.

During this period, in particular, the post-doc visitors played an important role in the scientific development of the graduate students, including their knowledge of the English language. Most of them worked on projects close to my own interests, but, following English rather than American practice, I gave little supervision to the post-docs, in retrospect perhaps too little. Kaniuga and, as a graduate student, Der Vartanian, studied NADH dehydrogenase [84] and succinate dehydrogenase, respectively. Joseph Tager, who had originally planned to spend 2 months with us, decided to emigrate from South Africa and I was able to appoint him to the staff. He played an increasingly important role, first in collaborating with me in the studies with oligomycin that will be reported below, and in supervising Ed de Haan and Rob Charles in their studies of the mechanism of glutamate oxidation and citrulline formation.

The unexpected finding that Jan Links' private collection of antibiotics contained oligomycin, which had been reported

by Lardy in 1958 as an inhibitor of oxidative phosphorylation, prompted me to ask Frans Huijing further to study its
action as part of his undergraduate research project. We
presented the results of this study in a paper published [85]
in the *Journal of the Japanese Biochemical Society* in honour
of the 70th birthday of Professor K. Kodama. The main conclusion was that oligomycin prevents reaction of the dinitro-
phenol-sensitive intermediate of oxidative phosphorylation
with phosphate. At the same time, Alan Snoswell [86]
showed that oligomycin does not inhibit the reduction of
NAD+ by succinate driven by the oxidation of succinate by
oxygen. These two papers, together with similar work by
Lars Ernster in Stockholm at about the same time, established one of the main tenets of my hypothesis of the mechanism of oxidative phosphorylation, the existence of a non-
phosphorylated high-energy intermediate of oxidative phosphorylation. It did not, however, give any information concerning the nature of this intermediate, which became the
main bone of contention in the succeeding years. Further
examination of the mechanism of energy-driven reduction of
NAD+ and its link to the reductive synthesis of glutamate
from 2-oxoglutarate and ammonia became the major research activity of Joseph Tager and myself over the next few
years [87].

As well as studying the effect of substituted guanidines,
Richard Guillory [88] introduced Dio-9 as a new tool in the
armoury of bioenergetics, that was later to prove to be a useful oligomycin-like inhibitor of photophosphorylation.

Coen Hemker, in work that, in my opinion, did not get the
attention it deserved, undertook a systematic study of the
effect of lipid solubility and pK on the uncoupling activity,
measured by stimulation of the latent ATPase in mitochondria of a homologous series of 2,6-dinitro-4-alkylphenols
[89]. He concluded that the uncoupling activity is directly
determined, not by the amount of phenol added to the me-

dium, but by the amount dissolved in the mitochondrial lipid.

Minnaert's work on cytochrome c oxidase was taken up by Bob van Gelder. By titration with NADH, in the presence of phenazine methosulphate, with or without inhibitors, he confirmed other work that indicated that the molecule contains four electron acceptors – two haems and two copper centres – and he was able accurately to determine the extinction coefficient at different wavelengths [90]. The validity of this work was shown recently, when unexpectedly it was found that the molecule contains three atoms of copper, not two [91]. However, two of these atoms are in a single centre that accepts only one electron. After defending his thesis in the autumn of 1966, van Gelder made two visits to Madison as post-doc with Helmert Beinert, where they made the important discovery of high-spin cytochrome a_3 at half reduction that has been described in Beinert's contribution to an earlier volume in this series [91]. Shortly after returning from his first visit, van Gelder was appointed Lector in Medical Chemistry with the responsibility of teaching chemistry to first-year students in medicine and dentistry, a task that I undertook myself when van Gelder was in the USA. He led an active research group in the laboratory until his tragic death from an accident in 1995.

Following up my interest in protein synthesis in mitochondria, arising from Harry Hulsmans' work, I asked Ab Kroon, another medical graduate, to make this his topic for his thesis. Ab confirmed other recent work showing that isolated mitochondria are able to incorporate amino acids into protein and that this system differs markedly from the ribosomal system, notably in its insensitivity to ribonuclease and sensitivity to chloramphenicol. He made a special study of the energy requirements for the incorporation. He found the incorporation to be inhibited by actinomycin D, a specific inhibitor of DNA-dependent RNA synthesis, implicating DNA in the process [92]. However, it proved difficult to obtain unequivocal evidence that mitochondria contain DNA. In agreement with previous work, he found that the mitochondrial preparations contain DNA, but it could not be excluded that this was due to contamination with nuclear DNA. Kroon gave special attention to this question in his thesis, which he defended in February 1966, by which time Nass and co-workers had published an electron micrograph showing a nucleic acid-like fibre in mitochondria, but chemical confirmation was still lacking. At about this time, Piet Borst, who had just returned from New York where he obtained the experience in nucleic acid chemistry, turned his attention to this problem. The final upshot of this was the demonstration by Borst and co-workers [93], in 1966, that mitochondria contain a specific circular DNA. This opened up a whole new field for the laboratory – the biogenesis of mitochondria.

Mitochondrial protein synthesis, which had got a bad name from the fraudulent activities mentioned above, suffered another attack when it was claimed that the incorporation of amino acids was, in reality, into contaminating bacteria. This was finally settled by demonstrating unaltered protein synthesis in sterile mitochondria [94].

Monne van den Bergh turned to the study of oxidative phosphorylation in the mitochondria of fly thoracic muscle, that Lewis and I had started in Cambridge. He was able to prepare exceptionally intact mitochondria with high respiratory rates with the natural substrates, pyruvate and α-glycerol phosphate, and with complete impermeability to other substrates [95]. After defending his thesis at the end of 1962, van den Bergh spent a year in Al Lehninger's laboratory and, on his return, studied fatty acid oxidation in mitochondria. In 1964, he filled a newly created additional post of Lector of Biochemistry in our laboratory. In June 1968, he was appointed Professor of Veterinary Biochemistry at the University of Utrecht.

Another line of research that was opened in the early 1960s was the study of the properties and mechanism of action of glyceraldehydephosphate dehydrogenase, which was the subject of four successive doctorate theses, by Ab Hilvers (1964), Jan de Vijlder (1970), Wim Boers (1976) and Ruud Scheek (1980). There were several reasons for my interest in this enzyme. First, it was the best-studied example of substrate-linked oxidative phosphorylation, which was a model for my hypothesis of respiratory-chain phosphorylation. Secondly, it was hoped that a study of the nature of the firmly bound NAD^+, known to be present in the crystallised enzyme, would give a clue to the nature of Purvis' 'extra NAD^+'. Thirdly, since it was a tetramer, it provided a good model for the study of inter-subunit interactions which, at that time, I considered to play an important role in the conservation of energy in the respiratory chain. Indeed, the most important result of this research was Jan de Vijlder's finding [96], simultaneously with and independently of Dan Koshland, of negative co-operativity of NAD^+ binding to the enzyme, the first example of negative co-operativity. Although this project yielded few other results of lasting significance, it taught us a lot about some of the pitfalls of NAD^+ chemistry, and provided important training in enzymology for the graduate students and undergraduates involved. After obtaining his doctorate, Jan de Vijlder joined the Department of Paediatrics and is now Professor in the Medical Faculty in Amsterdam.

What was to prove the most important development in the study of the mechanism of oxidative phosphorylation in this period was the publication by Peter Mitchell in 1961 of the chemiosmotic hypothesis. As I have described elsewhere [97], from the beginning I took this hypothesis very seriously and gave it pride of place in an extensive review [44] that I submitted early in 1964 and revised at the proof stage in 1966 to take account of Mitchell's new experiments. However, my early enthusiasm evaporated when Mitchell, in or-

der to explain that twice as many protons were ejected as
envisaged in his original hypothesis, proposed a 'looped' respiratory chain which I found difficult to reconcile with what
I knew about the respiratory chain [98].

At about this time, the 'chemical' hypothesis seemed to receive support from studies by Vignais and others showing
that mitochondria pre-incubated in the presence of substrate
and oxygen, but in the absence of ADP and phosphate, produced a 'burst' of ATP on the subsequent addition of ADP
and phosphate. It was concluded that the high-energy intermediate (A~C in my hypothesis) was built up during the
pre-incubation. However, as mentioned above, Karel van
Dam [75, 99] showed that the 'burst' could be completely accounted for, independent of whatever mechanism of oxidative phosphorylation was favoured, by the rapid oxidation of
NADH formed during the pre-incubation period. The importance of this finding was that it showed that the amount of
A~C, if it existed, was of an order of magnitude less than the
concentration of nicotinamide nucleotides or of ubiquinone in
the mitochondria. This was the death knell both of Purvis'
'extra NAD' and of quinone compounds as intermediates of
oxidative phosphorylation.

During a meeting in Warsaw in 1965, I argued that van
Dam's findings also created difficulties for the chemiosmotic
hypothesis, since it was far less than what I understood to
be the magnitude of $\Delta\mu_{H+}$ built up in the absence of acceptor.
This was further argued in a paper [100] which included experiments that seemed to cause difficulties in the interpretation of Mitchell's 'oxygen pulse' experiments. Apart from an
angry response from Mitchell, the points raised in this paper
were, still in my opinion, not adequately addressed.

Although not essential for the chemical hypothesis, it was
not unlikely that the reaction of A~C with ADP and phosphate to give ATP would proceed via an ~P intermediate. I
gave Bertus Kemp the task of seeking this elusive compound. As a starting point, I asked him to look at atracty-

loside, which Pierre Vignais had reported to inhibit the transfer of phosphate between the hypothetical –P compound and ADP. Following up a report by Bruni that atractyloside prevents the binding of adenine nucleotides to mitochondria, Bertus found that, in fact, atractyloside has no effect on oxidative phosphorylation *per se,* but prevents the interaction of externally added ADP with intramitochondrial ADP [101]. I was sitting near the front when Vignais presented his work at the International Congress of Biochemistry in New York in 1964, and managed to catch the Chairman's eye at the beginning of the discussion period in order to report Kemp's results. Martin Klingenberg, frantically waving his hand at the back of the hall, then reported similar results. I suppose that we can claim priority by a matter of seconds, but, in fact, Martin went further than we had by proposing that atractyloside acts on a specific adenine nucleotide carrier. A strong competition developed in the next few years between Vignais, Kemp and Klingenberg in the study of the properties of this carrier. Martin Klingenberg pursued this further than the others and in beautiful work has established the structure and mechanism of action of this and related carriers.

Cultures of single beating heart cells, prepared by Isaac Harary, provided a beautiful demonstration [102], that I used frequently in lectures, of the fundamentals of bioenergetics. Dinitrophenol stops the beating, but this is restored by ATP. Neither oligomycin nor iodoacetate alone inhibited the beating rate, but the combination of the two inhibitors do so. These experiments show clearly that ATP, derived from either glycolysis or respiration, is required. Unlike flagella movement, for example, the beating of the heart can not be driven by $\Delta\mu_{H^+}$.

Sometime in the early 1960s, I was told that the building on Jonas Daniel Meijerplein was to be abolished in connection with building the approach roads to a new tunnel to be built under the Amsterdam harbour. The Medical Faculty planned to move a number of the pre-clinical laboratories into a new building adjacent to the Academic Hospital in Amsterdam. I was loath completely to move away from the other laboratories in the Science Faculty and managed to persuade the latter to accommodate me into their building plans and to allocate space on the Plantage Muidergracht made available by tearing down an old physics laboratory. At the same time, I offered to the Medical Faculty to house the Section on Medical Enzymology in their new building which was called the Jan Swammerdam Institute. Much of the early 1960s was occupied with planning the new laboratories. Remembering that when my predecessor, B.C.P. Jansen, was brought back from Batavia to be the first Professor of Physiological Chemistry in Amsterdam, he had been promised a new building, which, because of the depression and the war never eventuated, I proposed the new building be called the B.C.P. Jansen Institute. In addition to the space for the laboratory itself, the building contained five lecture and seminar rooms for the use of other departments in the science faculty. I was made Director of the whole institute with a total gross area of 35 600 m². Monne van den Bergh organised the move into the new institute early in 1967 and claims that he did an experiment the next day (I was away skiing!). The new institute was officially opened on 7 June 1967, by the brand-new Minister of Education, Veringa, who a few years later was responsible for a new Higher Education Act that caused great disruption in the universities.

Hülsmann had moved into the Jan Swammerdam Institute a few years before, which eased the crowding in the old laboratory, but we were bursting at the seams by the time we moved. A large grant for new equipment opened up new possibilities for research.

The second decade (1965–1975)

Karel van Dam defended his doctorate in 1966 and shortly afterwards left for a post-doctorate year in Britton Chance's laboratory in Philadelphia. On his return to Amsterdam, he was appointed Lector in place of van den Bergh, who had moved to the Chair in Utrecht. With his students, Johannes Fiolet, Gerrit Veeneman, Tilly Bakker-Grunwald and Ruud Kraayenhof, Karel took up the study of photophosphorylation, which had been introduced in our laboratory by Keith Rienits, a former Canberra colleague of mine, during a sabbatical year in 1966. Ruud Kraayenhof later became Professor of Plant Physiology at the Free University in Amsterdam.

Joseph Tagerwas appointed Lector in the Medical Faculty in 1967 and Ed de Haan, who later succeeded Joseph as Lector, when the latter was appointed Professor in the Science Faculty, and Fred Meijer joined his group. Their work on glutamate oxidation expanded to the study of anion carriers in mitochondria.

Bertus Kemp stayed on after completing his thesis in 1968, taking over the lectures to dental students from Borst, when the latter was appointed Professor of Medical Enzymology after Hülsmann moved to the Erasmus University at Rotterdam. Bertus, who was later appointed Professor in the dental faculty, remained attached to my group until his untimely death in 1985.

In 1968, two graduate students joined my group, who played an important role in the development of the research into the 1970s, and at the time of writing occupy senior positions in the laboratory. Simon Albracht inherited from Bob van Gelder the facilities in the laboratory for measuring electron-spin resonance spectra and, after a post-doctorate stage with Vängård in Göteborg, became a leading exponent of this technique for the study of the metal components of the respiratory chain, particularly the iron-sulphur proteins.

Later, he was largely instrumental, together with Thauer, in establishing the presence of nickel in a group of hydrogenases.

Van Dam's findings did not exclude the possibility that the cytochromes, which are present in much smaller concentration than NAD, might be directly involved in energy conservation. However, in line with Paul Boyer's 'conformation' variation of the chemical hypothesis, I now pictured a 'high-energy' conformation of the apo-proteins, rather than an liganded form of the prosthetic group, as the intermediate of oxidative phosphorylation. I drew on the recent work of Changeux on allostery and the new structures being published by Max Perutz as a model [103].

Jan Berden undertook to solve what I called the cytochrome b paradox, namely that it appeared not to behave as a simple electron carrier. He was not helped in this by some experiments of my own, started during a summer visit to Britton Chance's laboratory in Philadelphia, that I incorrectly interpreted as showing the existence of a high-energy form of cytochrome b. Märten Wikström, who spent a year in our laboratory in 1970–1971, pointed out that my results could be explained by the oxygen-driven reduction of the cytochrome, something that I found very difficult to understand. Wikström and Berden [104] published an important paper in which they proposed a mechanism to explain this phenomenon. Without being able to explain, either to my colleagues or myself why, I was unhappy about their explanation. As I have explained elsewhere [6], my unhappiness was probably caused by my sub-conscious telling me that it did not explain Deul and Thorn's 'double kill' experiments (see above), which my conscious mind had forgotten. This was fully explained by Mitchell's Q cycle. Wikström and Berden deserve credit for proposing half the Q cycle, I deserve blame for not leading them to the discovery of the complete cycle. Of permanent value was Berden's calculation [105] from fluorescent quenching methods that the antimycin-binding site (now known to be the Q_{in}-binding site) is situated 2.4 nm from the haem of cytochrome b. I claim to be the first to realise, when the amino acid sequence was published by Sanger' s group, that cytochrome b is a single protein with two haems [106].

The laboratory was enriched in the late 1960s by a number of gifted visitors from abroad. In addition to Wikström, there were S. Muraoka from Japan, In-Young Lee from Korea, Jadwiga Bryla from Poland, Ruben Vallejos from Argentina and C.S. (Samson) Tsou from Taiwan. Among those joining Borst, Leslie Grivell from London deserves special mention, since he remained in Amsterdam and became Professor of Molecular Biology. Muraoka [107] discovered a 'cross-over point'[43] between cytochrome a_3 and oxygen, which was the first proof that energy is conserved in the reaction with oxygen, a conclusion that was later established by Wikström's demonstration of a proton pump in cytochrome oxidase.

I was struck by a report by Bert Pressman indicating that respiring mitochondria could build up what was called a 'phosphate potential'[44] which I realised was much higher than was easily accommodated by the generally accepted location of phosphorylation site 2 in the respiratory chain, independently of the mechanism of oxidative phosphorylation. Our studies [108], which included a re-investigation of the value for the free energy of hydrolysis of ATPase [109],[45] confirmed the high value of the phosphate potential.

---

43 A 'cross-over' point was defined by Britton Chance as the point in the respiratory chain, before which components became more reduced, and after which they became more oxidised on adding ADP to 'resting' mitochondria in the presence of substrate.

44 Equal to ΔG_0 + 1.36 log ([ATP]/[ADP][P$_i$]), where ΔG_0 is the standard free energy change in the hydrolysis of ATP.

45 Our value is still widely used. Krab [110] has recently corrected the values that should be used in the presence of magnesium.

During the late 1960s, we also directed our investigation of oxidative phosphorylation from the other end, so to speak, by studying the enzyme responsible for the last step in the synthesis of ATP, the ATPase. The role of this enzyme, originally known as coupling factor F_1, had been established some time earlier by Racker. In our laboratory, Ruben Vallejos isolated an especially active form of the coupling factor [111], which was shown to be a complex of F_1 with another protein, identified by van Dam and his students as the previously discovered 'oligomycin-sensitivity conferring protein' (OSCP), now known as one of the subunits of the multisubunit ATP synthase. It remains of some interest (and I think that it has been forgotten) that a complex of F_1 with OSCP can be easily isolated.

The F_1 ATPase became a major topic of research in my group in the early 1970s when Jan Rosing and David Harris[46] joined as graduate students. The most important finding was that F_1, as usually isolated, contains firmly bound ATP and ADP, more of the former than the latter [112]. I immediately realized that, if, by a conformation change in the ATPase protein, the operation of the respiratory chain would cause this firmly bound ATP to be dissociated from the protein, this would be the final step in oxidative phosphorylation. I developed this further in a Plenary Lecture to the International Congress of Biochemistry held in Stockholm that summer [103]. Independently and at about the same time, Paul Boyer, for different and sound reasons, had the same idea, which has now been firmly established as being correct. For me, the final proof came with Harvey Penefsky's demonstration that protein-bound ATP is spontaneously synthesised when ADP and phosphate are added to pure F_1.

---

46 David Harris spent 2 years in Amsterdam working for his D.Phil. Degree in Oxford, as part of a collaboration with George Radda.

Reorganization of Netherlands universities as a result of student pressure

By the beginning of 1969, the laboratory (whose name had changed from Physiological Chemistry to Biochemistry) was housed in two new buildings, occupying the whole of the B.C.P. Jansen Institute, located near the chemistry laboratories, and half a floor of the Jan Swammerdam Institute, located a few miles away alongside the Academic Hospital. It had grown enormously since 1955. In addition to myself, Borst was a full Professor and there were three Lectors (Tager, van Gelder and van Dam), 4 other senior academic staff (Pandit-Hovenkamp, Kroon, Kemp and de Haan), 28 graduate students, 23 part-time demonstrators and 56 members of the technical and administrative staff. In addition, there were a large number of doctoraal students working full-time, alongside the graduate students, on their research projects. Despite its size, it remained a workable entity, although there was inevitably some division, caused partly by geography and partly by the nature of the research, between those working on the two sites. Piet Borst was active in keeping this division within bounds.

As Director, I was legally responsible for the entire laboratory, but I naturally delegated and all matters concerning the laboratory, including appointments, promotions and allocation of funds between the various research groups were discussed at weekly meetings of the heads of the research groups. Various attempts to appoint an experienced manager from outside to help me failed and, by the beginning of 1969, I had entrusted this task to a senior technician, Henk Cornelissen, whom I had inherited from Jansen. He set up an effective financial control system, which, at that time was quite inadequate at the university level.

The system of governance at Dutch universities, like that in other continental countries, had changed little since the 19th century. The faculty, which made all the important decisions (or made recommendations which were nearly always accepted by a higher authority) consisted only of the full Professors, of whom there was typically only one per discipline. The most striking difference with the Anglo-Saxon system, with which I was familiar, was the lack of participation of senior non-Professorial academic staff in decision-making. Towards the end of the 1960s, some of the more politically active of the latter made representations to redress this situation. However, this reasonable aim was soon swamped by the wave of student unrest in European universities starting in 1968 in Paris. A number of our biochemistry students took a leading role in this movement in The Netherlands. Laboratories were encouraged by the university authorities to work out 'democratic' administrative structures. A committee of our laboratory worked out two possible solutions that was put to a secret ballot among the entire laboratory community. The result of this ballot, as first reported to me in my office, was a narrow victory for the more radical solution, which I indicated was not acceptable to me. Shortly afterwards, I was told that a recount had taken place and that now the less objectionable alternative had received the majority. I have often wondered about the 'recount'. In the event, the agreed structure was superseded by a new law, introduced in 1970 by a panicky Minister of Education, that transferred my responsibility to a governing body of 36 members of the laboratory, including all 18 members of the permanent academic staff and six representatives each of the technical and administrative staff, graduate students and undergraduates, the Chairman to be elected by the committee, 'for at least 1 year', as the new law stated. I was elected Chairman, but only for 1 year, but with the possibility of re-election each year. The election for only 1 year rather offended me, but I later realised that the closest thing to re-election for life is a yearly renewable election. I now give credit to the students understanding that better than I did. In the event, I had to take positive action when I decided to step down after my 60th birthday.

The 'democratic' structure was ill-advised. It was cumbersome and time-consuming and caused much disruption in many departments, and the end of some distinguished careers. That it worked reasonably well in the case of our department ('vakgroep' as it was now called) is due to the good sense of the members of the committee.

I found this period emotionally very upsetting and seriously considered leaving The Netherlands. I had already turned down two offers, one in 1958, to succeed Sir Rudolph Peters at Babraham near Cambridge, and one in 1963, to succeed Trikojus at the University of Melbourne when he retired in a few years. Although both offers were, for personal reasons, most attractive, I had refused them, both because of the obligation I had to my students in Amsterdam, particularly in 1958, and because of the exciting prospects of building a school that Amsterdam offered. Now, however, I felt that the most exciting period for me were over, that the future of the laboratory was secure in the hands of Piet Borst and others, and that I could move on with a clear conscience. Towards the end of 1969, I applied for the Chair of Biophysics at the John Curtin School of Medical Research of the Australian National University, which had been vacated by A.G. Ogston on his appointment as Master of Trinity College at Oxford. Nostalgia for Canberra played a part in what was I suppose a quixotic decision. In any case, I was not appointed. I have only applied for three jobs in my life, all to Canberra. In the first of these, I was successful, in the other two, both to the Australian National University, I was unsuccessful.

The new Higher Education Act threatened the unique structure of the Department that I had built up. Unlike other universities, there was a single Department of Biochemistry, linked to both the Science and Medical Faculties.

My original appointment was in the Medical Faculty, but in 1959, I was invited to join the Science Faculty. By the time of the new Act, two additional Chairs or Lectorates had been created in the Science Faculty (van Dam and Tager) and three in the Medical Faculty (Borst, van Gelder and de Haan), with other staff roughly equally divided between the two faculties. However, the Department was run as a single entity, which gave great flexibility in the allocation of teaching, research or administrative duties. The new Act required that the democratically constituted 'vakgroepen' be installed by a single faculty. I took this up with the Ministry who suggested that I propose a separate Faculty of Biochemistry. I thought that this was an excellent opportunity to develop biochemistry in The Netherlands, but could not get my colleagues in the other universities to agree. I still think that this was a chance missed. However, I was able to resolve this difficulty by persuading the two faculties in Amsterdam to sign an agreement by which the Medical Faculty created the vakgroep, but in which the rights of the Science Faculty were safe-guarded.[47]

Fortunately, by 1970, there were four strong groups besides my own in the laboratory carrying out independent research and it was possible to devise a workable structure, with five 'sub-vakgroepen' based on the main research activities. That for medical enzymology and molecular biology, under Piet Borst, was exceptionally active, attracting many

---

47 Soon after I retired, another amendment to the Higher Education Act necessitated that the agreement be renegotiated, which the Medical Faulty refused to do, with the result that two 'vakgroepen' for biochemistry were created. However, although this meant a division of the Department for teaching, the members of the two 'vakgroepen' resolved to remain together for research purposes in the form of an Institute for Biochemical Research, to which they honoured me by giving it my name. The Institute was founded in 1989. Later the Department of Microbiology joined the Institute, but, I am saddened to learn while writing this section, that as a result of reorganisation of the Medical Faculty (which has *de facto* been taken over by the Academic Hospital), the Biochemistry Department in the Medical Faculty has withdrawn from the Institute.

post-docs from other countries as well as able graduate students from within the country. Four visitors from England – Les Grivell (see above), Dick Flavell who discovered the introns in the haemoglobin gene, Alan Fairlamb and (now Sir) Alec Jeffries who, after returning to England developed DNA analysis for forensic purposes, deserve special mention. Piet Borst's own field developed from mitochondrial biogenesis to the study of gene expression, especially in trypanosomes. It was the work of his group, more than any other, in the increasingly important field of molecular biology that greatly enhanced the reputation of the laboratory. Joseph Tager's sub-vakgroep studied inborn errors of metabolism, particularly the lysosomal diseases, and mathematical models of metabolic control. Particularly the latter, which appealed to a bright group of students, also widened the scope of research in the laboratory. Van Dam's group was strengthened by Pieter Postma, who after a year in the USA developed a strong group on the uptake of sugars by bacteria, and Hans Westerhoff. Van Gelder's group, which included Ton Muijsers and Ron Wever, continued the study of cytochrome oxidase. Ron later extended this to peroxidases, particularly sea-weed chloroperoxidase in which he discovered that vanadium is an essential metal.

The last decade in Amsterdam 1975–1985

The research activities of my group, which included Bertus Kemp, Simon Albracht and Jan Berden, in the last decade of my Amsterdam period, were centred around four themes: (i) the Q cycle; (ii) the mitochondrial ATPase (ATP synthase); (iii) mechanism of oxidative phosphorylation; and (iv) hydrogenase and NADH dehydrogenase.

I was slow in realising the importance of Mitchell's proposal of the Q cycle, even when Mitchell wrote me a long letter in which he expressed the hope that it might resolve the nature of the 'Slater factor'. I can only suppose that I was too

heavily engaged in other activities to give his paper, or letter, the attention it deserved. It was only when Bernie
Trumpower mentioned in a lecture in the summer of 1979
that, after extraction of the Rieske iron-sulphur protein, antimycin prevents the reduction of cytochrome *b* that I realized (and said so in the discussion of Bernie's paper) that the
Rieske protein might be the BAL-labile factor and that, if
this were so, the Q cycle explained the 'double-kill' experiment (see also Ref. [6]). My graduate student, Simon de
Vries, soon proved that this was the case [113]. I now became an enthusiastic supporter of the Q cycle [114], which
explained many of the apparent anomalies around phosphorylation site 2 as well as the 'cytochrome *b* paradox'. Much
effort was expended by a number of able graduate students
(Gerrit van Ark, Simon de Vries, Carla Marres and Qin-shi
Zhu (from C.L. Tsou's laboratory in Beijing)), co-supervised
by Jan Berden, to decide between the various possible ver-
sions of the Q cycle suggested by Mitchell and, in particular,
to establish how the two turns of the cycle necessary for the
oxidation of one molecule of ubiquinol, by two molecules of
cytochrome *c*, are co-ordinated. Simon de Vries [115] established, for the first time, the existence of the hypothetical
$Q^{-\bullet}$(out) postulated as a transient intermediate in the Q cycle. Although the double cycle that we proposed turned out
to be over-elaborate, I think that we can claim that we did
contribute to the general acceptance of the Q cycle.

Under the leadership of Bertus Kemp and Jan Berden, we
spent much effort in establishing to our satisfaction and
convincing others, that F_1 contains 6 adenine nucleotide-
binding sites. The introduction by Bertus Kemp [116] of 8-
azido-ATP and -ADP (and later by Vignais of the 2-azido
analogues) played a decisive role in this. We concluded that
two of these sites located on the β-subunits are catalytic
sites, two are ADP-binding sites located on the interface of
the α and β subunits, and the other two have a structural
role. Further work by Jan Berden and his students, since I

left Amsterdam have, in our view, confirmed this conclusion, which is consistent with the three-dimensional structure of the enzyme recently published by John Walker, although the latter, as well as Paul Boyer, still favours three catalytic sites.

Although I was still not convinced by the mid 1970s that the Mitchell hypothesis is a correct description of the mechanism of oxidative phosphorylation, I had by that time accepted that it was the best working hypothesis and, in a memorandum to my co-workers and colleagues, encouraged them to plan their experiments on this basis. The opening sentence of the concluding remarks of my contribution to a review in the *Annual Reviews of Biochemistry* [117], written jointly with five leading workers in the field, was:

> The feasibility of the essential features of the Mitchell chemiosmotic hypothesis is now firmly supported experimentally.

Our experience with the Q cycle finally convinced me of the correctness and importance of the uni-directional trans-membrane electron transfer inherent in Mitchell's 'loops' as the origin of the membrane potential. I accepted that in mitochondria, at least, the energy of the respiratory chain is conserved as this membrane potential, that could be utilised directly for the electrogenic events, such as the uptake of cations into the mitochondria or the exchange of intramitochondrial ATP^{4-} for external ADP^{3-} or of intramitochondrial aspartate for external glutamate. However, although I was prepared to accept the likelihood that protons are involved in the synthesis of ATP, I had difficulty in accepting that protons are extruded from the membrane by the operation of the respiratory chain and are then drawn back again through the ATPase to make ATP from ADP and phosphate. To my mind, $\Delta\mu_{H^+}$ does not behave kinetically as would be expected of an intermediate of oxidative phosphorylation. For this reason, I favoured the alternative view, postulated by R.J.P. Williams that, if protons are involved, they are directly transferred from the electron-transferring proteins to the ATP synthase within the membrane. In a very thorough theoretical study, Karel van Dam's brilliant graduate student, Hans Westerhoff,[48] proposed a membrane chemiosmotic mosaic to overcome what we saw as the difficulties for the conventional chemiosmotic theory. As I stated during the formal 'opposition' at the conferring of his degree, I found it difficult to picture mosaic chemiosmosis in terms of the known physical structure of the membrane and of the proteins that it contained, and proposed, instead, that the energy transfer occurs by direct contact between the respiratory enzymes and the ATPase.. Work by Hackenbrock had shown that there is a rapid translational diffusion of the proteins in the plane of the membrane. I first presented what I called the 'collision hypothesis', together with experiments supporting the hypothesis, carried out by a graduate student, Marga Herweijer, at a symposium in Beijing in 1984 that was not published until 1987, although it appeared earlier in a review article [118]. After my retirement, I wrote a review [119] and made a number of contributions to symposia reviewing the evidence in favour of the hypothesis. I cannot say that it has caught on, although I have yet to see any explanation reconciling the conventional chemiosmotic theory with the experimental evidence that I submit is inconsistent with conventional theory (or alternatively demonstrating that the evidence is incorrect). I am also encouraged by recently published evidence [120] supporting the view that the lateral diffusion of protons along membranes can provide a direct link between sources and sinks involved in chemiosmotic coupling.

My interest in hydrogenase arose from the apparent energy crisis of the early 1970s. In common with other bioenergeticists, I considered the possibility of constructing a

48 Now Professor of Plant Physiology at the Free University of Amsterdam.

bio-reactor consisting of photochemical system II and hydrogenase linked by an electron carrier. Under the auspices of ZWO, a number of groups in The Netherlands formed what became known as the 'hydrogenase project'. I asked Simon Albracht, who was already experienced with the use of EPR for the study of iron-sulphur proteins, to join this project. Although it has not (yet?) contributed to the solution of the energy crisis, this project had important scientific spin-offs, leading to the discovery in Amsterdam and Marburg of nickel [121], and in Wageningen of a new type of iron-sulphur cluster [122] in hydrogenase.

Extra-university activities

Biochimica et Biophysica Acta

My association with *Biochimica et Biophysica Acta* (BBA) began in 1957, first as Managing Editor for *Preliminary Notes* and *Short Communications* and, from 1964 to 1982, Managing Editor for the whole journal. This was a major task, which I have described elsewhere [123].

Nomenclature

My activities with BBA led to my first international job as Secretary, from 1959 to 1964, of the Committee on Biochemical Nomenclature of the International Union of Pure and Applied Chemistry (IUPAC), later joint with the International Union of Biochemistry (IUB). I remained a member of the joint committee until my appointment as IUB Treasurer in 1971. Membership of this committee was fruitful, more for the friendships that were formed with, among others, Otto Hoffman-Ostenhof, Bill Klyne and Jo Fruton, than for the intellectual activity. A high point was getting David Green, Karl Folkers and Isler to agree, in paper, on the nomenclature of ubiquinone – coenzyme Q, although I note

that Karl and David's students, as well as the pharmaceutical industry, do not follow the internationally accepted nomenclature. A lot of time was wasted in territorial arguments between the two unions.

A related activity was my membership, from 1961 to 1981,
of the IUB Committee of Editors of Biochemical Journals
(President from 1969 to 1971).

EMBO and EMBL

In 1965, I was elected a member of EMBO and was a member of its Fund Committee from 1973 to 1978, being its
Chairman for the last 4 years. This I found a most rewarding and interesting job, since it showed me what the bright-
est of the young molecular biologists were doing. Early in the
development of the European Molecular Biology Laboratory
(EMBL), I succeeded Arthur Rörsch as Head of The Netherlands delegations to the European Molecular Biology Conference (EMBC), which finances EMBO, and the EMBL
Council. From 1977 to 1980, during which the building
housing the laboratory was officially opened, I had my turn
as President of the Council of the latter body. In this function, I was Chairman of the Search Committee for the Director of the laboratory to succeed John Kendrew. I believe that
this laboratory has been a great success, for which both John
Kendrew and his successor, Lennart Philipson, in their own
very different ways, deserve great credit. Both had to contend with a Council that, in my opinion, dealt too much in
the micro-details of the running of the laboratory. Most
delegations were led by a departmental official, sometimes
quite a junior one. The Netherlands delegation was an exception in this respect, in that I was Head of the delegation,
but, since I was representing The Netherlands government,
I had to follow instructions from my co-delegate from the
Ministry of Education, even when I had had no input in
drawing up these instructions. When the Ministry decided to

make the departmental official the Head of the delegation, I resigned on a question of principle and asked the Royal Netherlands Academy of Science to take up the issue with the Minister. When they failed to do so, because they had more important matters to take up with the Minister, I felt that they had let me down.

The Royal Netherlands Academy of Science

I was very surprised but delighted to receive a telephone call from Westenbrink, in the spring of 1964, telling me that I had been elected a member of the Academy. Not long afterwards, he called me again with the sad news that he had to go to hospital and asked me to take over as Editor of BBA. When he died in the summer of 1964, I was called upon to carry on many of the roles, both nationally and internationally, that Westenbrink had fulfilled with much distinction. I became Chairman (from 1964 to 1975) of the newly formed Commission of Biochemistry and Biophysics of the Academy, which included non-Academy members (indeed, I was now the only biochemist in the Academy) and was instrumental in creating a new Section for these disciplines, which made it easier to nominate future members (I had been elected into the Section for Chemistry). Shortly before Westenbrink's death, it was announced by the Academy that Mr Alfred Heineken, the Head of the brewery company, had donated funds to establish a Prize in Biochemistry and Biophysics, to be given once in 3 years, in honour of his father, Dr H.P. Heineken. I became Secretary of the Selection Committee and drew up the guidelines for the selection of the prize-winner, one of which was that Nobel Prize winners be excluded. However, the committee has been successful in foreshadowing Stockholm on no less than three occasions. I remained Secretary until 1976, when it became necessary for me to resign when someone kindly proposed me for the prize. I got a lot of pleasure from this job, especially the prize-giving ceremonies and the subsequent dinners with the Heineken family, and was sorry to have to give it up, especially as I was not the successful candidate! The Dr H.P. Heineken Prize is now one of the most important prizes in biochemistry and has been joined by analogous prizes, named after Alfred Heineken, in other subjects. I was particularly pleased when Piet Borst won the Dr H.P. Heineken Prize in 1992.

I always enjoyed the monthly meetings of the Academy in its beautiful old building – the Trippenhuis – in the centre of old Amsterdam, a short bicycle ride from my laboratory.[49] Indeed, I found the Academy a refuge of eliteship from the over-democratised outside world!

The Netherlands Organization of Pure Scientific Research (ZWO)

In common with my contemporaries all over the world, I had to do my share towards the evaluation of applications for government research grants. In retrospect, I think that I did more than my share and that, on the whole, this was not the best use I could have made of my time. I was a member of the ZWO Council for 15 years (1970–1985), and of the small central executive for 6 years (1970–1976). Particularly membership of the latter body was very time-consuming, since it finally adjudicated all grant applications, not only in natural sciences and medicine, but also in the humanities. Moreover, it was a time when the government, and therefore ZWO, was becoming more and more involved in scientific policy. A related activity, although not directly connected with ZWO, was my chairmanship, in 1980–1982, of a government committee, reporting directly to the Minister of Scientific Policy,

---

49 I recall the surprise of my colleagues in the USSR when I told them that, in contrast to the privileges of an Academy member in that country (car, dascha, extra salary, etc.), my only privilege as an Academy member in The Netherlands was being given a key to the bicycle shed.

on the position and future of biochemistry in The Netherlands. Although the use of citation analysis was at that time controversial, our report was generally well accepted both by the biochemical community and the government.

International Union of Biochemistry (IUB)

For most of my time in Amsterdam, and for 6 years after my retirement, I was concerned in some way or other with IUB, first in connection with nomenclature and biochemical journals and from 1964, when I was elected to the Council, with its governance, from 1971 to 1979 as Treasurer, 1985–1988 as President-elect and 1988–1991 as President. I now claim the unique record of having attended every International Congress of Biochemistry (and Molecular Biology).

China

My interest in China stems from my friendship with Chen-lu Tsou[50] at Cambridge. After he returned to China in 1951, we had little contact, although I met his chief, Yin-lai Wang, at the International Congress of Biochemistry in Brussels, in 1955. I then met Tsou briefly during the International Congress in Biochemistry in Moscow in 1961, and had a long talk with him during the FEBS Meeting in Warsaw in 1966, where he and Wang reported on the total chemical synthesis of insulin. The only news during the long period of the cultural revolution was a misleadingly optimistic report in 1973 from Emil Smith referred to by Tsou in his contribution to this series.[51]

When Taiwan was admitted to IUB in 1961, China withdrew from the Union with members of the Council being accused of being 'lackeys to American imperialism'. An altera-

---

50 See his recollections published in another volume of this series [29].
51 *Loc. cit.*, p. 368.

tion to the statutes making 'scientific communities' members, rather than countries, was unsuccessful in bringing them back into the fold. In the mid 1970s, I made an unsuccessful application for a visa to China to discuss the problem. However, by making use of an exchange programme between The Netherlands and Chinese Academies, I was able to visit China in the spring of 1979, with Chen-lu as my host. This, the first of ten visits to China (including Taiwan), was a most fascinating and enjoyable experience. It also gave the opportunity of opening up discussions with Wang and Tsou, and officials of the Academy, which, after two visits to Taiwan together with Bill Whelan, the General Secretary of IUB, finally led to the return of China to the Union. The negotiations have been described elsewhere [124] and are also referred to by Tsou [29].

In 1983, I was invited to become a member of an International Advisory Panel for the Chinese University Development Project of the World Bank. This involved extensive, not always very comfortable, visits to universities all over China, both before and after my retirement from my Chair in Amsterdam. I found it fascinating work and enjoyed the company of my fellow members on the Panel and its Chinese equivalent body. After completion of our work for this project, I spent two very cold weeks, in the winter of 1988, in Beijing, making a preliminary assessment for a new World Bank project focusing on research, in which the Chinese Academy of Science was also involved. This was put in abeyance after the events in Tiananmen Square the following summer. My last visit to China was in 1993 to attend a symposium in honour of Tsou's 70th birthday.

Retirement

When I was appointed, the retiring age for Professors was 70, but, not long before I was due to retire, Parliament reduced this to 65, even for those already appointed. The new law became operative shortly before my 67th birthday in January, 1984, but I was able to obtain a special dispensation until 1 September 1985, which allowed me to participate, while still in function, in the International Congress of Biochemistry in Amsterdam, of which I was Honorary President. I decided not to ask to be able to stay on as emeritus Professor in the laboratory after I retired, partly because, although this was common in Anglo-Saxon countries, this was not so in The Netherlands. I felt also that, after retirement, one should not even give the appearance of looking over the neck of one's successor(s). The Netherlands being such a small country, it seemed better to return either to Australia or England. Marion's preference for England where she has many relations and our daughter settling in England made that country the natural choice. My enthusiasm for sailing led further to the choice of Lymington in Hampshire and Professor M. Akhtar, Head of the Biochemistry Department at the University of Southampton was kind enough to propose that I be appointed an Honorary Professor of the university, which in 1993 honoured me with an Honorary Doctorate. I enjoy the contact that I have with members of this very active department.

Although I have had no research group since retiring from my Amsterdam Chair, I have been kept busy, first with keeping in touch with graduate students who still had to finish their theses in Amsterdam, as well as with the IUB, China, BBA (I am still Honorary Executive Editor) and participating in scientific meetings. The availability at the time of my retirement of reasonably priced, user-friendly, Word Processors has been a godsend for emeritus Professors. Although I no longer participate in the advances in bioenergetics, I keep in touch with developments, now with the aid of the Internet. It has been exciting in the last year to see the three-dimensional structure of three of the enzymes with which I have been closely concerned – cytochrome oxidase, hydrogenase and mitochondrial ATPase.

I still also sail a lot in the summer and, until this year, ski in the winter.

Concluding remarks

I found this article quite difficult – even traumatic – to write, since I found myself frequently overcome by nostalgia for far away places and for people some of whom are no longer with us.

It has turned out otherwise than I intended. Somewhat to my surprise, I find that considerably more than half of the narrative relates to the period before I moved to Amsterdam. This may also surprise others who know me only from my Amsterdam period, but I hope that my experiences during the early days of biochemistry will be of interest to some. In reviewing my career, I am once again conscious of how much I owe to so many with whom I have come into contact – my parents, wife, teachers, colleagues, technicians and students. Many are mentioned by name, but, in the interest of preventing this already long story becoming still longer, I have had to omit mention of outstanding contributions to the research activities of the laboratory made by many students and visitors from abroad, which do not fit into the topics I have selected, perhaps rather arbitrarily. I have also given completely inadequate coverage of the work of the groups of Piet Borst, Joseph Tager, Karel van Dam and Bob van Gelder, after they became independent. This was done deliberately, since this belongs to their memoirs. To all of them my deeply felt thanks.

References

1   Cunningham, M.M. and Slater, E.C., Aust. J. Exp. Biol. Med. Sci. 179 (1939) 457–464.
2   Brown, G.M., Curtis, R.G., Davies, W., Dopheide, T.A.A., Hawthorne, D.G., Hlubucek, J.R., Holmes, B.M., Kefford, J.F., Osborne,

J.L., Robertson, A.V. and Slater, E.C., Aust. J. Chem. 21 (1968) 483–489.
3   Jansen, B.C.P., Recl. Trav. Chim. Pays-Bas 55 (1936) 1046–1052.
4   Slater, E.C., Aust. J. Exp. Biol. Med. Sci. 19 (1941) 29–32.
5   Slater, E.C. and Rial, E.J., Med. J. Aust. 29 (1) (1942) 3–12.
6   Slater, E.C., in: A History of Biochemistry. Selected Topics in the History of Biochemistry: Personal Recollections. II. Comprehensive Biochemistry, Vol. 36 (Semenza, G., ed.), Elsevier, Amsterdam, 1985, pp. 197–253.
7   Slater, E.C. and Morell, D.B., Biochem. J. 40 (1946) 644–652.
8   Morell, D.B. and Slater, E.C., Biochem. J. 40 (1946) 652–657.
9   Slater, E.C., Nature 157 (1946) 803–806.
10 Kratzing, C.C. and Slater, E.C., Biochem. J. 47 (1950) 24–35.
11 Freeman, J., A Passion for Physics, Adam Hilger, Bristol, 1991, 229 pp.
12 Herbertson, B., The Biochemist 14 (2) (1992) 4–10.
13 Slater, E.C., Biochem. J. 45 (1949) 130–142.
14 Slater, E.C., in: Oxidases and Related Redox Systems (King, T.E, Mason, H.S. and Morrison, M., eds.), Liss, New York, 1988, pp. 51–64.
15 Slater, E.C., Nature 161 (1948) 405–406.
16 Slater, E.C., Biochem. J. 45 (1949) 1–7.
17 Slater, E.C., Biochem. J. 45 (1949) 8–13.
18 Slater, E.C., Biochem. J. 45 (1949) 14–30.
19 Slater, E.C., Biochem. J. 44 (1949) 305–318.
20 Slater, E.C., Nature 165 (1950) 674–676.
21 Slater, E.C., Biochem. J. 46 (1950) 484–503.
22 Slater, E.C., Pasteur 18 (1969) 90–92.
23 Slater, E.C., Nature 163 (1949) 532–523.
24 Williams, T.I., Howard Florey – Penicillin and After, Oxford University Press, 1984, pp. 282–293.
25 Slater, E.C., in: Structure and Function of Energy-Transducing Membranes (van Dam, K. and van Gelder, B.F., eds.), Elsevier–North Holland, Amsterdam, 1977, pp. 329–339.
26 Bernardi, P. and Azzone, G.F., Biochim. Biophys. Acta 679 (1982) 19–27.
27 Slater, E.C., Nature 166 (1950) 982–984.
28 Chance, B., Nature 169 (1952) 215–221.
29 Tsou, C.L., in: A History of Biochemistry. Selected Topics in the History of Biochemistry: Personal Recollections. III. Comprehensive Biochemistry, Vol. 37 (Semenza, G. and Jaenicke, R., eds.), Elsevier, Amsterdam, 1990, pp. 349–386.

30 Slater, E.C. and Bonner, W.D., Jr., Biochem. J. 52 (1952) 185–196.
31 Slater, E.C., in: The Physical Chemistry of Enzymes, Discussions of the Faraday Society 20 (1955) 231–240.
32 Dixon, M. and Webb, E.C., Enzymes, Longmans, Green and Co., London, 1958.
33 Slater, E.C. and Bonner, W.D., A Citation Classic Commentary. Current Contents: Agriculture, Biology and Environmental Sciences, Vol. 20, No. 50, 1989, p. 16.
34 Slater, E.C. and Cleland, K.W., Nature 170 (1952) 118.
35 Slater, E.C. and Cleland, K.W., Biochem. J. 55 (1953) 566–580.
36 Cleland, K.W. and Slater, E.C., Q. J. Microbiol. Sci. 94 (1953) 329–346.
37 Slater, E.C. and Cleland, K.W., Biochem. J. 33 (1953) 557–565.
38 Mitchell, P., Symposium of the Society for Experimental Biology, Vol. 8, University Press, Cambridge, 1954, pp. 254–261.
39 Cleland, K.W., Nature 170 (1953) 497.
40 Cleland, K.W. and Slater, E.C., Biochem. J. 53 (1953) 547–556.
41 Slater, E.C., Biochem. J. 53 (1951) 157–167.
42 Slater, E.C., Biochem. J. 53 (1953) 521–530.
43 Slater, E.C. and Holton, F.A., Biochem. J. 55 (1953) 530–544.
44 Slater, E.C., in: Biochemical Reaction Mechanisms. Comprehensive Biochemistry, Vol. 14 (Florkin, M. and Stotz, E.M., eds.), Elsevier, Amsterdam, 1966, pp. 327–396.
45 Slater, E.C. and Holton, F.A., Biochem. J. 56 (1954) 28–40.
46 Brand, M., The Biochemist, 16 (4) (1994) 20–24.
47 Slater, E.C., Nature 174 (1954) 1143–1144.
48 Slater, E.C., Biochem. J. 59 (1955) 392–405.
49 Lewis, S.E. and Slater, E.C., Biochem. J. 58 (1954) 207–217.
50 Slater, E.C. and Lewis, S.E., Biochem. J. 58 (1954) 337–345.
51 Slater, E.C., Nature 172 (1953) 975–978.
52 Weber, B.H., Biosci. Rep. 11 (1991) 577–617.
53 Tissières, A. and Slater, E.C., Nature 176 (1955) 736–737.
54 Slater, E.C., Trends Biochem. Sci. 2 (1977) 138–139.
55 Slater, E.C., Annu. Rev. Biochem. 22 (1953) 17–56.
56 Slater, E.C., in: Biologie und Wirkung der Fermente. 4. Colloqium der Gesellschaft für Physiol. Chem., Springer Verlag, Berlin, 1953, pp. 64–88.
57 Bouman, J. and Slater, E.C., Nature 177 (1956) 1181–1182.
58 Bouman, J., Slater, E.C., Rudney, H. and Links, J., Biochim. Biophys. Acta 29 (1958) 456–457.
59 Slater, E.C., Proc. 4th Int. Congr. Biochem. 9 (1958) 316–344.

60 Slater, E.C., Rudney, H., Bouman, J. and Links, J., Biochim. Biophys. Acta 47 (1961) 497–514.
61 Colpa-Boonstra, J. and Slater, E.C., Biochim. Biophys. Acta 27 (1958) 122–133.
62 Tissières, A., Hovenkamp, H.G. and Slater, E.C., Biochim. Biophys. Acta 25 (1957) 336–347.
63 Deul, D.H. and Thorn, M.B., Biochim. Biophys. Acta 59 (1962) 426–436.
64 Myers, D.K. and Slater, E.C., Biochem. J. 67 (1957) 558–579.
65 Hemker, H.C. and Hülsmann, W.C., Biochim. Biophys. Acta 48 (1961) 221–223.
66 Hülsmann, W.C. and Slater, E.C., Nature 180 (1957) 372–374.
67 Hülsmann, W.C., Elliott, W.B. and Slater, E.C., Biochim. Biophys. Acta 27 (1958) 664–665.
68 Borst, P. and Loos, J.A., Recl. Trav. Chim. Pays-Bas 78 (1959) 874–875.
69 Borst, P., Biochim. Biophys. Acta 57 (1962) 270–282.
70 Slater, E.C., Biochim. Biophys. Acta 1000 (1989) 323–326.
71 Borst, P. and Slater, E.C., Biochim. Biophys. Acta 41 (1960) 170–171.
72 Greengard, P., Minnaert, K., Slater, E.C. and Betel, I., Biochem. J. 73 (1959) 637–646.
73 Ernster, L., FASEB J. 7 (1993) 1520–1524.
74 Purvis, J.L., Nature 182 (1958) 711–712.
75 van Dam, K., Biochim. Biophys. Acta 92 (1964) 181–183.
76 Straub, J.P. and Colpa-Boonstra, J.P., Biochim. Biophys. Acta 60 (1962) 650–652.
77 Veeger, C. and Massey, V., Biochem. Biophys. Acta 64 (1962) 83–100.
78 Minnaert, K., Biochim. Biophys. Acta 50 (1961) 23–41.
79 Hulsmans, H.A.M., Biochim. Biophys. Acta 54 (1961) 1–14.
80 Oei, T.L., Acta Genet. 11 (1961) 205–216.
81 Klein Obbink, H.J., Acta Genet. 15 (1965) 21–32.
82 Hülsmann, W.C., Oei, T.L. and van Creveld, S., Lancet (1961) 581–583.
83 Nunnikhoven, R., Biochim. Biophys. Acta 28 (1958) 108–119.
84 Kaniuga, Z., Biochim. Biophys. Acta 73 (1963) 550–564.
85 Huijing, F. and Slater, E.C., J. Biochem. 49 (1961) 493–501.
86 Snoswell, A., Biochim. Biophys. Acta 52 (1961) 216–218; 60 (1962) 143–157.
87 Slater, E.C. and Tager, J.M., Biochim. Biophys. Acta 77 (1963) 276–300.

88 Guillory, R.J., Biochim. Biophys. Acta 89 (1964) 197–207.
89 Hemker, H.C., Biochim. Biophys. Acta 63 (1962) 46–54.
90 van Gelder, B.F., Biochim. Biophys. Acta 118 (1966) 36–46.
91 Beinert, H., in: A History of Biochemistry. Selected Topics in the History of Biochemistry: Personal recollections. IV. Comprehensive Biochemistry, Vol. 38 (Slater, E.C., Jaenicke, R. and Semenza, G., eds.), Elsevier, Amsterdam, 1995, pp. 193–258.
92 Kroon, A.M., Biochim. Biophys. Acta 108 (1965) 275–284.
93 van Bruggen, E.P.J., Borst, P., Ruttenberg, G.J.C.M., Gruber, M. and Kroon, A.M., Biochim. Biophys. Acta 119 (1966) 437–439.
94 Kroon, A.M., Saccone, C. and Botman, M.J. Biochim. Biophys. Acta 142 (1967) 552–554.
95 van den Bergh, S.G. and Slater, E.C., Biochem. J. 82 (1962) 362–371.
96 de Vijlder, J.J.M. and Slater, E.C., Biochim. Biophys. Acta 167 (1968) 23–34.
97 Slater, E.C., Biograph. Mem. R. Soc. 40 (1994) 283–305.
98 Slater, E.C., Eur. J. Biochem., 1 (1967) 317–326.
99 van Dam, K., Biochim. Biophys. Acta 128 (1966) 337–343.
100 Tager, J.M., Veldsema-Currie, R.D. and Slater, E.C., Nature 212 (1966) 376–379.
101 Kemp, A., Jr., and Slater, E.C., Biochim. Biophys. Acta 92 (1964) 178–180.
102 Harary, I. and Slater, E.C., Biochim. Biophys. Acta 99 (1965) 227–233.
103 Slater, E.C., in: Dynamics of Energy-transducing Membranes (Ernster, L., Estabrook, R.W. and Slater, E.C., eds.), Elsevier, Amsterdam, 1974, pp. 1–20.
104 Wikström, M.K.F. and Berden, J.A., Biochim. Biophys. Acta 283 (1972) 403–420.
105 Berden, J.A. and Slater, E.C., Biochim. Biophys. Acta 256 (1972) 19–215.
106 Slater, E.C., in: Chemiosmotic Proton Circuits in Biological Membranes (Skulachev, V.D. and Hinkle, P., eds.), Addison-Wesley, Reading, MA, 1981, pp. 69–104.
107 Muraoka, S. and Slater, E.C., Biochim. Biophys. Acta 180 (1969) 227–236.
108 Slater, E.C., Rosing, J. and Mol, A., Biochim. Biophys. Acta, 292 (1975) 534–555.
109 Rosing, J. and Slater, E.C., Biochim. Biophys. Acta 267 (1972) 275–290.

110 Krab, K, and van Wezel, J., Biochim. Biophys. Acta 1098 (1992) 172–176.
111 Vallejos, R.H., van den Bergh, S.G. and Slater, E.C., Biochim. Biophys. Acta 153 (1968) 509–520.
112 Harris, D.A., Rosing, J., van den Stadt, R.J. and Slater, E.C., Biochim. Biophys. Acta 314 (1973) 149–153.
113 Slater, E.C. and De Vries, S., Nature 288 (1980) 717–718.
114 Slater, E.C., Trends Biochem. Sci. 8 (1983) 239–242.
115 de Vries, S., Albracht, S.P.J., Berden, J.A. and Slater, E.C., J. Biol. Chem. 256 (1981) 11996–11998.
116 Wagenvoord, R.J., van der Kraan, I. and Kemp, A., Biochim. Biophys. Acta 460 (1977) 17–24.
117 Boyer, P.D., Chance, B., Ernster, L., Mitchell, P., Racker, E. and Slater, E.C., Annu. Rev. Biochem. 46 (1977) 1015–1026.
118 Slater, E.C., Berden, J.A. and Herweijer, M.A., Biochim. Biophys. Acta 811 (1985) 217–231.
119 Slater, E.C., Eur. J. Biochem. 166 (1987) 489–504.
120 Heberle, J., Riesle, J., Thiedemann, G., Oesterhellt, D. and Dencher, N.A., Nature 370 (1994) 379–384.
121 Albracht, S.P.J., Graf, E.-G. and Thauer, R.K., FEBS Lett. 140 (1985) 311–313.
122 Hagen, W.R., van Berkel-Arts, A., Krüse-Wolters, K.M., Dunham, W.R. and Veeger, C., FEBS Lett. 201 (1986) 158–162.
123 Slater, E.C., Biochimica et Biofysica Acta, The Story of a Biochemical Journal, Elsevier, Amsterdam, 1986, 122 pp.
124 Slater, E.C. and Whelan, W.J., Trends Biochem. Sci. 5 (2) (1980) III–V.

G. Semenza and R. Jaenicke (Eds.)
Selected Topics in the History of Biochemistry: Personal Recollections, V
(Comprehensive Biochemistry Vol. 40) © 1997 Elsevier Science B.V.

Chapter 4

A Lifetime Journey with Photosynthesis

ALEXANDER A. KRASNOVSKY

Laboratory of Photobiochemistry, A.N. Bakh Institute of Biochemistry,
Lenynsky prospect, 33, Moscow 117071, Russian Federation

Preface

On 16 May 1993, my father, academician Alexander Abramovitch Krasnovsky, passed away in the hospital of the USSR Academy of Science. This happened 3 months before his 80th birthday. In his files I found letters from Professor R. Jaenicke, the editor of *Comprehensive Biochemistry*, inviting my father to write recollections, a copy of a letter with his positive reply and a manuscript of recollections in Russian. This work was apparently not finished. Probably, he wanted to finish it in time for his birthday, on 26 August. The manuscript consisted of many separate pieces typed on a type-writer. Some of these pieces were obviously corrected several times and looked like a connected narrative, others were first drafts. There was no general plan, list of literature or figures. I felt that my duty was to complete this paper. Working on it, I did not want to add anything from myself. My desire was to express my father's thoughts as accurately as possible. It was easy to arrange the biography section according to the actual chronological sequence of events. To arrange the scientific part, I followed the order in which the materials in the original manuscript were presented. In unclear places I used plans of his previous reviews. Some-

Alexander Abramovich Krasnovsky in his office at the A.N. Bakh Institute of Biochemistry, Moscow, 1984.

times when the narrative was incomplete, I added the extracts from these reviews. The most helpful were papers [1–3], especially the recollections published in the *Historical Corner of Photosynthesis Research* [3]. The references and figures have also been taken from these reviews. In my work I used comments from my mother, Fanya Lvovna Kromysheva, and my father's coworkers and colleagues, N.V. Karapetian, A.V. Umrikhina and others. I am very grateful to E.F. Litvina and L. Orr for their kind help in translating the manuscript into English.

Now this work is finished. I have taken into account all materials, notes and comments indicated in the original manuscript. It was a sad and emotionally saturated work which allowed me to continue for one more year, our endless talks about life and science. Reading the final version of the recollections, I see that the most important events of my father's life have been adequately reflected. If he was alive, he would definitely add more details, names and scientific results. However, sudden death interrupted this work. Now his recollections are a part of the history of science. I believe that this paper will be of interest for my father's friends and colleagues and useful for young people who start their scientific career.

Alexander A. Krasnovsky, Jr.
27 July 1994

Introduction

When I began to write these recollections, at the invitation of *Comprehensive Biochemistry*, I had to answer the question whether it was necessary at all. I lived during the greater part of the 20th century, in the epoch of wars, revolutions and social cataclysms. The historical background of my generation has been many times described in literature. I see no reason to write about it again. Fortunately, I was not in-

volved in political storms and also was not a victim of repri-
sals.

As in the whole history of mankind, science was developing in spite of worldly events. Sometimes, authorities
stopped supporting scientific investigations, sometimes they
generously supported them, but the spring of science was
flowing slowly, then, was turning into a powerful river – in
spite of all changes, during war or peace, starvation and
prosperity, sometimes, in spite of everything – as in the legend of Archimedes.

I never liked abstract theoretical speculations and mathematical schemes. I always wished to describe the studied
phenomena by simple models, which could be investigated
by relatively simple experiments. Maybe, this was a result of
my education as a chemist which was based on the idea that
all phenomena are built upon free playing molecules. Biochemical systems are complicated by the existence of ordered structures which determine their function and regulation.

To create the models I needed experimental results as a
permanent food for thought. At some stages my studies required having assistants. As the group of my coworkers was
growing, I had to face the problems of human interrelations,
which much differ from the scientific problems and sometimes cannot be easily solved.

Scientific work is not always a straight road, you often
end up in a blind alley that you would rather forget. Joys
and sorrows on that path can not be objectively described.
Nevertheless, I will try to describe the events which took
place many years ago as some kind of a logical process
(which did not exist in reality), and try to write a connected
narrative.

Childhood and primary education

I was born in Odessa, in 1913. The strongest recollection of my childhood was the sea, now stormy, now calm, changing colors, the smell of rotting algae on the shore and ships on the horizon. Revolution – red flags, Marseillaise, Liberty, Equality and Fraternity, then, Civil War – starvation, typhus, frequent changes of authorities: life became very hard. In 1921, my mother and I escaped Odessa and went by train in a 'teplushka' (a heated carriage for goods) to Moscow through the whirlwinds of the Civil War.

In Moscow we stayed with our relatives; it was warm and crowded there. Food was sold by ration-cards, one could even get white bread. For the first time I saw a French baguette, renamed a 'city loaf', during the fight against cosmopolitanism. In the requisitioned mansions, municipal (communal) flats were arranged. We got one room in such a flat, in a wooden mansion with columns, that was built soon after the War of 1812 against Napoleon.

In front of the window there was a wonderful garden with lilacs in bloom and poplar trees. In winter, problems with heating arouse. Big 'Holland' furnaces, intended for the big noble house, could not heat small rooms of municipal flats. So we all were using small stoves called 'burjuika', the pipes of which were connected to the big chimney of the old furnaces. For cooking we used little kerosene ovens with wicks, 'primus' and 'kerosinka', which were more safe.

Firewood was sold near the Kievsky Railway station, the measure of its amount was 'sajen' (fathom). Birch wood was considered the best, giving more heat. This wood was brought on horse-driven carts (lomovaya telega) to the yard of our house, where we sawed and chopped the wood and put it into the wood-sheds. From the perspective of modern people, all this seems rather unreal, but I did not find it inconvenient, this was the natural way of living. I began feeling uncomfortable later in 1938, when I got married and brought my wife into this single room where my mother, aunt and grandmother lived.

Our communal flat was a cross section of the society of that time, housing people with loud noble names, intellectuals, bureaucrats, workers, alcoholics and street girls. I lived more than 40 years there and got a flat of my own only in 1963, in the epoch of Khrushchev, when I was a Doctor of Science, Professor and the father of a 20-year-old son. All this is typical of the life of my generation, and is not easily understood by the young.

I studied at school number seven in Krivoarbatsky lane. The school was located in the building of the former Khvostovskaya Girls Gymnasium, from which it inherited highly qualified teachers, excellent physical and chemical laboratories, a gymnastic hall, and even a small 'zoo' in the school yard where rabbits and guinea-pigs led a rather dull existence. Opposite the school, there was a large empty piece of land which in winter was flooded with water and transformed into a skating-rink. In the late 1920s, a house was built there by the famous architect Mel'nikov, which now represents an example of the constructivism of that time.

School education included 7 years of general studies, and after that 2 or 3 years of the so-called 'specialized courses' resulting in qualification as a chemist-technician, land surveyor, constructor, or accountant. Thus, a young man of 17 or 18 could either work and earn money, or pass the exams to enter a higher educational institution. The latter was not easy, as the preferential admission was for the children of the workers and peasantry.

Primary education in our school was rather peculiar: no home work, no grammar at all, many excursions to museums and historical places, and no grades. Grades were considered creating inequality and harmful to a child's psyche. This order lasted for the first 5 years after the Civil war. Later, a more formal system was adopted which has been generally accepted (in Russia) until now. The undoubted merit of this system for kids was having plenty of time for games and free reading not connected with the obligatory order. Usually, each school had some dominating subject, ours majored in chemistry. That meant that we spent relatively more time in theoretical and practical chemistry classes. We studied two foreign languages, German and French, although methods of teaching were rather poor. Soon French was canceled. As for German, we knew it rather badly by the time we left school. We had a brilliant teacher of literature, Georgii Ivanovitch Fomin, who made us love Russian literature. We also had excellent teachers of mathematics, drawing and physical training.

I must say that reading books was a real cult in our school. We used to swallow an enormous amount of literature without any special choice, and often attended the outdoor book sales near Kitaygorodsky Wall and on Tverskoy Boulevard. There, one could find interesting literature on all subjects. It was the epoch of NEP (New Economical Policy) when people were selling off their own books and some were confiscated from private libraries.

It is worth noting that at that time, the campaign for abolishing illiteracy had been actively carried out, and we were involved in teaching basic reading skills and the four rules of arithmetic to street cleaners (dvorniks) from the neighboring houses and members of their families.

In 1923, a radio-range beacon was built by the famous constructor Shukhov; it is used now for TV. At that time, we all had been infected by the 'radio-fan virus'. All boys were involved in radio-receiver construction. To make them, one needed a self-induction coil, the construction of which required so called 'ring wire' which was not available for sale. To get it, not completely legal actions were applied. Capacitors were home-made. In order to make them, we used tin foil wrappers for sweets and paraffin paper designed for medical purposes. The detector crystals were also home-made. Leaden shavings were fused together with sulfur in a tube. Then, the tube was cooled, broken with a hammer and pieces of alloy, sulfurous lead (called 'halenite') were obtained. The second electrode was a copper wire with a bevel cut. It was very difficult to find the active point between copper wire and halenite. The hardest problem was to get a telephone receiver. When all parts were orderly arranged, then, what luck!, you could hear call-signals from Comintern Station at longwave lengths.

As I have already mentioned, it was the epoch of NEP, so the demand for radio parts stimulated production. In Arbat, a lot of small stores appeared selling wire, halenite, telephone receivers, and the first Soviet electrode lamp – a triode called 'micro'. As a radio amateur, I made good progress constructing quite a decent three-valve receiver with one step high frequency amplification, detector and one step low-frequency amplifier. When atmospheric conditions were favorable, it was possible to listen to Moscow and the nearest European stations. Unfortunately, that was the end of my radio hobby as I had to begin working.

Professional education in chemistry

After the seven-year school, I participated in the Chemistry Courses which were given in the 10th Nansen School in the Merzlyakovsky lane. There, I met Fanya Kromysheva, my future wife. We had a very serious training in chemistry. It was the time of the First Five-Year Plan, according to which, we were trained for the Bobrikov Chemical Enterprise near Tula. However, by the time we had finished our training, this Enterprise had not yet been built. In 1931, I began to work at the Butyrsky aniline-dye factory in Moscow. The work there was really interesting: we had to learn new methods of analysis and synthesis, and apply them to the actual factory work. In addition, we young chemists, taught workers some basic notions of chemistry which helped them to be aware of the processes going on in the various reactors.

As I remember, the staff, both engineers and workers, were very competent and worked with enthusiasm and high efficiency, although the salary was rather low. Unfortunately, that could not be said about the administration, because the director and the engineer-in-chief were recruited from the so called 'vydvyzenzev' (promotion reserve), whose major qualifications were membership in the Communist Party.

The factory was privileged to send a limited number of workers to the D.I. Mendeleev Institute of Chemical Technology for additional education, without leaving their jobs at the factory. In 1933, I was admitted as a second-year student to this institute, because the courses had given me a very substantial training in chemistry and other main subjects. I believe studies combined with work in the same field of specialization are most useful for those who are strong enough for it. To be able to study, I had to work night shifts and attend the institute in the mornings. However, the knowledge we got did not remain sheer abstraction, as it was used in everyday work, and, hence, was well remembered.

The training at the Mendeleev Institute had a narrow specialization which corresponded to the field we worked in. As a student I witnessed the informal character of relations between professors and students. This was a part of the pre-revolutionary tradition of the Russian Higher Education. A student was a colleague for the professor, a junior, a rookie, but still, a colleague. It was quite natural for a professor to go with a student to the beer-house on Miusskaya Square and spend time there together.

My diploma thesis was devoted to the method for synthesis of the aluminum-calcium complex of alizarin (dioxyanthraquinone), a red dye, called 'kraplak'. I had to work out the technique for synthesis and to defend the procedure for production. In 1937, after the defense of my thesis, I got a diploma of Chemical Engineer and was accepted for postgraduate training at the Mendeleev Institute. So, I left the Butyrsky factory which had become like a home for me. In

1938, I married my former classmate Fanya Kromisheva, who has been my wife and friend ever since. At that time, she had graduated from the Chemistry Department of the M.V. Lomonosov Moscow State University and started to work on applied problems at the Moscow Institute of Steel and Alloys.

The head of the department, Professor Vassily Kiselyov, proposed that I choose for my post-graduate training any theme in the field of chemistry and technology of pigments and dyes. I had chosen the problem of the photosensitizing action of titanium dioxide which was beginning to be employed as a white pigment at that time. This compound had a considerable drawback: it decomposed the surface layer of the material it covered. We managed to show that this phenomenon was connected with the ability of TiO_2 to photosensitize oxidative processes.

Studies of titanium dioxide, which is a typical semiconductor, required knowledge of physics of semiconductors and physical and organic chemistry as well. I remember that this area was at the initial stage of development: the idea of 'excitons', was proposed just at that time by a Leningrad physicist, Ya Frenkel. As I worked mostly among technology engineers, I felt that communication with researchers working in fundamental science was needed.

In specialized literature, I found that photosensitization processes were investigated by A.N. Terenin at the State Optical Institute in Leningrad. I went to Leningrad, found the Optical Institute at Birzhevye Linii, and was almost lost in the numerous corridors of this ancient building. Walking around, I asked everybody whom I met where I could find Terenin. Finally I came to a small room where I saw a young man wearing protective glasses who was soldering glass pieces. In despair I asked him too where was Terenin, and the young man answered: 'I am Terenin. What do you want?' We got acquainted, and living in different towns, began an active correspondence. A.N. Terenin had been just elected

Corresponding Member of the Academy of Science. He was quite young, thirty nine, as I remember. He had no family, science was his sole interest. Terenin used to inform me of the most important literature and to criticize the experimental data I presented in my letters. It was surprising how he could spare so much time for a young man like me who did not have any fundamental education in physics.

I felt the necessity to improve my education, mainly in physical and organic chemistry and physics. The conditions for this were favorable in the Mendeleev Institute. Nobody interfered with my experimental work and we had no difficulties in getting chemicals (like we are having now). The Institute library was excellent. I started to teach at the Institute of Professional Improvement of the Technical and Engineering Staff, where I prepared and gave a full-year lecture course of physical chemistry and later a full-year course of organic chemistry. It was really of greatest use to me, but it is not clear whether it was equally useful for my students.

At that time, I became keenly interested in the Trautz-Perren hypothesis which suggested that activation of all chemical reactions, dark and photochemical as well, occurred at the expense of different forms of radiation. Later, this concept was found to be wrong, but at the time it seemed very attractive, as it seemed to provide an explanation for the activation of chemical reactions of all types.

For my experiments, I managed to construct a small spectrophotometer with changeable light filters and a selenium photoelement. My experience as a radio amateur was useful. Using this set-up, I measured photosensitized peroxide accumulation. The method was based on the formation of ferrous ions as a result of the reaction of peroxides with ferric ions. Ferrous ions were detected by their reaction with potassium thiocyanate which led to red coloring of the solutions.

A system containing TiO_2 was illuminated by an ultraviolet lamp which we got with difficulties from the Polyclinics. From the experiments, it seemed clear that the coating was destroyed as a result of TiO_2-photosensitized oxidation. The technological task was now to inhibit this photochemical activity. This was achieved by adsorption of ions of the transition metals Co, Ni or Fe. However, more important was the theoretical problem of photochemical activity of semiconductors, the presence of microimpurities and their specific crystalline structure. Using X-ray structural analysis, I managed to observe that titanium oxide samples with the anatase structure showed higher photochemical activity than those with ruthyl structure.

We were also interested in peroxide accumulation during the photochemical oxidation of isomeric xylols. It was shown that the ortho-isomer was the most photochemically active compound. The mathematical analysis of the kinetic data led to the conclusion that the chain propagation mechanism, and the mechanism based on consecutive reactions, yield similar equations for peroxide accumulation. This strongly disappointed me, and for a long time I lost interest in the application of mathematical methods for the analysis of chemical processes.

In 1940, I was awarded the Candidate Degree (PhD), my examiners were A.N. Terenin and N.P. Peskov, the chief of the Division of Physical and Colloid Chemistry of the Mendeleev Institute. After that I got a position as an assistant (assistant professor) in that division. Teaching and fundamental research were not much supported by the state, and my salary was very low.

War and the beginning of the biochemical career

At that time there was war in Europe, and everyone felt that the war was approaching the USSR. In 1941, when the war crossed the USSR border, I was directed to the military chemical enterprise in Siberia. There, my son, Alexander was born. In 1943, while in Moscow on a business trip, I met

A.N. Terenin who was also working in military chemistry. He had come to Moscow to take part in the Annual Meeting of the USSR Academy of Sciences where he presented a lecture 'Light and Chemical Processes'. It was clear that the end of the war was not too far. A.N. Terenin thought that the enormous scientific potential formed within the military-industrial complex must serve towards the development of fundamental science. Both of us felt strongly that, for the survival of mankind, the main problem was not an arms race, but efficient solar energy conversion in photosynthesis. A.N. Terenin invited me for a doctorate position in a program which was going to be organized at the USSR Academy of Sciences.

In 1944, the government decision regarding the 'preparation of cadres of the highest qualification for the USSR Academy of Sciences' was officially announced. According to this decision, specialists were directed for post-graduate and doctoral training to the Academy of Sciences, from the army and the military-industrial complex. The same year, I was admitted to the Doctoral Program according to the recommendation of academician A.N. Terenin. W.A. Engelhardt,[1] who was the director of the Institute of Biochemistry, took me into his laboratory. His wife, Militsa Nikolaevna, let me use her Warburg apparatus, which was an extraordinary generosity at that time.

Using this apparatus, I started to investigate the oxidation of ascorbic acid, catalyzed by copper ions and other copper-containing compounds. I learned about this reaction from the red volumes of the 'Cold Spring Harbor Symposia', (Volume 1939). For many years, they were available in the Institute library, being a real gold mine of biochemical information. The problem I was actually interested in was the

---

1 One of the founders of molecular biology in the USSR, discoverer of oxidative phosphorylation, member of the USSR Academy of Science since 1953, see: W.A. Engelhardt. Life and Science. Annu. Rev. Biochem. 1982, 51: 1–19.

effect of light on the catalytic oxidation of ascorbic acid. To observe it, I illuminated vessels of the Warburg apparatus using a table lamp which tested the patience of Militsa Nikolaevna.

W.A. Engelhardt decided to improve my knowledge of biochemical problems. For that purpose he asked me to translate several articles from *Currents in Biochemical Research*, a book edited by D. Green. The translation was published in Russian by the 'Inostrannaya Literatura' Publisher in 1948. It contained ideas about mechanisms of gene behavior which were considered as sheer heresy in the USSR at the time when T.D. Lysenko was coming into power. It was at the end of 1948 when the session of VASCHNIL (All Union Academy of Agricultural Sciences and Forestry) took place which was a horrible blow to genetics and physico-chemical biology in our country.

Following Engelhardt's advice, I had attended the seminars in the Institute of P.L. Kapitsa[2] for a few years. I remember that Kapitsa invited Engelhardt to report his studies of the mechanism of muscular contraction. To my surprise, there were a lot of people present, not only biologists, but also chemists, physicists and even scholars from the humanities as well. The Kapitsa seminars were famous for their broad horizons and the freedom to advance new ideas. P.L. Kapitsa, himself, was a very tough leader of the discussions. Engelhardt had a really hard time because physicists did not know biochemistry and were mostly interested in the thermodynamic and physical mechanisms of muscular contraction and relaxation. During the discussions, Engelhardt showed an outstanding knowledge of biochemistry, chemistry and physics. I had the impression that both sides profited enormously from the exchange of ideas. Among other lecturers, I remember Rapoport (a fearless opponent of

[2] Famous physicist, one the founders of low-temperature physics, Nobel Prize winner, 1978, member of the USSR Academy of Science since 1939.

Lysenko) who talked at the seminar just after the rout of genetics by the session of VASCHNIL.

I had worked for about a year in Engelhardt's laboratory, when I got a separate room in the basement and, later, a better one in the main building of the Biochemistry Institute. In 1945, a small research group was organized under Terenin's supervision. My first associate was Galina P. Brin who was invited by A.N. Terenin, on the recommendation of academician Kasanskii, director of the Moscow Institute of the Fine Chemical Technology, from where G.P. Brin had graduated. V.B. Evstigneev joined us a year later. Consisting of three scientists, our group was renamed the 'Laboratory of Photobiochemistry'.

The first thing we needed for our experiments was a spectrophotometer. We tried to make it by ourselves, using an old pre-war monochromator and sodium photoelements occasionally found in the store of the Institute of Biochemistry. However, soon V.N. Bukin, the Head of the Laboratory of Vitamins at our Institute, told us that the Institute of Vitamins had received two Beckman DU spectrophotometers which did not work but might still be usable. V.B. Evstigneev and I went to the Institute of Vitamins and acquired one of these spectrophotometers, in exchange for an analytical scale. I doubt that it was really a fair exchange.

The spectrophotometer we obtained had evidently been dropped during transport, as its slit was damaged, but we managed to repair it within a short time. The use of this spectrophotometer allowed us a wide variety of interesting experiments on aerobic systems. Application of the Thunberg tubes (modified for placement into the spectrophotometer) made it possible to carry out photochemical experiments under anaerobic conditions. In addition, we managed to use the spectrophotometer also as a spectrofluorimeter, by applying an external light source for excitation. This spectrophotometer appeared to be the only modern apparatus in all the Institutes of the Biological Section of the Academy of

The first days of the photobiochemistry group: Alexander Nikolaevich Terenin (sitting), Alexander Abramovich Krasnovsky and Galina Petrovna Brin in the A.N. Bakh Institute of Biochemistry, Moscow, 1946.

Science because people from other institutes of the Biological Section used to come to us for their spectrophotometric measurements in the visible and ultra-violet regions where light absorption by proteins and nucleic acids is observed.

Photochemistry of chlorophyll

We were interested in mechanisms of chlorophyll participation in photosynthesis. It was proposed by several researchers in the past, that reversible transformations of chlorophyll might underlie its action. This idea was most clearly expressed by K. Timiriazev [4]. He showed that chlorophyll could be reduced by metallic zinc, and suggested that a pigment, protophylline, which he found in etiolated leaves, was the product of chlorophyll reduction. It was established later that protophylline was a precursor in the chlorophyll biosynthesis. Efforts of different researchers to reveal reversible chlorophyll transformations were described in details by E. Rabinovich in his book, *Photosynthesis* [5–7]. It was clear that photosynthesis was a redox process. Our goal was to observe experimentally reversible photochemical redox transformations of chlorophyll in the simplest models, i.e., in solutions of the pigment.

When we started our experiments, we were convinced that chlorophyll was too complicated for us. Therefore, we began with studies of the spectral and photosensitizing action of highly hydrophobic synthetic phthalocyanins and their magnesium and copper complexes, which we received from the Research Institute of Semiproducts and Dyes (NIOPIK). The experiments were carried out in aqueous suspensions and solutions of phthalocyanines in organic solvents. Phthalocyanines turned out to be very similar to chlorophyll, comparing their absorption spectra and photochemical properties. Water suspensions of all phthalocyanines actively photosensitized the oxidation of ascorbic acid, regardless of the nature of the central metal atom. In organic solvents, the most active complex was magnesium-phtalocyanin, with its bright red fluorescence. We had shown that this phtalocyanin has the ability to effect partly reversible photoreduction and photooxidation [8, 9]. This encouraged us to begin studies of chlorophyll.

Reversible chlorophyll photoreduction

In order to work with chlorophyll, we had to solve numerous technical problems that required great experimental efforts. The first problem was that chemically pure solvents were not available at that time. G.P. Brin made a 3 m high rectification column with small glass rings inside. Using that column, we could obtain pure diethyl ether, pyridine, acetone, and alcohol by distillation of technological solvents. In the case of ether, the removal of peroxides was needed prior to distillation. That was achieved by the infusion of ferric sulfate. Chlorophyll was obtained from dry nettle leaves by the procedure of F.P. Zscheile and C.L. Comar, with final chromatographic separation on sugar columns. Stuffing the sugar columns was a special art: the column was stuffed with small sugar portions using a copper pestle; the intensity of the 'blows' affected the density of the column and the degree of separation of leaf pigments. After much practice, we managed to obtain chlorophyll a and chlorophyll b, with the absorption spectra previously described in literature.

In these experiments, as it is often the case, much was governed by chance. According to L. Pauling's concept of the resonance of structures, we expected to see formation of chlorophyll anion-radicals in basic media. As an electron donor, we used ascorbic acid, which, we knew, was a perfect substrate for sensitized oxidation. It was clear also that experiments should be carried out in the absence of oxygen, which was an active electron acceptor.

Even the first experiments were successful. After illuminating chlorophyll solutions in pyridine containing ascorbic acid, a red intermediate was observed with the absorption maximum at 520 nm. It reacted reversibly in the dark with the regeneration of the initial green color of chlorophyll [10]. This was the first reversible chlorophyll photoreaction revealed experimentally.[3] Later on, we observed such photochemical reactions with all the available chlorophyll analogs and derivatives. For example, during pheophytin photoreduction, a similar intermediate was formed. In photoreduction of protochlorophyll (obtained from the inner coats of pumpkin seeds), an intermediate with the absorption maximum at 470 nm predominated. Photoreduction of porphyrins, depending on the pH of the medium, led to the formation of various intermediates with absorption maxima in both long-wavelength and short-wavelength regions of the spectrum [11]. These reactions attracted the attention of researchers and were extensively studied in many laboratories [12–14].

Reversible chlorophyll photooxidation

First experiments devoted to reversible chlorophyll oxidation were reported by Rabinovich and Weiss who used ferrous ions as oxidants [16]. In our first studies, pigment photooxidation by air oxygen or benzoyl peroxide was analyzed. In these reactions, intermediate products were formed which did not have distinct absorption maxima in the UV-VIS region. The component most resistant against photooxidation was pheophytin; chlorophyll had lower stability, the most labile were bacteriochlorophylls a and b. After addition of a reductant, ascorbic acid, the initial pigments were partly regenerated from the intermediates. This phenomenon was most pronounced during a short illumination of bacteriochlorophyll solutions [9, 15]. Subsequently, we managed to observe completely reversible photooxidation of chlorophyll,

---

3 This reaction is known today as the Krasnovsky reaction.

bacteriochlorophyll and their analogs in viscous alcohol-glycerol mixtures at −70°C using quinones (including ubiquinones) as reducing agents. Low temperature and high viscosity of the solutions prevented fast back reactions of the photooxidation products. After thawing, complete regeneration of the initial pigments was observed [16]. The absorption spectra of the photooxidized intermediates that we obtained under stationary illumination of the solutions, were similar to the absorption spectra of chlorophyll radicals detected using the technique of time-resolved flash-photolysis [17].

Chase for free radicals

From the very beginning of our studies, we have always assumed that the primary photoprocesses involve formation of the pigment triplet state and then a pair of cation and anion radicals (Fig. 1). In order to reveal free radicals experimentally, we first applied the method of initiation of methyl methacrylate polymerization. In those experiments, chlorophyll or its analogs, and the electron donor, ascorbic acid, were dissolved in methyl methacrylate, and then the system was illuminated with red light absorbed by the pigments. After several minutes, the solution became solid as polymerization of methyl methacrylate occurred. This phenomenon was noticed by N. Uri in 1952 (see Refs. in [18]). We studied it in detail and proposed that polymerization was initiated by a pigment anion-radical formed under light. However, as traces of oxygen were always present in solutions, polymerization could also be initiated by the superoxide radical formed as a result of oxygenation of the ascorbic acid [18].

In order to resolve the two alternative mechanisms, ESR measurements were needed because this is the most efficient approach to detect free radicals. In 1958, we were kindly permitted to use ESR spectrographs in the laboratory of Professor Bresler at the Institute of Macromolecular Com-

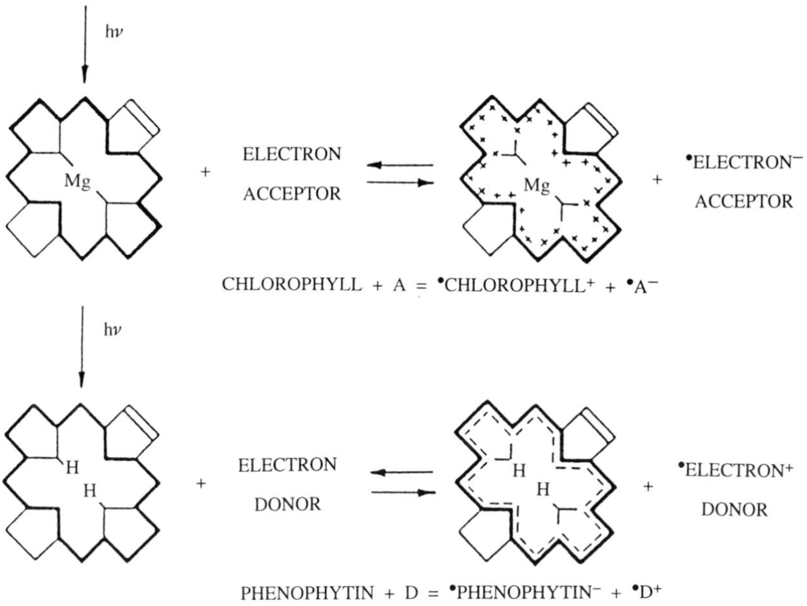

Fig. 1. Primary steps of the pigment-photosensitized charge separation.

pounds in Leningrad, and the laboratory of L.E. Blumenfeld and A.E. Kalmanson at the Institute of Chemical Physics in Moscow. Unfortunately, neither apparatus allowed illumination of the samples directly in the cavity. When solutions containing chlorophyll and ascorbic acid were placed into the cavity after illumination, only a very weak ESR signal was observed which was difficult to analyze.

We continued these experiments on the ESR spectrometer built by N.N. Bubnov in the V.V. Voevodskii laboratory at the Institute of Chemical Physics. This apparatus had a hole in the cavity which allowed illumination of our samples. Under illumination, well resolved ESR signals of free radicals were detected. The spectrum of this signal resembled that observed under dark oxidation of ascorbic acid in piperidine.

We proposed that a radical of monodehydroascorbic acid was detected. A similar observation was reported in 1958 by B. Commoner at the IVth International Biochemical Congress in Vienna. After further detailed analysis, we came to the conclusion that free radicals observed in the solutions containing both chlorophyll and ascorbate correspond to monodehydroascorbic acid alone. As formation of the radical pair was inevitable, it was reasonable to assume that chlorophyll radicals were also present in the system, but we could not detect them [19].

After 1960, we continued our ESR studies in collaboration with Dr. L.P. Kaushin and his associates. The signals of the triplet state of chlorophyll and its analogs were detected in frozen pigment solutions at liquid nitrogen temperature [20].

In 1967, we bought an ESR spectrometer from the Institute of Chemical Physics. It was the first model manufactured in the workshop of that institute for sale. This set-up has been used for all subsequent experiments in our laboratory with free radical formation. Samples were placed in the cavity and illuminated with a xenon lamp, their temperature could be changed from +50 to −196°C. This allowed us to find suitable conditions for the observation of the ESR signals originating from the pigment triplet states, ascorbic acid, quinone and chlorophyll radicals.

The experiments with the p-benzoquinone-containing systems were especially successful. By varying the temperature, we could observe the predominant accumulation of either the semiquinone or the pigment signal. The most distinct results were obtained in the systems containing bacteriochlorophyll and quinone. For detailed reviews of our data and the studies of other laboratories, see [21, 22].

Chlorophyll-photosensitized electron transfer

In the late 1940s, Melvin Calvin's research group discovered that CO_2 reduction was a cyclic process which required a photochemically generated reductant in the system. It was assumed that reduced pyridine nucleotides might be such reductants [23]. Stimulated by this idea, we investigated the system ascorbic acid–chlorophyll–nicotinamide adenine dinucleotide (NAD). Pure NAD was isolated from yeast by G.P. Brin. Under the action of red light absorbed by chlorophyll, we observed a maximum at 340 nm in the different absorption spectra corresponding to NADH formation [24]. Unfortunately, we could not carry out such experiments with chloroplasts as we did not know how to isolate them from plants. Two years later (1951), in the laboratories of D. Arnon, H. Gaffron and S. Ochoa, the formation of NADH and NADPH in the illuminated chloroplast was first observed, opening the way to elucidation of the photosynthetic electron transfer chain.

Later on, investigation of photochemical electron transfer reactions in different systems, by varying electron donors (D), pigment-sensitizers and electron acceptors (A), led us to the conclusion that chlorophylls, pheophytins and their analogs serve as uphill electron carriers [11, 25, 26]. In connection with this, it is worth noticing that the use of methyl viologen as an electron acceptor was a major step forward in studying photosensitized electron transfer processes. The redox potential of this acceptor is close to the hydrogen electrode. In addition, methyl viologen forms a characteristic blue intermediate with an absorption maximum at 640 nm after one-electron reduction. Using methyl viologen, we managed to observe quite a few interesting photochemical phenomena, studying, for example, electron-transfer reactions in heterogeneous systems, as well as detergent micelles and liposomes.

For instance, the chlorophyll-photosensitized reduction of methyl viologen by thiourea was observed by A. Luganskaya in the medium from which oxygen had not been removed. We were able to show that under illumination, the oxygen concentration was strongly decreased as a result of photosensitized oxygenation of thiourea. Methyl viologen was reduced by the photoproducts of thiourea oxygenation which appeared to be stronger reducing agents than thiourea itself [27]. In order to create models similar to the chloroplast membrane, we incorporated chlorophyll, pheophytin, bacteriochlorophyll and bacteriopheophytin into the liposomal lipid phase. Similar systems were studied by M. Calvin, G. Tollin, H.T. Tien and others. A peculiarity of our experiments was the introduction of ascorbic acid inside the liposome vesicles, and methyl viologen in the exterior water phase. Upon illuminating liposome suspensions, we observed transmembrane electron transfer. Reduced methyl viologen was accumulated under the action of light. When light was switched off, a slow backward reaction took place. In the absence of oxygen, the back reaction was found to depend on the permeability of the liposomal membranes to the back flow of electrons from the reduced methyl viologen to the products of ascorbic acid photooxidation. The highest quantum yield of the methyl viologen reduction (up to 30%) was observed when pheophytin and bacteriopheophytin were used as photosensitizers. Introduction of bacterial hydrogenase in this system allowed us to observe the evolution of molecular hydrogen, which appeared as a result of electron transfer from a donor through methyl viologen to hydrogenase (Fig. 2) [28–30].

Flash photolysis experiments were performed in collaboration with the Institute of Chemical Physics, showing that in the case of bacteriopheophytin, the primary event is reduction of the pigment triplet state, whereas the electron transfer from the reduced pigment to methyl viologen is the subsequent step. As was found much later, the transmembrane potential greatly affected the charge separation [31]. Liposomes are a promising model of charge separation in photosynthesis and further studies of photoinduced electron transfer in this model seems worthwhile.

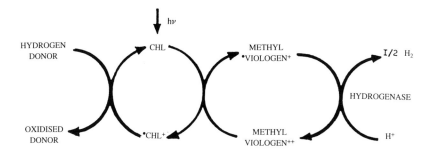

Fig. 2. Scheme of the photosensitized reduction of methyl viologen coupled to hydrogenase.

The state of the photosynthetic pigments in living cells

In 1950, with the initial idea to find photochemical reactions of phycobilins, I went to the Sevastopol Biological Station of the Academy of Science, which had been restored after its destruction during the war. The staff of the Station were several enthusiastic scientists, headed by Professor Vodyanitsky, who actively invited researchers of the Academy of Sciences to come for work. Living conditions at the Station were hard at that time. I and the ichthyologist and future academician Svetovidov lived and worked in one of the laboratories of the Station. Our food consisted predominantly of bullheads which Professor Svetovidov used to fish right from the wall of the Station, and which I boiled in the laboratory's Erlenmeyer flasks.

I was given a boat which I used for voyages in the harbor waters and for collecting various species of red algae from the stones. *Calitamnium Ribosum* was the best for my experiments. It lost color in distilled water with the formation of aqueous solutions of phycoerythrin and phycocyanine. The next step was the purification of these pigments. As an adsorbent for the columns, I used a tooth powder which I was buying in large amounts in the pharmacy store, very suspicious to the store personnel. Unfortunately, there was some 'flavor' component in the tooth powder which could only be removed by repeated washing with water, leading to a significant loss of the powder itself. Phycobilin pigments were readily separated on the tooth powder columns, and easily taken off by slightly alkaline solutions of sodium phosphate.

Thus, I acquired a lot of phycobilins in solutions. Unfortunately, later it appeared that they were rotting when kept under aerobic conditions. As a result, I brought to Moscow my phycobilin solutions with disgusting odors. However, after additional chromatography, I obtained beautiful solutions with extremely intensive luminescence. Our subsequent analyses have shown that phycobilins are incapable of significant reduction and oxidation. A weak photochemical activity, photosensitization of ascorbic acid oxidation, was observed after protein denaturation [32, 33]. This confirmed the hypothesis that phycobilins are light-harvesters of solar radiation, transmitting the excitation energy to chlorophylls. Probably, evolution has led to the use of phycobilin pigments in the light-harvesting antenna, leaving to chlorophyll the main role as the participant in electron transfer [34].

All our experiments with isolated pigments were most satisfying from the point of view of photochemistry. However, they could not solve the problem of the involvement of these pigments and their interesting spectral properties in photosynthesis, in vivo. Therefore, it was evident that our next goal had to be the elucidation of the state of the pigments in the photosensitizing organisms and the mechanism of their reversible transformations in the intact photosensitizing structures.

Regarding the state of chlorophyll in situ, it has been known from the intensive research in the period 1870–1920, that the absorption spectra of chlorophyll in plants differ from those in chlorophyll solutions, resembling the spectra of the pigment in colloidal aqueous systems. This led D.I.

Ivanovsky (the discoverer of tobacco mosaic virus) and the German chemists R. Willstätter and A. Stoll, to propose that chlorophyll is in a colloid state in plants. K.A. Timiryazev wrote about 'solid' chlorophyll; in contrast, V.N. Ljubimenko proposed a concept (which is now generally accepted) that chlorophyll is connected with proteins. E.C. Wassink, in 1939, was able to show that bacteriochlorophyll a forms three spectral forms in bacterial cells. The occurrence of these forms Wassink explained by different pigment-protein interactions [5].

In the beginning of our studies (1948), we confirmed the observation of early workers that chlorophyll in plants is spectrally different from chlorophyll in solutions. Using a young leaf of *Tradescancia* infiltrated with water, we observed a fine structure of the absorption spectrum with maxima at 670 and 678 nm. Further experiments made in collaboration with L.M. Kosobutskaya, led us to the conclusion that different forms of chlorophyll and its analogs exist in cells [35]. Thus, it was of a great interest to find a sufficiently transparent photosensitizing organism and study the fine structure of the absorption spectra using a spectrophotometer with high resolution.

When I worked at the Sevastopol Station, my attention was attracted by the red alga *Phyllophora*. Its thallus was transparent due to a great amount of agar. Thus, it presented an excellent object for spectral studies. By the way, it was practically impossible to extract phycoerythrin from it. Several years later, I learned that near Odessa, not far from the seashore, whole fields (Zernov's field) of Phyllophora were found, and commercial extraction of agar was organized.

A.N. Terenin told me that an old professor of physics at the Odessa University, A. Kirillov, had a spectrophotometer in his laboratory. This news made us happy. The coincidence of the object of studies with a spectrophotometer in the same geographical point stimulated us to undertake a trip to

Odessa, where Terenin and I arrived in 1955. Professor Kirillov was most hospitable and friendly and we were provided with *Phyllophora* in enormous quantities. In *Phyllophora* thalluses we observed two main systems of absorption maxima in the red absorption range – at 675–678 and 685–690 nm. Sometimes, a less distinct maximum was observed at 665–670 nm. This work was reported in the first volume of the *Biofizika Journal* in 1956 [36].

C.S. French and coworkers developed a derivative spectrophotometer to study chlorophyll absorption spectra. By applying derivative spectroscopy, they revealed several chlorophyll spectral forms in plant materials [37]. Later, a combined analysis of derivative absorption, fluorescence and fluorescence excitation spectra showed the existence of more than ten spectral forms of chlorophyll in the photosynthetic apparatus, with the absorption maxima varying from 660 to 740 nm [38]. We proposed that the long wavelength shift of the red maximum of chlorophyll absorption of leaves (as compared to that of the dissolved pigment) was due mainly to different types of pigment aggregation caused by the high concentration of chlorophyll in chloroplasts and interactions between the adjacent molecules. Similar ideas were developed in the laboratory of J.J. Katz [39].

In our experiments, the absorption spectra of the aggregated pigments, obtained as solid films by evaporation of pigment solutions, were investigated [40, 41]. Later, fluorescence of these films was studied [42, 43]. Usually, the absorption and fluorescence maxima of the solid films were shifted towards longer wavelengths, as compared to those of pigment solutions. Most peculiar was the behavior of solid films of bacteriochlorophyll *c* (bacterioviridin) and bacteriochlorophyll *a*. As a rule, bacteriochlorophyll *a* exhibited three absorption maxima in the near infrared, very close to the maxima observed by E.C. Wassink and associates for cells of purple bacteria [44]. Wassink attributed those maxima to various pigment-protein complexes, because bacteriochlorophyll a isolated from the cells had only one absorption maximum at 770 nm. However, in our experiments similar spectra were observed in the absence of proteins as a result of pigment-pigment interaction in the aggregated forms.

The absorption and fluorescence spectra of bacteriochlorophyll c (bacterioviridin) films were also surprisingly similar to the spectra of green bacteria [40–43]. The principal absorption maximum was at about 745 nm, instead of the maximum at 660–670 nm for isolated bacteriochlorophyll c in solution (Fig. 3). To our delight, it has been shown in subsequent papers of several groups, that the antenna complexes (chlorosomes) of green photosynthetic bacteria really consist mostly of aggregated bacteriochlorophyll and the protein concentration is very low [45, 46].

We supposed that aggregated pigment forms should ap-

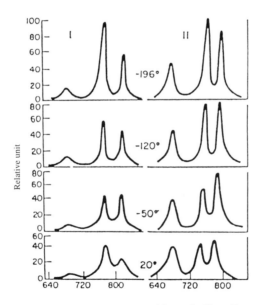

Fig. 3. Fluorescence spectra of bacteriochlorophyll c *(bacterioviridin) in cells of sulfur bacteria (I) and in solid films (II) at different temperatures.*

pear gradually in the process of pigment biosynthesis: first the monomeric forms should appear and subsequently, with their accumulation, pigment aggregation should occur. In order to investigate this idea, we (my post-graduate student F.F. Litvin and I) started studies on the spectral properties of chlorophyll fluorescence during the greening of etiolated leaves. For these measurements, we used the technique of fast freezing of leaves by their placement in liquid nitrogen. At liquid nitrogen temperature, biochemical processes are stopped so that pigment properties at fixed intermediate states of the greening process can be observed. We found that at the initial stages of greening, chlorophyll(-ide) with fluorescence maxima at 670–690 nm, dominated, and with further greening and increased chlorophyll accumulation, fluorescence maxima shifted to 730 nm, as a consequence of aggregated pigment forms [47].

We also revealed the transformation of protochlorophyll (-ide), with a fluorescence maximum at 655 nm, to chloro-phyll(-ide) with fluorescence maxima at 670–680 nm, apart from 'inactive' protochlorophyll(-ide) with maxima at 630–635 nm [47]. These fluorescence spectra correlated with the findings of K. Shibata who studied absorption spectra of greening etiolated leaves [48]. Litvin et al. discovered chlo-rophyll precursors also in mature green leaves [49]. Recently, our studies (together with N.N. Lebedev and P. Shif-fel) confirmed the findings of Litvin et al. Using sensitive fluorescence equipment, we observed in mature leaves a se-quence of protochlorophyllide photoreduction intermediates similar to those observed in etiolated leaves [50]. In experi-ments made in collaboration with M.I. Bystrova and F. Lang, we were able to show that protochlorophyll(-ide) ag-gregates formed in solid pigment films are spectrally similar to three protochlorophyll(-ide) forms in etiolated leaves [41, and Refs therein]. Thus, we managed to observe the pres-ence of various chlorophyll forms: aggregated and mono-meric, in model systems, plants and bacteria. However, in the meantime, experiments of many research groups showed a far more complicated picture of the biogenesis and spatial organization of the chlorophyll-protein complexes of the light harvesting antenna and reaction centers [51].

Reaction centers

In the 1930s, an assumption was advanced by R. Emerson and W. Arnold that several hundreds of chlorophyll molecules supply one reaction center in the photosynthetic apparatus. By measuring photoinduced difference absorption spectra of photosynthetic bacteria, L.N.M. Duysens found the absorption bands corresponding to the active centers and antenna bacteriochlorophyll [52]. These observations corroborated our ideas with respect to the occurrence of photochemically active and inactive chlorophyll in chloroplasts [35, 40].

In extending our studies of the organization of chlorophylls in situ, we tried to find reversible chlorophyll transformations in algal cells or in isolated chloroplasts. Investigations to tackle this problem were started in our laboratory by N. V. Karapetyan (1957) who spent much effort constructing a very sensitive difference spectrophotometer. Unfortunately, using this set-up we could not observe absorption changes of chlorophyll. However, we found a strong increase of chlorophyll luminescence upon illuminating chloroplasts and algal cells. This phenomenon was subsequently studied in detail with the result that the intensity of fluorescence is determined by the redox state of the photosystem II (PS-II) reaction centers [53, and Refs therein]. In particular, an induction of chlorophyll fluorescence was observed in chloroplasts when a highly reductive medium was created [53]. Here, quinone was fully reduced so that it was highly suggestive to propose the existence of an electron carrier with oxidoreduction potential near the hydrogen electrode. It was shown later that this carrier was pheophytin.

When R. Clayton and coworkers isolated reaction centers from chromatophores of photosynthetic bacteria, they showed that the reaction centers contain bacteriopheophytin, along with bacteriochlorophyll [54]. Later on, W. Parson's and L. Dutton's research groups applied picosecond spectroscopy and showed that bacteriopheophytin is involved into photoinduced electron transfer [55, 56]. These discoveries strongly enhanced our long-term interest in the photochemistry of pheophytins. For instance, they suggested that the reversible photoreduction of bacteriopheophytin we observed in 1951 for bacteriopheophytin solutions [15], might be relevant in connection with more recent work on the photosynthetic reaction centers.

In 1970, at the Institute of Photosynthesis in Pushchino (Moscow Region), I organized a research group for the study of photosynthetic reaction centers. The members of this group were, among others, my former postgraduate students, Vladimir Shuvalov and Vyacheslav Klimov. In the 1970s, I was invited by the late academician L. Mandelstam, the Director of the Institute of Spectroscopy in Troitsk, to give a report on spectral problems of photosynthesis. After my lecture we discussed the possibility of a collaboration in this field. I was informed that the picosecond technique was available in the group of Yu.A. Matveetz and I asked V.A. Shuvalov if he would like to start joint studies with the Matveetz group. Shuvalov very actively started these experiments and elucidated the sequence of electron transfer processes in the reaction centers isolated from *Rhodopseudomonas viridis* at picosecond time resolution. It was shown that bacteriochlorophyll photooxidation is coupled to the photoreduction of bacteriopheophytin and then further on to ubiquinone [57]. Under the stationary illumination of chromatophores, the photoreduction of bacteriopheophytin was detected by Klimov et al. [58]. Thus, no doubt, bacteriopheophytin participated in the minichain of electron transfer in the reaction centers. Naturally, we asked the question: to what extent can one observe pheophytin participation not only in bacterial reaction centers, but also in that of PS I and PS II?

Reviewing our photochemical experiments, I noted the similarity in the behavior of chlorophyll and pheophytin: both were capable of reversible photooxidation and reduction. The difference was that chlorophyll underwent more readily reversible photooxidation, whereas pheophytin showed photoreduction. It seemed that pheophytin must be somehow involved in photosynthesis. Many researchers had previously reported that during pigment extraction from green leaves, there were always small amounts of pheophytin detectable. Perhaps, rather than being a purification artifact, pheophytin was significant in electron transfer?

Already, in our experiments in 1968, in collaboration with M.G. Shaposhnikova, we applied the fluorimetric method of pheophytin determination under conditions excluding pheophytinization (low temperatures, etc.). However, even then, a small amount of pheophytin (making up 1–1.5% of the chlorophyll content) was always present [59]. The puzzle was solved later by the research group in Pushchino when pheophytin was found to be a component of the reaction centers of PS II. The experimental evidence of this was obtained when membrane particles enriched in PS II reaction centers were studied. Klimov et al. [58] managed to observe photoreduction of pheophytin coupled to photooxidation of reaction center chlorophyll. This concept is now generally accepted (Fig. 4).

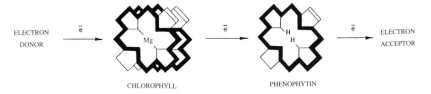

Fig. 4. Photoinduced electron transfer between chlorophyll and pheophytin.

The situation was different in the case of PS I, where the reaction center pigment – P700 was discovered by B. Kok (1957) [60]. Photoreduction of pheophytin was not observed in PS I. According to the picosecond absorption measurements, in this case, the role of pheophytin was played, in this case, by a unique form of chlorophyll [61, 62].

Thus, an ability of chlorophyll and pheophytin to engage in reversible oxidative-reductive phototransformation, discovered in the early work of our laboratory, is actually realized in the photosynthetic reaction centers. Pheophytin was found to be an electron carrier in the reaction centers of both photosynthetic bacteria and PS II of higher photosynthetic organisms. This led to the hypothesis of the evolutionary proximity of both types of reaction centers. The bacterial reaction center has been studied in great detail. Its high resolution three-dimensional structure by X-ray crystallography has been decoded by Huber, Deisenhofer and Michel [63], work for which the authors were awarded the Nobel Prize in Chemistry of 1988. The differences between the reaction centers are obvious: only the redox potential of the reaction center of PS II allows oxidation of water to molecular oxygen. So far, the details of the mechanism remain enigmatic, leaving one of the mysteries of photosynthesis still unresolved.

Inorganic models of reaction centers

The experience of my early work with inorganic semiconductor-photosensitizers helped me to introduce them into photosynthesis research as model reaction centers. Here, under the action of light, an electron is transferred from the valence to the conduction band, leaving behind an electron vacancy, a hole. Apparently, migration of electron and hole to different active centers of the semiconductor surface, determines the possibility of charge separation and realization of the redox processes.

In 1961, we conducted an experiment that we called an 'inorganic model of the Hill reaction' [64]. We used TiO_2, ZnO, and WO_3 illuminated with the 365 nm mercury emission band. In water suspensions containing semiconductor particles and electron acceptors (ferric compounds or p-benzoquinone), we observed oxygen photogeneration, although, with a low quantum yield (1–3%) (Fig. 5). Both formation of ferrous compounds and the release of hydrogen ions corresponded stoichiometrically to the oxygen produced [65]. In the experiments conducted in collaboration with researchers at the Institute of Chemical Physics (using ^{18}O-labeled water), it was shown that oxygen was actually generated from water molecules [66].

Under aerobic conditions, when methyl viologen was used as an electron acceptor and ascorbate, dithiothreitol and other compounds as electron donors, we observed a photosensitized reduction of methyl viologen. In the presence of an appropriate catalyst (bacterial hydrogenase), the evolution of

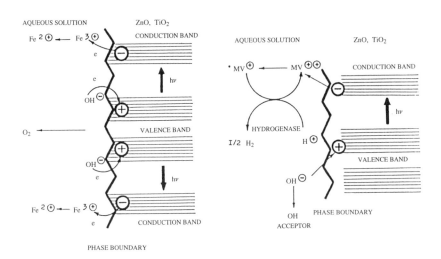

Fig. 5. Hypothetical scheme of photosensitized oxygen and hydrogen evolution on the particles of TiO_2 and ZnO.

molecular hydrogen was observed (Fig. 5) [67]. In these experiments, methyl viologen served as the intermediate electron carrier from semiconductor particles to hydrogenase. Recently, we observed coupling of TiO$_2$ particles with living *Clostridium* cells. In this case, reversibly photoreduced methyl viologen diffused through cell walls to the hydrogenase localized inside the cells (Fig. 6) [68].

In experiments performed by M.A. Shlyk et al. [69], the electron transfer from the semiconductors (TiO$_2$ and CdS) to hydrogenase, was shown, without methyl viologen as intermediate electron carrier. When TiO$_2$ was used, the quantum yield of electron transfer to hydrogenase achieved 20%. Tris-buffer, sulfhydryl compounds or formate served as electron donors in these reactions [69]. A number of systems of hydrogen photoproduction with the use of inorganic catalysts has been described by M. Gratzel [70].

It seems attractive to use molecular nitrogen as electron acceptor in reactions of this type. G.N. Schrauser and T.D. Guth experimentally observed dinitrogen photoreduction on the surface of TiO$_2$ [71]. Using the system TiO$_2$–N$_2$–acetaldehyde, photosensitized amino acid formation was observed. The quantum yield of these reactions was extremely low, about 10^{-6} in the latter case [72]. The use of semicon-

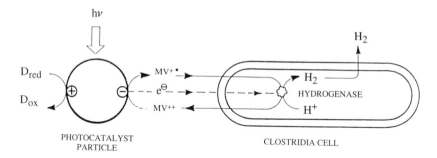

Fig. 6. Scheme for hydrogen photoproduction in the coupling of mineral photocatalyst to Clostridium *cells. MV, methyl viologen; D$_{red}$ and D$_{ox}$, reduced and oxidized electron donor, respectively.*

ductors in the investigation of nitrogen photoreduction is a very important direction of study, undoubtedly promising for technical applications. The above described systems are not structural models of the reaction centers. They only used the principles of photochemical charge separation for the conversion and storage of the energy of light quanta.

Involvement in problems of the origin of life

In 1924, A.I. Oparin wrote a short book, *Origin of Life on the Earth*. According to Oparin's concept, organic substances, including protein-like compounds, were formed abiotically from inorganic components of the Earth's atmosphere under the action of electrical discharges or UV radiation. These organic substances formed in the 'primary broth', the phase-isolated structures, so-called 'probionts', which were the prototypes of the livings cells. Further evolution was a consequence of natural selection of the primary cells. This book had a wide influence in the world and its ideas were supported by many researchers.

In 1957, A.I. Oparin organized, in Moscow, the first international symposium which was devoted exclusively to the origin of life. The symposium attracted many prominent scientists, not only biologists, but also physicists and chemists. The main sensations of the symposium were papers dealing with the abiotic formation of amino acids from mixtures of ammonia, carbon dioxide, water and other inorganic compounds under electric discharge and UV radiation. Papers with such data were presented by S. Miller and the Nobel Prize winner H. Urey, A.G. Pasynskii, T.E. Pavlovskaya and A.N. Terenin.

Participation at this meeting enhanced my interest in the problem of the origin of life. My lecture at the symposium was devoted to the development of photocatalytic systems in organisms. According to Oparin's hypothesis, the primary cells were heterotrophic. However, the existence of life on our planet is undoubtedly connected with plant photosynthesis. What were the primary forms of photosynthesis and how did they develop on Earth? The most ancient photosynthetic organisms we know of are cyanobacteria or blue-green algae which are found in Pre-cambrian deposits formed three-billion years ago. However, cyanobacteria are highly developed organisms with the present day's complex pigment system, containing chlorophylls, phycobilins, carotenoids, and they are capable of photosynthesis of organic substances from CO_2 and water, oxygen release and assimilation of molecular nitrogen. Elucidation of the metabolism of cyanobacteria is undoubtedly interesting from the viewpoint of evolutionary biochemistry, but does not answer questions regarding the origin of photosynthesis.

In connection with this, the photoreactions of inorganic catalysts and semiconductors might be of interest as models of primary photobiological processes. It may be assumed that the inorganic components of the Earth's crust are primary photosensitizers of chemical processes. The diversity of the photocatalytic reactions involving electron transfer, formation of hydrogen, oxygen and various reduced and oxidized compounds, suggests a possible use of the inorganic photosensitizers in primary cells. In the existing organisms, no inorganic sensitizers have been found; instead evolution has preferred porphyrin derivatives. However, it cannot be excluded that those primary organisms could have been a blind alley of evolution.

Another direction of research, which seemed promising to us from the evolutionary point of view, was photoactivation of coenzymes involved in dark enzymatic reactions. We studied cytochromes, flavins, pyridine nucleotides and porphyrins. Cytochromes and iron-porphyrins, possess very low photochemical activity. However, illumination of the oxidized forms of cytochrome c, under anaerobic conditions, led to their partial reduction by electron donors. Also, oxidation of reduced cytochromes by oxygen has been observed [73].

The high photochemical activity of flavins was well known prior to our studies. We confirmed these observations and, in particular, showed that deazaflavin, isolated from methanogenic bacteria, is an extremely active photosensitizer of redox reactions [74].

The most interesting results were obtained with pyridine nucleotides. We found that reduced pyridine nucleotides, NADH and NADPH, when excited at their absorption maxima (340 nm), were capable of reducing a number of compounds, including those with redox potentials exceeding the potential of the hydrogen electrode. Coupling with hydrogenase caused release of molecular hydrogen. The ESR measurements revealed formation of free radicals by photoactivated NADH and NADPH in solution and in the solid state. Photoactivation of hydrogen production has also been observed in *Clostridium* cells, as the consequence of NADH photoactivation [75–77]. One may assume that photoactivation of coenzymes has been used in ancient organisms as a primitive prototype of photosynthesis.

The possibility of the abiotic formation of porphyrins from simple organic compounds was shown by Rotemund in the 1930s. He observed porphyrin formation in reactions of pyrrol with formaldehyde. We established that this reaction indeed causes the formation of various porphyrin compounds – porphyrins, chlorins and bacteriochlorins. The introduction of inorganic photocatalysts in the system allowed us to observe, in some cases, the photoactivation of porphin synthesis [78]. The natural pathway of chlorophyll biosynthesis in plants has been decoded by S. Granick; it includes intermediate porphyrin formation. Based on the idea that phylogeny and ontogeny are correlated, one might postulate that the primary sensitizers were simple porphyrins, and only much later, in the process of evolution, chlorophyll and its derivatives appeared. In this context, we were able to show that simple porphyrins, such as porphin and hematoporphyrin, readily undergo reversible photoreduction and photooxidation, and serve as photosensitizers of electron transfer reactions.

In search of the evolution of photosynthetic apparatus, one may turn to the study of the cosmos, but no traces of porphyrins have been found on other planets of the solar system. Hope remains that in some regions of Mars, microorganisms may exist, similar to the terrestrial organisms living under extreme conditions, as for example, in the Antarctic ices. Porphyrins found in samples of soil from the Moon turned out to be of abiotic origin: they were artifacts formed in exhaust gases of space rockets.

An idea to organize an international interdisciplinary society of physicists, chemists and biologists, for the promotion of studies of the physico-chemical basis of the origin of life, had been discussed for a long time after the Moscow symposium in 1957. In 1970, the society (International Society on the Origin of Life, ISSOL) was founded during an international symposium in Pont-a-Mussone (France). The symposium organizer was Professor R. Buvet. A.I. Oparin was unanimously elected the President. I was elected as one of the members of the Executive Committee and in 1977, I was elected Vice-president of the society.

Teaching at the Moscow University

In 1953, academician A.N. Terenin, and I, entered the old building of M.V. Lomonosov Moscow State University at Mokhovaya Street. In a room in the basement, we met a broad-shouldered handsome man with lively eyes, Boris Nikolaevitch Tarusov. We got acquainted and Tarusov told us that he was charged by the University Rector I.G. Petrovsky with organizing the first Chair of Biological Physics in our country. It was planned that it would be integrated into the Department of Biology with the goal to train specialists familiar with effects of radioactivity on living organisms. However, Tarusov believed that the goals of the

Chair should be less specialized. In his opinion, biophysics was physical chemistry in biology, and the Chair should deal with the action of a wide variety of physical factors, including light, to the living systems. Terenin and I were invited to organize a photobiology branch within the Department.

What was the Chair like at that time? In the first years, B.N. Tarusov invited about 30 students for postgraduate training. The number of undergraduate students was much less at that time. They were supposed to defend their Candidate (PhD) Theses in different areas of biophysics which they were free to select at their own discretion. B.N. Tarusov never oppressed the initiatives of young scientists; the atmosphere of free creativity predominated. However, it was an urgent problem to find qualified supervisors for the postgraduates. Terenin and I chose three post-graduates for training in photobiophysics: Yu.A. Vladimirov, S.V. Konev, and F.F. Litvin. We offered them a proposal to develop methods of luminescence measurements. Yu.A. Vladimirov and S.V. Konev began to study the luminescence of aromatic amino acids in proteins. F.F. Litvin was more interested in photosynthesis and began to study chlorophyll luminescence in leaves at liquid nitrogen temperature. They started their research using photographic techniques for detection of fluorescence spectra and later managed to build a very sensitive fluorimeter based on the use of photomultiplier tubes.

From the first year of work at the Moscow University on, I offered a course of lectures on photobiology (photobiophysics and photobiochemistry), which had not previously been given. The aim of the course was to explain the molecular principles underlying the effect of light on living organisms. Students showed keen interest in these lectures. They were attended by students from the Chairs of Biophysics, Plant Physiology, and Microbiology of the Biology Department, and the Departments of Physics and Chemistry. B.N. Tarusov was much interested in photobiological problems himself, as he was involved in investigations of the superweak spontaneous luminescence of living cells and tissues, and therefore he also joined the course. We enjoyed discussing problems of mutual interest. Teaching was always attractive to me, giving me deep satisfaction, and moreover, an opportunity to attract young people for the undergraduate and postgraduate training in my own research group. The education in photobiology at the Chair of Biophysics promoted the development of this science in the country considerably. Since 1954, about sixty students obtained their Candidate (PhD) and Doctor of Science Degrees under my supervision. Most of them passed photobiology training at the Chair of Biophysics of the Moscow University, many have become leading scientists in the field, studying various problems of photobiophysics. This has been a source of pride and satisfaction for me throughout my professional life.

Epilogue

As I mentioned, I started my scientific career as a Chemical Engineer, in 1937. My Candidate Thesis (1940) was devoted to photosensitization properties of titanium dioxide. I obtained a Doctor of Science Degree in 1948, for studies of photosynthetic pigments; my dissertation was entitled *The Study of Photochemical Reactions in Photosynthesis*. My first steps in fundamental science were made in collaboration with outstanding scientists such as A.N. Terenin, V.A. Engelhardt, and many other talented people. With their help, I have been privileged to become integrated into the network of national and international scientific traditions which they, themselves, took over from their teachers and colleagues. I hope that I managed to pass this relay to younger scientists who worked with me as students or collaborators. Now, many of them have organized their own laboratories and Chairs and raise new teachers and researchers.

Starting from the 1960s, scientific meetings devoted to mechanisms of photosynthesis and photobiology always attracted numerous participants from many universities and research institutes of the Soviet Union. In 1962, I was elected the corresponding member and in 1976, the full member of the USSR Academy of Sciences. In 1991, my colleagues and I were awarded the USSR State Prize in Fundamental Science. I consider these awards as recognition of the validity of our research direction.[4]

We always knew that we were a part of the international scientific community. However, personal contacts with foreign colleagues have become possible only since the late 1950s, during the time of Khrushchev reforms. Scientific programs for the country members of the Council for Mutual Economic Aid (CMEA) were created. We were involved in the program on photosynthesis which was initially headed by A.N. Terenin. This program included scientists from Bulgaria (Profs. M. Popov and I.T. Iordanov), Hungary (Profs. A. Faludi-Daniel, L. Szalay and F. Lang), Poland (Profs. W. Zelawsky, S. Wieckowski and D. Frackowiak), East Germany (Prof. P. Hoffmann), and Czechoslovakia (Dr. Z. Sestak). Participants from the Soviet Union were from leading laboratories of the USSR Academy of Sciences, Academies of Belorussia and Ukraine, Moscow State University and Universities of other towns. A common program of the fundamental studies existed. Annual symposia were organized in all member countries of the CMEA. This collaboration was definitely useful. It helped to increase the quality of the research and train new specialists in photosynthesis. Prof. W. Hendrich (Poland), F. Lang and E. Lehoczki (Hungary), P. Shiffel and M. Durhan (Czechoslovakia) and I. Djelepova (Bulgaria) worked in our research group.

At the time of the USSR friendship with China, we had undergraduate and postgraduate students from there also. I

---

4 A.A. Krasnovsky was also a member of foreign academies and international societies.

remember that before the cultural revolution in China, 50% of all undergraduate students at the Chair of Biophysics were Chinese. Interesting research has been carried out in our laboratory by H.I. Tan.

As far as I can remember, the International Symposium on the Origin of Life, organized by A.I. Oparin in 1957, was the first conference which attracted many prominent scientists from USA, Germany, Japan and other countries to Moscow. Among others, we met Nobel Prize winners L. Pauling and M. Calvin there.

In 1958, for the first time in my life, I had the opportunity to go abroad as a member of the Soviet delegation to the International Biochemical Congress held in Vienna. In 1959, I was invited as a plenary lecturer at the International Botanical Congress in Montreal. In 1961, Professor H. Tamyia and I organized a symposium on photosynthesis at the International Biochemical Congress held in Moscow. In 1968, I was invited as a plenary lecturer at the first International Congress on Photosynthesis held in Freudenstadt. I mention here only my earliest participation in international scientific events as they left lasting impressions on me. At these meetings, I met many distinguished colleagues: E. Rabinovitch, M. Calvin, S. French, M. Gibbs, D. Arnon, A. Moyse, L. Bogorad, H. Gaffron, B. Kok, Govindjee, M. Kamen, H. Witt, A. Benson and others. Personal contacts with colleagues greatly enhanced my scientific imagination and accelerated our progress in research.

Being interested in the various problems I have discussed above, I now consider that my main scientific interest lies in the creation of an efficient model system that would allow conversion of solar energy based on the principles of photosynthesis. The problem is to find photosensitizers capable of efficient charge separation under the action of solar radiation, and to avoid back reactions of electrons and holes in the functioning of the artificial electron transfer chains. I think that the search of photobiochemical systems, capable not only of solar energy conversion, but also of long-term energy storage, presents a scientific issue extremely important for the future of mankind, in order to solve the problems of energy and food that the world faces today and will face even more drastically in the future.

References

1   A.A. Krasnovsky, Annu. Rev. Plant Physiol., 11 (1960) 363–410.
2   A.A. Krasnovsky, Transformation of Solar Energy in Photosynthesis. Molecular Mechanisms. A lecture presented at the XXIXth Annual Memorial Session devoted to academician A.N. Bakh, Nauka, Moscow, 1974.
3   A.A. Krasnovsky, Photosynth. Res., 33 (1992) 177–193.
4   K.A. Timiryazev, Proc. R. Soc. London, ser. B., 72 (1903) 424.
5   E.I. Rabinovich, Photosynthesis and Related Processes, Vol. 1., Interscience Publ., New York, 1945.
6   E.I. Rabinovich, Photosynthesis and Related Processes, Vol. 2.1., Interscience Publ., New York, 1951.
7   E.I. Rabinovich, Photosynthesis and Related Processes, Vol. 2.2., Interscience Publ., New York, 1956.
8   A.A. Krasnovsky and G.P. Brin, Dokl. Akad. Nauk SSSR (Biochemistry), 53 (1946) 447–450.
9   A.A. Krasnovsky and G.P. Brin, Dokl. Akad. Nauk SSSR (Biochemistry), 58 (1947) 1087–1090.
10 A.A. Krasnovsky, Dokl. Akad. Nauk SSSR (Biochemistry), 60 (1948) 421–424.
11 A.A. Krasnovsky. In: B.S. Neporent (ed.), Elementary Photoprocesses in Molecules, Plenum Press, New York, 1968, pp. 163–183.
12 T.T. Bannister, Plant Physiol., 34 (1959) 246–254.
13 G.R. Seely and A. Folkmanis, J. Am. Chem. Soc., 86 (1964) 2763–2770.
14 H. Scheer and J.J. Katz, Proc. Natl. Acad. Sci. USA, 71 (1974) 1626–1629.
15 A.A. Krasnovsky and K.K. Voynovskaya, Dokl. Akad. Nauk SSSR (Biochemistry), 81 (1951) 879–882.
16 A.A. Krasnovsky and N.N. Drozdova, Dokl. Akad. Nauk SSSR (Biochemistry), 150 (1963) 1378–1381.
17 A.K. Chibisov, Photochem. Photobiol., 10 (1969) 331–347.
18 A.A. Krasnovsky and A.V. Umrikhina, Dokl. Akad. Nauk SSSR (Biochemistry), 104 (1955) 882–885.

19 N.N. Bubnov, A.A. Krasnovsky, A.V. Umrikhina, V.F. Tzepalov and V.Ya. Shlyapintokh, Biofizika, 5 (1960) 121–126.
20 G.T. Rikhireva, L.A. Sybeldina, Z.P. Gribova, B.S. Marinov, L.P. Kaushin and A.A. Krasnovsky, Dokl. AN SSSR (Biophysics), 181 (1968) 1485–1488.
21 A.V. Umrikhina, N.V. Bublichenko and A.A. Krasnovsky, Biofizika, 18 (1973) 565–568.
22 A.A. Krasnovsky, A.V. Umrikhina and N.V. Bublichenko. In: A.A. Krasnovsky (ed.), Spectroscopy of Molecular Phototransformations, 'Nauka' Leningrad, 1977, pp. 106–131.
23 M. Calvin, Photosynthesis Res., 21 (1989) 3–16.
24 A.A. Krasnovsky and G.P. Brin, Dokl. Akad. Nauk SSSR (Biochemistry), 67 (1949) 325–328.
25 V.B. Evstigneev. In: B.S. Neporent (ed.), Elementary Photoprocesses in Molecules, Plenum Press, New York, 1968, pp. 184–200.
26 A.A. Krasnovsky, Biophys. J., 12 (1972) 749–763.
27 A.A. Krasnovsky and A.N. Luganskaya, Dokl. Akad. Nauk SSSR (Biochemistry), 183 (1968) 1441–1444.
28 A.A. Krasnovsky, A.N. Semenova and V.V. Nikandrov, Dokl. Akad. Nauk SSSR (Biochemistry), 262 (1982) 469–472.
29 A.N. Semenova, Ya.V. Barannikova, V.V. Nikandrov and A.A. Krasnovsky, Biological Membranes (in Russian), 4 (1987) 448–557.
30 A.N. Semenova, V.V. Nikandrov and A.A. Krasnovsky, J. Photochem. Photobiol. B, 1 (1987) 85–91.
31 V.V. Nadtochenko, V.V. Lavrentyev, I.R. Rubtsov, N.N. Denisov, Ya. V. Barannikova and A.A. Krasnovsky, Photochem. Photobiol., 53 (1991) 261–269.
32 A.A. Krasnovsky, V.B. Evstigneev, G.P. Brin and V.A Gavrilova, Dokl. AN SSSR (Biochemistry), 82 (1952) 947–950.
33 L.G. Erokhina and A.A. Krasnovsky, Molek. Biol. (in Russian), 2 (1968) 550.
34 C.S. Yochum and L.R.J. Blinks, J. Gen. Physiol., 38 (1954) 1.
35 A.A. Krasnovsky and L.M. Kosobutskaya, Dokl. AN SSSR (Biochemistry), 91 (1953) 343–346.
36 A.A. Krasnovsky, E.A. Nesterovskaya and A.B. Goldenberg, Biofizika, 1 (1956) 328–333.
37 C.S. French, J.S. Brown and M.C. Lawrence, Plant Physiol., 49 (1972) 421–429.
38 F.F. Litvin and V.A. Sineschekov. In: Govindjee (ed.), Bioenergetics of Photosynthesis, Academic Press, New York, 1975, pp. 619–661.
39 J.J. Katz, Photosynth. Res., 26 (1990) 143–160.

40   A.A. Krasnovsky, Voynovskaya and L.M. Kosobutskaya, Dokl. AN SSSR (Biochemistry), 85 (1952) 389–392.
41   A.A. Krasnovsky and Bystrova, Biosystems, 12 (1980) 181–194.
42   A.A. Krasnovsky, Yu.E. Erokhin and Hun Yui Zun, Dokl. AN SSSR (Biochemistry), 143 (1962) 456–459.
43   A.A. Krasnovsky, Yu.E. Erokhin and B.A. Gulyaev, Dokl. AN SSSR (Biochemistry), 152 (1963) 1231–1234.
44   E.C. Wassink, E. Katz and R. Dorrestein, Enzymologia, 7 (1939) 113.
45   R.E. Blankenship, D.C. Brune, B.P. Witmershaus. In: Stevens, Jr. and D.A. Bryant (eds.), Light Energy Transduction in Photosynthesis: Higher Plants and Bacterial Models, Am. Soc. Plant Physiol., 1988, pp. 32–46.
46   A.R. Holtzwarth, K. Griebenow and K. Shaffner, J. Photochem. Photobiol. A: Chem., 65 (1992) 61–71.
47   F.F. Litvin and A.A. Krasnovsky, Dokl. AN SSSR (Biophysics), 117 (1957) 106–109.
48   K. Shibata, J. Biochem. (Tokyo), 44, 3 (1957) 147–173.
49   F.F. Litvin, G.T. Rikhireva and A.A. Krasnovsky, Dokl. AN SSSR (Biophysics), 127 (1959) 699–701.
50   N.N. Lebedev, P. Siffel and A.A. Krasnovsky, Photosynthetica, 19 (1985) 183–185.
51   O.B. Belyaeva and F.F. Litvin, Photobiosynthesis of Chlorophyll, Moscow University Publ., Moscow, 1989.
52   L.N.M. Duysens, Thesis, Utrecht, 1952.
53   N.V. Karapetyan, V.V. Klimov, I.N. Krakhmaleva and A.A. Krasnovsky, Dokl. AN SSSR (Biochemistry), 201 (1971) 1244–1247.
54   R.K. Clayton, Photosynth. Res., 19 (1988) 207–224.
55   M.G. Rockley, W.W. Windsor, R.I. Cogdell and W.W. Parson, Proc. Natl. Acad. Sci. USA, 71 (1975) 2251–2253.
56   K.J. Kaufmann, P.L. Dutton, T.L. Netzel, J.S. Leigh and P.M. Rentzepis, Science, 188 (1975) 1301–1303.
57   V.A. Shuvalov, A.V. Klevanik, A.V. Sharkov, Ya.A. Matveetz and P.G. Krukov, FEBS Lett., 91 (1978) 135–139.
58   V.V. Klimov, V.A. Shuvalov, I.N. Krakmaleva, N.V. Karapetyan and A.A. Krasnovsky, Biokhimia, 41 (1976) 1435–1441.
59   A.A. Krasnovsky and M.G. Shaposhnikova, Fiziol. Rast., 17 (1970) 436–439.
60   B. Kok, Acta Bot. Nederl., 6 (1977) 316–336.
61   V.A. Shuvalov, B. Ke and Dolan, FEBS Lett., 100 (1979) 5–9.
62   J.M. Fenton, M.J. Pellin, Govindjee and K.J. Kaufmann, FEBS Lett, 100 (1979) 100–104.

63 J. Deisenhofer, O. Epp, K. Miki, K. Huber and H. Michel, Nature, 318 (1985) 618–624.
64 A.A. Krasnovsky and G.P. Brin, Dokl. AN SSSR (Biochemistry), 147 (1962) 656–659.
65 A.A. Krasnovsky and G.P. Brin. In: H. Metzner (ed.), Photosynthetic Oxygen Evolution, Academic Press, London, 1978, pp. 405–410.
66 G.V. Fomin, G.P. Brin, M.V. Genkin, A.K. Ljubimova, L.A. Blumenfeld and A.A. Krasnovsky, Dokl. AN SSSR (Biochemistry), 212 (1973) 424–427.
67 A.A. Krasnovsky, G.P. Brin and V.V. Nikandrov, Dokl. AN SSSR (Biochemistry), 229 (1976) 990–993.
68 A.A. Krasnovsky and V.V. Nikandrov, FEBS Lett., 219 (1987) 93–96.
69 M.A. Shlyk, V.V. Nikandrov, N.A. Zorin and A.A. Krasnovsky, Biokhimia, 54 (1989) 1598–1606.
70 M. Cratzel. In: A.M. Braun (ed.), Photochemical Conversions, Press Polythechnique Romandes, Lausanne, 1983.
71 G.N. Schrauser and T.D. Guth, J. Amer. Chem. Soc., 99 (1977) 7189–7193.
72 A.A. Krasnovsky, T.E. Pavlovskaya and A.N. Telegina, Dokl. AN SSSR (Biochemistry), 308 (1989) 1258–1261.
73 A.A. Krasnovsky and K.K. Voynovskaya, Biofizika, 1 (1956) 120–126.
74 V.V. Nikandrov, E.S. Panskhava and A.A. Krasnovsky, Photobiochem. Photobiophys., 13 (1986) 105–114.
75 A.A. Krasnovsky and G.P. Brin, Dokl. AN SSSR, 153 (1963) 721–724.
76 V.V. Nikandrov, G.P. Brin and A.A. Krasnovsky, Biokhimia, 43 (1978) 636–645.
77 I.V. Zhukova, V.V. Nikandrov and A.A. Krasnovsky, Biofizika, 25 (1980) 1095–1096.
78 S. Granick and D. Mauzerall. In: D.M. Greenberg (ed.), Chemical Pathways of Metabolism, Vol. 2, Academic Press, New York, 1961.

G. Semenza and R. Jaenicke (Eds.)
Selected Topics in the History of Biochemistry: Personal Recollections, V
(Comprehensive Biochemistry Vol. 40) © 1997 Elsevier Science B.V.

Chapter 5

Efraim Racker:
28 June 1913 to 9 September 1991

GOTTFRIED SCHATZ

Biozentrum der Univ. Basel, Dept. Biochemistry, Klingelbergstrasse 70,
CH-4056 Basel, Switzerland

When he entered our Vienna laboratory on a hot summers day in 1961, I was struck by his youthful stride that belied his white hair, his foreign-looking bow tie, and a curious tension in his face. My friends told me later that I had just seen Efraim Racker, one of the foremost biochemists of our time, and that this was his first visit to Vienna since he had fled this city more than 23 years ago.

In 1961, Racker's work on biological ATP production had made him one of the stars of bioenergetics, the branch of biochemistry dealing with energy conversion by living cells. I had already made up my mind that I wanted to do my post-doctoral work with him and asked him the next day whether he would accept me. He drew me aside to quiz me about my work, but interrupted me after my first few sentences by asking, 'How come you speak English so well?' Flattered, I explained in my German-accented English, that I had spent my last year of high school as an exchange student in the United States. His immediate riposte, 'How come you speak English so badly?' made us both laugh and started a lifelong

Efraim Racker and Severo Ochoa

bond between us that was severed only when he died 30 years later.

Racker was born on June 28, 1913, in the town of Neu-Sandez in Poland, to Jewish parents who moved to Vienna before his second birthday. His family was not wealthy, and settled in the second district of Vienna called Leopoldstadt, then largely an enclave of recent immigrants from the eastern reaches of the Austro-Hungarian Empire and beyond. Most of these immigrants were poor Jews who had hoped for a better life, but then faced anti-semitism, the deprivations of World War I, the social unrest that followed the collapse of the Empire in 1918, and the rise of fascism that culminated in the Nazi takeover of Austria in 1938. During these stormy years, Racker attended elementary school and high school. The formal atmosphere there did not appeal to him at all; he much preferred playing soccer or chess in the Augarten, the local park. Even as an old man, he loved competitive sports such as soccer, ping-pong or tennis, and was an excellent chess player, but he had only vague notions about the usual staple of classical schooling, such as the names of Greek goddesses or the Habsburg family tree. However, to the end of his life, he retained an intimate knowledge of the literature, music and art of the Vienna he had known. Young Efraim was fascinated by the public lectures of the writer and critic, Karl Kraus, knew many of the local musicians, and was profoundly influenced by the paintings of Egon Schiele. For his twelfth birthday, one of his aunts gave him a painting set which soon accompanied him on his regular forays into Augarten and began his lifelong passion for painting. Encouraged by the painter and art educator, Victor Löwenfeld, who had become his artistic mentor, Racker initially decided to become a painter. After finishing high school, he passed the highly competitive admission examination to the Vienna Academy of Art, but was once again turned off by the rigid and formal style of training he encountered. He soon left the academy to study medicine at the University of Vienna.

In the early 1930s, the fame of Vienna's Medical School had largely faded, and quite a few of the professors and students belonged to right-wing student organizations called *Burschenschaften.* Many of these *Burschenschaftler* later joined the Nazi movement. I got to know many of them during my own student days in postwar Austria. They were then aging but unrepenting, and I shudder when picturing Racker among them, for it was easy to see that he was Jewish. How many hidden scars did these years leave? Perhaps they explain why Racker always had such a strong gut reaction against power, arrogance and pompousness. This trait made him a great chairman and scientific adviser, but sometimes a difficult adversary of deans, provosts and other members of official hierarchies.

Although disappointed by medical school, Racker was captivated by the discoveries of Sigmund Freud, whose home and office in Berggasse 16 he passed on his daily walk to the university. Had Racker remained in Vienna, he would probably have become a psychiatrist. Indeed, his older brother, Heinrich, who later fled to South America, was a psychoanalyst. The two brothers were very close and Racker often told me of their many late-night discussions on Freud, Adler, and their often highly-strung disciples. Racker would have made an excellent psychiatrist; many of his postdoctoral fellows (including myself) had good reason to beware of his uncanny ability to read other people's minds; however, fate decided that his interest in the workings of the human mind should lead him to study brain metabolism and, later on, biological ATP production.

Racker's graduation from medical school almost coincided with Hitler's march into Austria. Racker wisely decided to leave while it was still possible and fled, via Denmark, to Great Britain where the biochemist, J. Hirsh Quastel, offered him a job at Cardiff City Mental Hospital in Wales. In

this rather remote place, the two of them tried to detect biochemical defects that could explain the mental abnormalities
of their patients. Racker apparently decided to tackle this
problem from the bottom up because his first publication
was humbly entitled *Histidine Detection and Estimation in
Urine* [1]; there was nothing humble about the opening sentence, however: 'During the last 10 years, many attempts
have been made to detect changes in the liver metabolism
of patients suffering from mental disorders'. The inexperi-
enced M.D. from Vienna clearly had set his aims high!
He also studied the effect of oxygen deprivation on the metabolism of tissue slices, but quickly realized that his hopes
of finding biochemical causes for mental diseases were
doomed as too little was known on the metabolism of normal
cells.

    But once again Racker's scientific career was buffeted by
the political gales that started to blow across Europe. When
Great Britain entered the war, Racker suddenly found himself an enemy alien whose experiments with human urine
near the strategically sensitive coast posed a security risk.
He lost his job at Cardiff and, together with many other
refugees from Nazi Germany, was interned on the Isle of
Man, where he practiced medicine for the first time in his
life. Although he enjoyed being a 'real' doctor, he soon decided to try his luck as a researcher in the United States, not
knowing that he was embarking on a 25 year odyssey. He
started out doing a brief stint as a research associate in the
Physiology Department of the University of Minnesota at
Minneapolis (1941–42), but then once again worked as a
physician in New York city's Harlem Hospital (1942–44). His
career as a biochemist started in earnest only in 1944, with
his appointment as staff member in the Microbiology Department of the New York University Medical School. He
often spoke with great fondness of this department and the
great scientific debt he owed to several of its members, particularly to Severo Ochoa, and Colin MacLeod, the depart-

ment chairman. It was during the 8 years in this department that he finally became a professional biochemist. Yet when he was offered the position of associate professor at Yale Medical School in 1952, he accepted and moved to New Haven. In 1954, his odyssey seemed to end when he accepted the position of chief of the Nutrition and Physiology Department at the Public Health Research Institute of the City of New York. Little did our modern Odysseus know that Manhattan would be but another way station in his wanderings and that he would reach his Ithaca only 12 years later.

It was at the Public Health Research Institute that I joined him as a postdoctoral fellow in the summer of 1964, and I will never forget my shock when I first saw this Mecca of bioenergetics. The institute was a decrepit and grimy building, wedged between a run-down police garage and a coal-fired power plant whose dusty emissions darkened the window panes and settled into every crevice of the laboratory. The address 'Foot of East 16th Street' should have warned me; it certainly warned cab drivers who often refused to go there. The laboratories had most of the required equipment, but many of the instruments were old and not well maintained. Cockroaches, many of ghastly dimensions, were everywhere, as floating corpses in buffer solutions, uninvited guests in lunch boxes, or electrocuted culprits in short-circuited electric equipment.

Yet I had one of the most exciting and productive years of my life in that building. It was brimming with talented, motivated people from all over the world, and 'our' fourth floor appeared to have the best of them. There were Racker's trusted colleagues, Maynard Pullman, Ray Wu and Harvey Penefsky, all of them already well known in their own right. Many of the postdoctoral fellows at that time, such as Ron Butow, Yasuo Kagawa, Howard Zalkin and Richard McCarty, later went on to distinguished careers. Two irreverent and bright graduate students, Peter Hinkle and Gladys Monroy, loved to disagree with Racker and injected spice into our lunch discussions. Lunches were taken together, in what was grandly called a lunch room, and usually consisted of a homemade sandwich and, for Racker, a small can of Hawaiian fruit punch, a mysterious concoction fortunately restricted to North America. There were visiting professors such as Michael Schramm; and there was Racker himself, 'Ef' to his senior colleagues. At that time he did not yet invite postdocs to address him by that nickname, but I did so anyway and he did not seem to mind. Having just turned fifty-one, Ef was a splendid leader, full of vigor, wit, enthusiasm, and self-confidence. Aided by one or two technicians, he managed to work at the bench nearly every day, yet keep all of us under close surveillance. Here in his lab, he was not tense at all; he was outgoing, relaxed, and clearly aware of the fact that he was at the height of his scientific powers. There certainly was no trace of a bow tie. In fact, the best that could be said of his dress was that it matched the building. Coming to work on Saturdays was de rigueur; offenders were received on Monday morning with a frosty 'How was your weekend?' and usually sinned no more. In order to avoid rush hour traffic, work started at about 1000 h and went on until 1900 h or longer. We were a proud and happy crowd, but we also understood why Ef had his detractors. He either liked you or he did not. If he did not, his quick mind and sharp tongue could leave long-lasting wounds. We learned that doing science was not only joyful exploration, but also a game of intellectual domination and that Ef played the game well. Maynard Pullman was welcome support when Ef's presence or impatience became too overpowering. Maynard was universally respected for his discoveries in the field of oxidative phosphorylation and was well liked for his warmth and balanced judgment. He could stand up to Ef for us, and often did. Harvey Penefsky kept more to himself, but was much in demand for precise scientific information and critical discussion.

Despite his grueling schedule, Ef always seemed to have time. He never closed the door to his office; he never secluded himself in the library to read the latest journals, and he never used his hours at the institute for working on a manuscript or the book he was then writing [2]. Yet he answered every letter and rarely missed deadlines. But, as if that were not enough, he also spent much of his time at home, painting and producing brilliant acrylics, which he then gave to his scientific friends and collaborators. In later years, he also sold his paintings for the benefit of the Edsall Fund, which he had set up to aid needy students. Although few of his friends knew this, he was also a voracious reader, who over the years amassed an extensive and varied personal library. He knew how to organize his time. He never wasted a minute. *Gemütlichkeit* was not for him.

I spent my first postdoctoral weeks, in 1964, reading most of Ef's previous publications. Upon moving to Minnesota in 1941, he had continued his search for a biochemical basis for brain diseases, by studying the effect of a polio virus infection on glycolysis in mouse brain. Right away, an exciting result: the virus inhibited glycolysis [3]. After his first move to New York city in 1942, more excitement: one could also inhibit glycolysis by adding a purified preparation of another neurotropic virus, directly to the brain homogenate. Then disappointment: the inhibition was caused by iron which contaminated the virus preparations [4]. At this point, most others would have given up. Indeed, the discovery of this artifact called into question Ef's previous papers on this topic and might well have stopped his scientific career before it had ever taken off. But Ef's ingenuity converted this defeat into his first scientific triumph. Undeterred, he went on to show that the inhibition could be overcome by glutathione, a ubiquitous cysteine-containing tripeptide whose role in metabolism was still poorly understood; however, there was good evidence that glutathione was an essential cofactor of the enzyme glyoxylase which converts glyoxal to glycolic acid. When Ef showed that this reaction proceeded through a carboxyl-S-glutathione intermediate, he had identified the first 'energy-rich' thioester of biological relevance. Glyoxylase was a rather esoteric enzyme, but this could not be said of the glycolytic enzyme triose-phosphate dehydrogenase which resembled glyoxylase in its sensitivity to compounds reacting with sulfhydryl groups. Could it be that triose-phosphate dehydrogenase worked through a similar thioester intermediate? The enzyme catalyzes the energy-yielding oxidation of an aldehyde (glyceraldehyde-3-phosphate) to a carboxylic acid and couples it to the energy-requiring formation of 1,3-diphosphoglyceric acid with inorganic phosphate as phosphoryl donor. Warburg had already proposed that this reaction proceeded by direct addition of inorganic phosphate to the aldehyde group and subsequent oxidation of the adduct to 1,3-diphosphoglyceric acid. Undeterred by Warburg's authority, Ef and his technician, Isidore Krimsky, showed convincingly that the aldehyde group reacted first with an enzyme-bound sulfhydryl group, that the resulting thio-hemiacetal was then oxidized to an energy-rich thioester, and that this thioester was 'phosphorylyzed' by inorganic phosphate to 1,3-diphosphoglyceric acid. Warburg first scoffed at what he called 'Racker's Umweg' (Racker's detour), but later had to concede that nature took the Umweg rather than the direct route. Although the reactive sulfhydryl group was later shown to belong to the enzyme itself, rather than to tightly bound glutathione, the elucidation of the mechanism by which a biological oxidation is coupled to ATP formation still ranks as one of the most important biochemical discoveries of all time. The simplicity of this reaction was so persuasive that for the next 20 years most biochemists were convinced that oxidative phosphorylation in mitochondria and bacteria had to obey the same principle. How wrong they were!

After moving to Yale in 1952, Ef continued his work on carbohydrate metabolism. He discovered and purified transketolase, a key enzyme of the pentose phosphate pathway. This finding, together with work by others such as Bernhard L. Horecker, eventually led to a detailed description of the entire pathway.

When Ef returned from Yale to New York city in 1954, in order to join the Public Health Research Institute, he first continued to work on the mechanism of glycolysis and the pentose phosphate pathway. By then, most of the steps of glycolysis were known, but there was disagreement on how the process was regulated. Why was glycolysis of intact cells inhibited by respiration ('Pasteur effect')?, why did glycolysis inhibit respiration ('Crabtree effect')? and why did most tumor cells, unlike normal cells, convert glucose to lactate even under aerobic conditions ('aerobic glycolysis')? The answers could only come from work with reconstituted in vitro systems. Together with Ray Wu and Shimon Gatt, Ef investigated the effect of respiring mitochondria, nucleotides and specific inhibitors on glycolysis catalyzed by a cytosolic extract from Ehrlich ascites tumor cells, and showed that glycolysis was dependent on the continuous regeneration of ADP and inorganic phosphate by an ATPase. The family of glycolytic enzymes thus included an ATPase, but these in vitro systems could not tell which of the many cellular ATPases was responsible for the regulation in intact cells.

This important finding did not receive the attention it deserved, probably because biochemists were then mesmerized by allostery and preferred to place the burden of glycolytic control solely on the shoulders of phosphofructokinase. Even today we do not fully understand how glycolysis is controlled in living cells and why most tumor cells exhibit aerobic glycolysis. Ef and Ray stopped working on this problem in the mid 1960s, but Ef returned to it during his final years.

I was always captivated by Ray Wu's quiet charm and professionalism. Watching him set up his complex enzyme systems, with intense concentration, taught me much about how to do a successful experiment. Wu rarely spoke up or contributed jokes during our lively lab discussions, but was a great help for us postdocs, and a close and trusted friend to Ef.

Ef's return to New York city, from Yale in 1954, had also not dimmed his interest in the pentose phosphate pathway. Together with Dan Couri, he reconstituted the pathway from purified components and showed that the reconstituted system catalyzed the complete oxidation of glucose-6-phosphate. Together with June Fessenden and others, Ef also continued to investigate the detailed mechanism of several enzymes of this pathway.

Soon after Ef had moved to the Public Health Research Institute, Maynard Pullman joined his department. Pullman had set his sights high: he wanted to go after the Holy Grail of bioenergetics, the mechanism of ATP synthesis in mitochondria and chloroplasts. For a start, he decided to isolate the mitochondrial enzymes which coupled the oxidation of nutrients to the synthesis of ATP from ADP and inorganic phosphate. Pullman was aware that this was a formidable undertaking and must have been very pleased when the gifted Harvey Penefsky joined him as a graduate student.

Following a procedure pioneered by David Green, they obtained fresh bovine hearts from a nearby slaughterhouse, disrupted them in a mechanical blender, and isolated from the resulting homogenate several grams of mitochondrial membrane fragments which still catalyzed oxidative phosphorylation. They wanted to use these 'submitochondrial particles' as their starting material for resolving, and ultimately reconstituting, the individual enzymes of respiration-driven ATP synthesis. Both knew that David Green at the University of Wisconsin, Paul Boyer at the University of Minnesota and Albert Lehninger at Johns Hopkins University were hot on their trail, and the race was on.

It was a long and frustrating race for several reasons. First, the structure of biological membranes was, at that time, unknown. Second, nobody knew how to assess the purity of a hydrophobic protein since SDS-polyacrylamide gels did not appear on the scene until 1967. Third, most biochemists assumed that ATP synthesis was coupled to respiration through a 'high-energy' intermediate of the type that functions in glycolytic ATP production. This intermediate (fondly called x-squiggle-y) was avidly sought, but never found. By the mid 1960s, the many futile attempts had led to frustration and heated controversies; however, most outsiders have later painted an exaggerated picture of the situation, perhaps because the mercurial temperament and acid humor of some of the leading mitochondriacs did not appeal to everyone. By today's standards, relationships between the competing laboratories remained civilized; experimental discrepancies were usually resolved by joint experiments and I cannot recall any instance where a laboratory withheld requested reagents or information from a competitor.

We postdocs loved to gossip about the relationships between the key players. Ef seemed to be closest to Britton Chance whose brilliance, boyish temperament and experimental skill matched his own. He genuinely liked Paul Boyer, Henry Lardy, Albert Lehninger, Bill Slater and Lars Ernster, and respected their scientific rigor and masterful grasp of biochemistry, but in the mid 1960s, none of them worked directly on the resolution of oxidative phosphorylation and scientific interactions were less frequent. Our gossip usually focused on David Green and his group. Green was imaginative, self-assured, flamboyant and quick with tongue. He usually disagreed with Ef, easily matched him as a debater, and was Ef's perennial foil. When Peter Mitchell emerged as a major figure in the field several years later, he and Ef developed a friendly rapport, yet it seemed to me that they were never quite at ease with each other.

But I am getting ahead of my story. At first, progress in Ef's department was amazingly fast. In order to solubilize the enzymes coupling respiration to ATP synthesis, the submitochondrial particles were vigorously shaken with tiny glass beads in a shaker, as originally described by the Australian biochemist, Peter M. Nossal. Mike Kandrach, our gifted and eccentric mechanic, soon built a monstrous US version, which he himself considered so dangerous that he screwed it to the floor of a separate room, operated it by remote control, and allowed nobody else to touch it. This contraption emitted a lugubrious rumble that shook the building and would have sent any contemporary Californian diving for the nearest earthquake shelter. When these tortured mitochondrial fragments were sedimented in an ultracentrifuge they still respired, but no longer synthesized ATP. Adding back the supernatant from the centrifugation sometimes restored ATP synthesis, as Pullman and Penefsky had hoped, but this effect was quite irreproducible, particularly after the first few attempted purification steps. The project seemed to be stuck; but, being the master biochemist that he was, Pullman systematically varied his experimental protocols until he found that the partially purified soluble fractions restored ATP synthesis nearly every time if they were stored at room temperature. In order to protect the garden of oxidative phosphorylation from unworthy intruders, God had guarded the entrance with a cold-labile enzyme! But the assay was still very tedious until Pullman began to suspect that the cleavage of ATP, which was catalyzed by the soluble fraction, might be just another activity of the mysterious factor that restored oxidative phosphorylation. He started to monitor purification by assaying ATPase activity which was much faster and easier than assaying restoration of oxidative phosphorylation, and from then on, progress was fast: when he and Penefsky purified the ATPase from the supernatant, it was indeed identical to the 'coupling factor' of oxidative phosphorylation. The splitting of ATP in a test tube was clearly a reversal of the reaction which the enzyme catalyzed in living cells. Since the ATPase was the first defined factor that coupled respiration to ATP synthesis, it was named Factor 1, or F_1. The first enzyme of oxidative phosphorylation had been identified and purified!

This fundamental discovery was published in 1960. It was greeted with universal admiration and hopes were high that the enzymology of oxidative phosphorylation would soon be understood. In a letter he wrote to me in Vienna early in 1963, Ef suggested various topics for my upcoming postdoctoral stay with him. He discouraged me from planning to work on oxidative phosphorylation with the words 'Progress on this front is now quite fast and by next year our interests may have shifted to other topics.' In the following years of slow progress, I could always enliven my seminars by showing a slide with this sentence in Ef's own handwriting.

But for a while, Ef's prophecy seemed to be correct. Together with Vida Vambutas, he purified a closely similar coupling factor from spinach chloroplasts. The purified chloroplast F_1 (termed CF_1) restored light-driven ATP synthesis to EDTA-treated chloroplast fragments, but did not cleave ATP unless it was gently treated with trypsin. This result confirmed the general expectation that photophosphorylation and oxidative phosphorylation functioned by a similar mechanism. In experiments that he first did himself and later together with Yasuo Kagawa, a gifted and hardworking postdoctoral fellow from Japan, Ef subfractionated submitochondrial particles with cholate and salt, and identified a membrane factor that anchored F_1 to the membrane and rendered it cold-stable and sensitive to the toxic antibiotic, oligomycin. As oligomycin was then considered the most specific inhibitor of oxidative phosphorylation, its lack of effect on the ATPase activity of F_1 had been the most serious argument against a role of F_1 in oxidative phosphorylation. The identification of a factor conferring oligomycin sensitivity on soluble F_1 not only silenced this criticism, but also paved the way towards resolving the membrane-embedded components of the oxidative phosphorylation machinery. Kagawa and Ef named their insoluble F_1-binding factor F_o

(in contrast to what is generally thought, the subscript does not signify zero, but the letter o, for oligomycin).

Kagawa presented his results to a packed audience at the 1965 Annual Federation Meeting in Atlantic City. Although he had been intensively coached by some of the American postdoctoral fellows in the lab, he struggled with the sounds and the grammar of the English language. He had particular difficulty pronouncing 'F_o', making it sound like the exhortation 'Ef, ho!' Still, all the experts in the audience realized that they were witnessing a landmark presentation. The response in Japan must have been similarly positive because Kagawa was soon offered an attractive full professorship at Jichi University, where he still works today.

Ef's work with Kagawa also showed that the characteristic knobs which lined the inner face of the mitochondrial inner membrane in electron micrographs were, in fact, F_1: the knobs disappeared when F_1 was stripped off, and reappeared when purified F_1 was added back. At that time, this was perhaps the clearest evidence for the molecular asymmetry of a biological membrane.

Buoyed by these advances, Ef's oxidative phosphorylation team now tried to isolate additional protein factors, by mistreating phosphorylating submitochondrial particles in various ways. The particles obtained with our Nossal-type glass bead shaker were termed N-particles. Treatment with ammonia solution yielded A-particles, intensive sonication, S-particles, and treatment with phospholipids, P-particles. All these particles were defective in coupling respiration to ATP synthesis and could be partly reconstituted by adding back other mitochondrial protein fractions which were termed F_2, F_3, and F_4. However, these fractions elicited only marginal effects which were often irreproducible. Today we know that these fractions were impure and cross contaminated with each other, that the factors under study had not always been completely stripped from the test particles, and that the non-linear response of the particles to added factors led many a hapless postdoc astray. Also, all efforts to reconstitute oxidative phosphorylation from completely solubilized submitochondrial particles proved unsuccessful.

By the beginning of 1966, the optimism in Ef's lab had faded somewhat and several of his postdoctoral fellows decided to chose other fields when they started their own laboratories elsewhere. As so often happens in a scientific career, this lull coincided with a wave of official recognition. In the spring of 1966, Ef was elected into the US National Academy of Sciences. A few months later, Robert Holley (who was soon to receive the Nobel Prize for his work on the structure of tRNA) and Robert Morison from Cornell University in Ithaca, New York, visited Ef and tried to persuade him to help create and lead the biochemistry department of a new biology unit at that campus. The innovative concept of this plan had received a generous grant from the Ford Foundation, and Ef's own position was to be endowed by one of the prestigious Einstein professorships, through which the State of New York hoped to attract outstanding scholars to its new university system.

Ef had of course received offers before, but this one had the right ring. It also came at just the right time. Although Ef had grown up in Vienna, he was never a 'city type' and he was beginning to loathe the inconvenience of working in New York city. In order to give his family the peace they needed, he lived in faraway Mount Vernon and the daily commuting by car was proving to be a burden. The challenge of creating a new research unit may have also excited him, but I suspect that his decision was really swung by his wife Franziska. He had known Franzi (Frances to most Americans) in Vienna, where she received her M.D. at about the same time as he. Born to an established Viennese family who lived in the rather fashionable ninth district, she, too, had emigrated via Great Britain to the United States. She had just obtained an advanced degree in public health at Harvard Medical School, when she and Ef met again in the

New World. They married in 1945. Intelligent, musical, practical and charming, she was everybody's favorite and an adopted grandmother for many of the postdocs' children. She was then, and still is, an active and successful physician. As ambitious and strong-willed as Ef himself, she was a perfect companion for him, by showing him unflinching devotion while guiding him with a firm and understanding hand. She loves nature and gardening (quite unlike Ef), and was attracted by the prospect of living in the green hills of rural Ithaca, and of being closer to her husband and her only child Ann, who was then 16 years old. After only a brief deliberation, Ef decided to take the plunge; in the fall of 1966, he moved his family and most of his laboratory to Ithaca.

The decision proved to be an excellent one. Cornell greeted Ef with open arms and gave him free rein in creating 'his' department. His reputation helped him to recruit outstanding senior scientists such as Quentin Gibson from Great Britain, Leon Heppel from NIH, André Jagendorf from Johns Hopkins University, as well as his trusted colleagues, Harvey Penefsky and Ray Wu. 'Young Turks' were added by the arrival of Stuart Edelstein, Peter Hinkle, Richard McCarty and David Wilson. While all of this was happening, I was back in Vienna trying to get reaccustomed to Europe; however, in 1968 I decided to cross the Atlantic once again and join Ef's new department as a faculty member. All of us were housed in Wing Hall which had been refurbished and expanded. Research funds and jobs were still plentiful and our spirits were high. We owned the world.

Ef proved to be a smooth administrator who coped effortlessly with the added burden of creating and running a big department. He had a talent for picking capable aides and letting them do things their own way. His teaching obligations were light. Because of his intuitive and idiosyncratic style of thinking, he preferred to teach seminar courses and the students loved him. For the first time in his life, his appointment brought him in close contact with undergraduate students and it proved to be an immediate and reciprocal love affair. These intoxicating 'hippie years' were hard on parents, but they were great for outgoing and unconventional academic teachers such as Ef. Even his sartorial negligence was a distinct advantage. In spite of his many new commitments, Ef continued to work at the bench and to supervise the postdoctoral fellows who now flocked to him in growing numbers. Some of the outstanding young people from this period include Günter Hauska, Richard Huganir, Baruch Kanner, Ladislav Kováč, Chris Miller, Maurice Montal, Nathan Nelson, Michael Newman, Jan Rydström, Dennis Stone, Bernie Trumpower and Charles Yocum. I became particularly close to Nathan Nelson, who often reminded me of Ef himself. His amazing productivity was sustained by the help from his smart wife Hannah. Both of them became lifelong friends of Ef and his family.

With such a high-caliber cast, the Cornell team soon scored its first successes. These successes were triggered by the team's growing conviction that oxidative phosphorylation was not mediated by a high-energy chemical intermediate, but by a transmembrane proton gradient, as Peter Mitchell had proposed a decade ago. Ef's conversion to Mitchell's ideas was triggered by his many discussions with the brilliant Peter Hinkle, who had done postdoctoral work with Mitchell in Great Britain and who now saw it his mission to save Ef's soul by converting him to chemiosmosis.

Ef loved to argue with Hinkle, but in the end his conversion probably came from two experiments in which he participated himself. Prompted by studies done by Hinkle, Ef showed that the oxidation rate of cytochrome c by cytochrome oxidase, reconstituted into liposomes, was controlled by a transmembrane proton gradient. This 'respiratory control' closely mimicked that seen with respiring intact mitochondria when ATP synthesis was prevented by lack of the phosphate acceptor, ADP. Second, he and Kagawa (who had returned to Ef's lab for an extended sabbatical) finally succeeded in reconstituting ATP-^{32}Pi exchange, a partial reaction of oxidative phosphorylation, from pure F_1 and solubilized F_0. This activity was only seen when the two components were reconstituted into a sealed liposome, and was lost when the liposomes were made leaky to protons.

Perhaps the most famous experiment from this period was done by Ef and Walther Stoeckenius, a German biologist working at the University of California at San Francisco. Stoeckenius had discovered bacteriorhodopsin, a purple chromoprotein from the archaebacterium *Halobacterium halobium* and, together with Dieter Oesterhelt, had found that this protein functioned as a light-driven proton pump in the bacterial plasma membrane. Why not incorporate this simple, pure proton pump into a liposome, together with the F_1F_0-ATPase from bovine heart? If Mitchell was right and the two components oriented themselves properly in the liposome membrane (which was the big if), then the protons pumped out by the illuminated bacteriorhodopsin should flow back through the F_1F_0-ATPase and generate ATP from ADP and inorganic phosphate. Stoeckenius visited Ef at Cornell to do this experiment with him, even though neither of them gave it much of a chance. The gods must have been pleased to see two elderly scientists working together at the bench; the experiment worked and convinced even the most obdurate skeptics that Mitchell's hypothesis was correct. Almost two decades after elucidating a key reaction of ATP formation in glycolysis, Ef had helped to nail down the mechanism of ATP formation in mitochondria, chloroplasts and the bacterial plasma membrane.

In the years that followed, Ef and his collaborators reconstituted an astonishing array of different membrane enzymes into liposomes, and established reconstitution as a powerful and generally applicable approach for unraveling the mechanism of pumps, transporters and receptors. Numerous prestigious honors and prizes came his way, such as the Warren Triennial Prize in 1974, the National Medal of

Science in 1976 and the Gairdner Award in 1980. But many biochemists were disappointed when the 1980 Nobel Prize for Chemistry went to Mitchell alone. As usual, the Nobel committee did not divulge its reasoning, but many of the leading biochemists were surprised, or even upset, by the fact that Ef had not shared the prize. Only Ef himself seemed to be little concerned. He continued to work at the bench as usual and painted more avidly than ever in his new enlarged studio which had been a birthday present from Franzi. He also enjoyed his new role as grandfather; he was never more at ease than when playing with children.

There was also renewed excitement in the laboratory, for his long and distinguished career seemed to be headed for a triumphant finale. Having been a key figure in unraveling the enzymology of glycolysis, the pentose phosphate pathway and oxidative phosphorylation, he had returned to the problem that had still defied him: the abnormal glycolysis in tumor cells. His previous studies had convinced him that glycolysis was controlled by cellular ATPases. Could one of these ATPases be hyperactive in tumor cells and thereby cause an abnormally high glycolytic rate?

It seemed like a stroke of luck that, just at this time, Mark Spector joined him as a graduate student. Intelligent, hard-working, and unusually skilled at the bench, Spector embodied many of the qualities Ef admired in others. A brilliant hypothesis was born and documented by a quick succession of experiments. A tumor gene encodes a protein kinase. This kinase phosphorylates and thereby activates another kinase. This second kinase activates a third, and so on until the last member of this 'kinase cascade' phosphorylates a subunit of the ATP-driven sodium-potassium pump in the plasma membrane; phosphorylation of the pump renders it less efficient, increasing its rate of ATP consumption, and thereby the rate of glycolysis. But in 1981, it became clear that Spector had fabricated his data [5], and it could not have happened at a worse time. The United States was just then going through one of its recurring crusades against human evil, and this time the evil was 'scientific misconduct.' Ef found himself in the center of a storm and it was then that his true stature showed. He immediately published retractions of the questionable papers, withdrew those still in print, and offered to resign from his various committees until the issue had been resolved. He set an example of courage and honesty to us and future scientists. It was no coincidence that the American Society of Biological Chemistry chose this moment to award him their prestigious Sober Memorial Lectureship. To him it must have been a dark time, but to me and many of his friends it was perhaps his finest hour.

It was in one of the difficult years that followed that he spent the only sabbatical of his life in my Basel laboratory. Once again I admired his quick mind, his enthusiasm, his experimental skill, his undimmed scientific curiosity and his truly astounding openness to new ideas. His books on bioenergetics and the social impact of science were much appreciated in Europe and he loved to discuss them with my students and postdocs. But a long day in the lab left him exhausted and his private talks with me often touched upon the dark sides of life. His lectures still impressed and captivated his audience, even though his deteriorating hearing made it difficult for him to handle the subsequent discussions.

Upon his return to Ithaca, his letters became more frequent and unusually warm and personal, but his suddenly erratic handwriting made me worry about his health. On 6 September 1991, he came home from a hard Saturday in the laboratory and was felled by a severe stroke. He never regained consciousness and died in Syracuse 3 days later.

What kind of man was Efraim Racker? Someone as brilliant, artistic and intuitive as he will always defy definition. Like many great scientists, Ef had several personalities and therefore elicited different responses in different people. He was one of the last great figures of biochemistry's heroic age. He embodied the artistic, even romantic approach to a field of science that has become increasingly dominated by organized collective efforts. He could be egocentric, insensitive, even overbearing, but those who knew him well will cherish the memory of his warmth, his immense intellectual range, his lack of prejudice, and his unshakable belief in the power of human reason and inventiveness. Once you were his friend, he was always on your side and forgave you everything. Few knew that he was sometimes haunted by depressions which showed in the many leafless trees of his paintings and his never-ending fascination with the secrets of the human mind.

Perhaps one would only understand the many layers of Ef's character if one could retrace his formative years in the Vienna of his youth. But the world of his youth has been brutally shattered forever and few are still with us to tell about it. Ef rarely talked about that time and maintained only infrequent contacts with relatives outside his immediate family. Upon his emigration he had even tried, with some success, to forget the German language and tended to stammer when being addressed in that language.

But every human life is mystery and best weighed by its influence on others. Efraim Racker showed me that scientific exploration and art are but two manifestations of the same powerful spirit that makes us human, brings us joy and gives us wings.

Acknowledgments

Many friends and colleagues have helped me retrace Ef's private and scientific life and improve this brief account. My special thanks go to Franziska Racker, Judy Caveney, Peter Hinkle, Harvey Penefsky, Maynard Pullman, Ray Wu, Carolyn Suzuki, Andreas Matouschek and Dennis Stone.

References

1   E. Racker. Histidine detection and estimation in urine. Biochem. J. 34 (1940) 89–96.
2   E. Racker. Mechanisms in Bioenergetics. New York: Academic Press, 1965.
3   E. Racker and H. Kabat. The metabolism of the central nervous system in experimental poliomyelitis. J. Exp. Med. 76 (1942) 579–585.
4   E. Racker and I. Krimsky. Inhibition of glycolysis in mouse brain homogenates by ferrous sulfate. Fed. Proc. 6 (1947) 431.
5   G.B. Kolata. Reevaluation of cancer data eagerly awaited. Science 214 (1981) 316–318.

Selected bibliography

1951
Mechanism of action of glyoxalase. J. Biol. Chem. 190: 685–696.

1952
With I. Krimsky. Mechanism of oxidation of aldehydes by glyceraldehyde-3-phosphate dehydrogenase. J. Biol. Chem. 198: 731–743.

1955
With G. de la Haba and I. Lecler. Crystalline transketolase from baker's yeast: isolation and properties. J. Biol. Chem. 214: 409–426.

1959
With S. Gatt. Regulatory mechanisms in carbohydrate metabolism II. Pasteur effect in reconstructed systems. J. Biol. Chem. 234: 1024–1028.
With R. Wu. Regulatory mechanisms in carbohydrate metabolism III. Limiting factors in glycolysis of ascites tumor cells. J. Biol. Chem. 234: 1029–1035.
With R. Wu. Regulatory mechanisms in carbohydrate metabolism IV. Pasteur effect and Crabtree effect in ascites tumor cells. J. Biol. Chem. 234: 1036–1041.
With D. Couri. The oxidative pentose phosphate cycle V. Complete oxidation of glucose-6-phosphate in a reconstructed system of the oxidative pentose phosphate cycle. Arch. Biochem. Biophys. 83: 195–205.

1960
With M.E. Pullman, H.S. Penefsky and A. Datta. Partial resolution of the enzymes catalyzing oxidative phosphorylation I. Purification and

properties of soluble dinitrophenol-stimulated adenosine triphos-phatase. J. Biol. Chem. 235: 3322–3329.

With H.S. Penefsky, M.E. Pullman and A. Datta. Partial resolution of the enzymes catalyzing oxidative phosphorylation II. Participation of soluble adenosine triphosphatase in oxidative phosphorylation. J. Biol. Chem. 235: 3330–3336.

1963

A mitochondrial factor conferring oligomycin sensitivity on soluble mito-chondrial ATPase. Biochem. Biophys. Res. Commun. 10: 435–439.

1965

Mechanisms in Bioenergetics. New York: Academic Press.

With V.K. Vambutas. Partial resolution of the enzymes catalyzing photo-phosphorylation I. Stimulation of photophosphorylation by a prepa-ration of latent Ca^{++} dependent adenosine triphosphatase from chlo-roplasts. J. Biol. Chem. 240: 2660–2667.

1966

With Y. Kagawa. Partial resolution of the enzymes catalyzing oxidative phosphorylation IX. Reconstruction of oligomycin-sensitive adeno-sine triphosphatase. J. Biol. Chem. 241: 2467–2474.

With Y. Kagawa. Partial resolution of the enzymes catalyzing oxidative phosphorylation X. Correlation of morphology and function in sub-mitochondrial particles. J. Biol. Chem. 241: 2475–2482.

1971

With Y. Kagawa. Partial resolution of the enzymes catalyzing oxidative phosphorylation XXV. Reconstitution of vesicles catalyzing ^{32}Pi-adenosine triphosphate exchange. J. Biol. Chem. 246: 5477–5497

1974

With W. Stoeckenius. Reconstitution of purple membrane vesicles cata-lyzing light-driven proton uptake and adenosine triphosphate forma-tion. J. Biol. Chem. 249: 662–663.

1975

With T-F. Chien and A. Kandrach. A cholate-dilution procedure for the reconstitution of the Ca^{++} pump, ^{32}Pi-ATP exchange and oxidative phosphorylation. FEBS Lett. 57: 14–18.

1976

A New Look at Mechanisms in Bioenergetics. New York: Academic Press.

G. Semenza and R. Jaenicke (Eds.)
Selected Topics in the History of Biochemistry: Personal Recollections, V
(Comprehensive Biochemistry Vol. 40) © 1997 Elsevier Science B.V.

Chapter 6

A Life with the Metals of Life*

BO G. MALMSTRÖM

Department of Biochemistry and Biophysics, Göteborg University,
Medicinaregatan 9C, S-413 90 Göteborg, Sweden

Introduction: the shaping of a bioinorganic chemist

A scientist, like any other human being (or biological crea-
ture, for that matter), is a product of nature and nurture. A
creative individual may be highly original and independent,
but even such a person is unavoidably influenced in intellectual development by contacts with other humans – parents,
teachers, other investigators, but also friends and acquaintances in other walks of life besides natural science. In this
essay, I would like to concentrate on descriptions of my interactions with other individuals, who have had profound
influences on my life as a biochemist. Of necessity, I cannot
do this without devoting considerable space to my own activities in research and related professional work.

During my 48 years of research, I have worked on a large
number of biochemical systems, but, with a few exceptions,
they have had one common denominator – their biological
function involves interaction with metal ions. This is, however, not as serious a restriction as it may seem to some, because metal ions play essential roles in much of the chemical
machinery of living organisms. This is evidenced, for exam-

---

* The text of this chapter was concluded in December 1995.

ple, by the fact that a very large fraction of all known enzymes are metalloenzymes, i.e., they contain strongly bound metal ions as prosthetic groups, or are activated by metal ions, such as Mg^{2+} or K^+. The study of the biological functions of metals is at the interface of chemistry and biology, and it has led to the emergence of a new chemical discipline, bioinorganic chemistry. In the preface to a recent textbook in this field, the authors state: 'Many critical [biological] processes require metal ions, including respiration, much of metabolism, nitrogen fixation, photosynthesis, development, nerve transmission, muscle contraction, signal transduction, and protection against toxic and mutagenic agents.' [1]. In other words, the study of the role of metals in biological systems covers a very large area of the field of biochemistry.

Early influences on my intellectual development

A gymnasium with a university curriculum

In 1942, I entered the gymnasium in Bromma, an upper middle class suburb of Stockholm. My father was managing editor of Sweden's second largest newspaper, *Svenska Dagbladet*, and it was undoubtedly my home environment with a lot of books and my parents association with people in intellectual professions, which had made me want to seek a career involving work in natural science or medicine already at that time. My view of science and scientists was very romantic, having been formed by popular science books, such as *Microbe Hunters* by Paul de Kruif, and American movies, like *The Story of Louis Pasteur*. Even if I soon realized that this picture was naive as well as erroneous, it definitely was very important in developing my desire to become a scientist. The decisive influence, however, came from the exceptional standard of teachers at my gymnasium.

Swedish gymnasiums in the 1940s were very exclusive schools. The teachers had generally been 'docent' at a university, a title given to those who received a high enough grade on their doctor's thesis. A few got a university appointment as docent, a position for research and teaching limited to 6 years. At the end of that period, if none of the few professorships had become vacant, the alternative was generally a job as a gymnasium teacher, called 'lektor'. The best researchers were commonly attracted to gymnasiums in the major cities, and the Stockholm area first of all. Bromma gymnasium had a particularly qualified group of lektors. The science lektors were all internationally recognized investigators, a statement which I can illustrate with an experience from 1951, when I was at the University of Minnesota. One day I was approached by a biology professor, who was going to Sweden on a sabbatical and he wanted some advice from me. I asked him to which university he was going and his answer was 'Bromma gymnasium'.

The Bromma chemistry lektor, Herman Rinde, made a particularly deep impression on me. In 1924, he had written the first publication on the ultracentrifuge, together with the Nobel laurate, The Svedberg (Svedberg and Rinde, 1924). Rinde's thesis was still used in the 1960s as a standard work by specialists on the analytical ultracentrifuge. For example, Matt Meselson at Harvard told me in 1972, that he had gotten the idea for cesium chloride gradient ultracentrifugation, used in his experiments demonstrating semi-conservative replication of DNA (Meselson and Stahl, 1958), from this thesis. Rinde was also the, so-called, faculty opponent when Arne Tiselius defended his thesis on electrophoresis in 1930, which, together with his contributions to chromatographic techniques, earned him the Nobel Prize for Chemistry in 1948.

It is in some ways surprising that Rinde had such a strong positive influence on me, because he used methods which would have had discouraging effects on many people. In fact, some of the students switched from the mathematics-science to the Latin-humanities line of the gymnasium as a result of comments like this, offered in response to a student's answer to a question: 'What a fool you make of yourself. Should you not take the Latin line instead and learn about Hannibal's march through the Alps?' His demands were colossal: 'Perfect knowledge of the textbook is the minimum requirement for a passing grade'. He also, however, gave much of his own time and energy to the pupils: writing and duplication of advanced material, not found in the textbook; 'voluntary' seminars at the end of school, 1 day per week, based on a classic book on thermodynamics (Lewis and Randall, 1923); 'voluntary' sessions of chemical problem solutions in the school auditorium during Easter vacation; 'voluntary' laboratory work during weekends and vacations (it was tradition that Rinde came to the laboratory at noon on Christmas Eve, gave us coffee and said that now we may go home). Due to protests from other teachers that the chemistry studies took all of our time, an investigation was started and showed that the demands for a passing chemistry grade at Bromma gymnasium were slightly higher than those for the first-term chemistry course at Stockholm University.

A book with a mission

Certainly, it was Rinde's teaching that kindled my keen interest in chemistry, but an important factor for my early decision to become a chemist was undoubtedly the fact that the chemistry studies had taken so much time, which I otherwise had felt to be wasted. At the time, I had no intention of becoming a biochemist, since, due to Rinde's influence, I was mainly attracted to inorganic and physical chemistry. On a summer vacation, however, I read a book which, no matter how pathetic this may sound, clearly changed my life in several major respects.

The book with such a tremendous influence on me was *Man, the Unknown*, by Alexis Carrel, who had won the Nobel Prize in Physiology or Medicine in 1912, for his work on organ transplantation. The book represented an attempt to present a synthesis of everything known in 1935, about Man, from a natural science point of view. Carrel also had a mission. His opinion was that a large part of the problems of our world were caused by our ability to understand and control the lifeless matter surrounding us with the aid of physics and chemistry, at the same time as we lacked any real understanding of the physical and chemical basis of our body and soul. Carrel was a pioneer environmentalist: he considered that physics and chemistry had changed our environment to an extent that we were no longer physically or mentally adapted to it. He advocated a massive support of the life sciences in a wide sense so, as he expressed it, biology could take the step from 'a lower form of science', the 'descriptive', to the 'abstract and quantitative' level, which physics and chemistry had already reached. (Unfortunately the book ended with a chapter on eugenics with clear features of fascism and a belief in Nietsche's 'Obermensch', but this should not lead one to discard the values of the main body of the text.)

Carrel's style suited my romantic concept of science, at the same time as his arguments convinced me intellectually. According to Carrel, what we needed was primarily a better understanding of life processes from a physical and chemical point of view; in other words, we must develop biochemistry and biophysics. As a result, I decided at the age of 17, to become a biochemist, a resolve from which I never later faltered. I told Rinde, my chemistry teacher, about my intention, which made him very disappointed. As was common among the chemists of the day, he regarded biochemistry as a lower form of chemistry. He was, of course, right in one sense, since biochemistry in the early 1940s was largely a descriptive science. After all though, it was towards a change in this very situation that I wanted to contribute.

From gymnasium to college studies in the United States

In 1946, the year that I finished the gymnasium, there was no direct curriculum to study to become a biochemist in Sweden. I knew that Arne Tiselius had a personal chair in Biochemistry in the science faculty of Uppsala University, and in the catalogue of Stockholm University I discovered that Karl Myrbäck held a chair in organic chemistry with biochemistry. It was, however, not until in the 1950s that they were allowed to grant a degree of 'filosofie licentiat', the closest correspondence to the present day Ph.D., in biochemistry. Thus, in the science faculty it was not possible to start research in biochemistry until one was close to 30 years old.

Another possibility was to study at the medical faculty, where biochemistry was already a dominating field of research, and I decided to do this. A contributing factor was the fact that a medical school professor, Hugo Theorell, was the Swedish biochemist with whose work I was most familiar, because he often contributed popular science articles to my father's newspaper. Consequently, in the summer of 1946, I applied to and was accepted as a medical student at the Karolinska Institute in Stockholm. Another development soon changed my plans, however.

Sweden, during the 1940s, was a neutral corner isolated from a war-ravished Europe. When the isolation was broken at the end of the war, in 1945, I had a longing to get out into a world which was now open. I was only 19 and thought that I could afford a year abroad. Thus, I applied for a scholarship for college studies from the Swedish–American Foundation and was given one to Muhlenberg College in Allentown, PA. On the *Gripsholm*, sailing for New York in August 1946, was also Hugo Theorell, to whom I wrote a letter asking for a meeting. He received me very kindly and gave me advice and information about biochemistry studies both in Sweden and in the United States.

Muhlenberg is a small liberal arts college, founded in 1848. It specialized (and still does) in pre-theological (it has ties to the Lutheran church) and pre-medical education. The latter has had the effect that the science departments, particularly biology and chemistry, have always been strong. I still treasure the notes from the inspiring biology lectures of John Shankweiler. The head of the Chemistry Department, George Brandes, was an excellent organic chemist, and he was joined during my second year by a young silicon organic chemist, Russel Smart. In 1963, when I spent 6 months in the US, my wife and I visited Muhlenberg and saw both Brandes and Smart. I also have lasting impressions of the mathematics teaching of Truman Koehler, who thought that I had a flair for mathematics and could not understand why I should want to devote myself to 'trivial applications in the physical sciences'. Another strong influence was that of the philosophy professor, Russel Stine, who kindled in me a lifelong interest in analytical philosophy, especially that of Bertrand Russel.

In 1993, I renewed my ties with Muhlenberg, when the present President, Arthur Taylor, elected me member of an International Visiting Committee. This meets every autumn, and in 1994 my wife and I also made a visit in May, when the College made me an honorary doctor. On my regular visits I see the professors that are still alive, for example, Russel Smart and my physical chemistry professor, Thomas Lloyd, who is still active in research. I have also made friends with several of the new members of the faculty and administration.

My first year at Muhlenberg, together with some of the advice given by Theorell, made me reconsider my study plans. I discovered that in US universities, there are no strict borders between the faculties; a student can freely combine medical school and science courses, something which is still largely impossible in Sweden. So I reasoned, why should I, who wanted to become a biochemist, learn to deliver babies or remove infected appendices? In the US, I could take the medical courses of interest to a biochemist, e.g., physiology, physiological chemistry and bacteriology, and combine them with studies in physics and chemistry. Thus, I decided to continue my education in the US.

The excellent standard of my gymnasium curriculum made it possible for me to acquire a B.S. Degree after five semesters at Muhlenberg. During my last term, I wrote to many universities in an attempt to be admitted as a graduate student. My first choice was Columbia University, and I got a polite but negative reply from Hans T. Clark, who was then head of the biochemistry department. Finally, economic realities decided for me, and in the summer of 1948, I moved to the University of Minnesota, which had offered me a graduate fellowship.

The graduate school years

The beginnings of research

In order to choose a topic for my thesis work, I went around and talked to the potential advisers about their research. I was then fascinated by the description Dave Glick gave of his studies in quantitative histochemistry. Glick had spent a period in the Carlsberg Laboratory in Copenhagen and had been introduced to histochemistry by Kaj Linderstrøm-Lang and Heinz Holter. Their histochemistry was not of the type where one stains microtome sections with 'specific' reagents and tries to estimate the amount of a given component from the intensity of the stain. Had it been, I would not have been interested; in fact, my very first publication (Glick et al., 1951) was a critique of such methods, based on a seminar I gave in my final year as a graduate student. Instead, what the Danish group had done was to scale down conventional analytical methods so that they could work with sample volumes as small as 10 μl containing substances to be analyzed in amounts of the order of nanograms. Glick had continued this type of work and applied his methods to the gastric mucosa and the adrenal gland, in particular.

In view of my interest in physical chemistry, my choice of adviser may seem curious. I tried, however, to combine histochemistry with my physical interests. In a course in biophysics, given by Otto Schmitt, I had worked with X-ray absorption, and I had hoped to be able to localize chemical elements in microtome sections by this method. To do this, one had to use monochromatic X-rays with wavelengths on each side of the absorption edge of the element to be determined. I decided to try to determine Zn, because to get X-rays of suitable wavelengths, one then had to bombard targets of gold and platinum, respectively, with electrons. Zn was known to be a component of carbonic anhydrase and was consequently of considerable biological interest.

My idea appealed very much to Glick, who had the mechanism of acid formation in the stomach as one of his main interests at the time. There was a controversy going on as to whether carbonic anhydrase was directly involved or not. To determine carbonic anhydrase activity on the histochemical level, Glick and I deemed unrealistic, but we saw here a possibility of estimating the carbonic anhydrase content indirectly by Zn analysis. Unfortunately, I soon discovered that the sensitivity of the X-ray absorption method was too low, and I ended up developing a microcolorimetric method with a sensitivity of about 1 ng (Malmström and Glick, 1951). For this purpose I built a colorimeter that could be used with about 5 μl of sample, and an extraction apparatus to extract the Zn from about 20 μl of aqueous phase into an equal volume of tetrachlorethylene containing the reagent, dithizone. With my method, I showed that the cells forming acid in the stomach, the parietal cells, have a higher concentration of zinc than the surrounding cells (Malmström and Glick, 1952).

In the summer of 1949, I did get to use an X-ray absorption method together with Arne Engström at the Karolinska Institute, who had been the pioneer in the applications of such methods to biology. My father had died in May of 1949, and I returned to Sweden for the funeral and stayed the whole summer to help my mother. The work with Engström did not involve elemental analysis but determination of the dry weight of gastric mucosa cells (Engström and Malmström, 1952). We found that the parietal cells have a particularly high water content and suggested that this favored rapid acid-base equilibria by allowing fast diffusion.

At the time I worked with Engström, Joe Neilands, later professor in Berkeley, was a postdoc with Hugo Theorell, whose lab was in the same building. We became good friends and have so remained. We were also united by a common interest in metalloenzymes, but we have never collaborated. Fifteen years later, we did write a review together, however (Malmström and Neilands, 1964).

Intellectual influences on a graduate student

Glick is a poet but also a superb writer of science. He gave me a very hard drilling in writing papers, and it is undoubtedly thanks to him that my composing of about 200 publications has been relatively painless. Apart from style, one thing he taught me is to be meticulously accurate in the description of experimental details. Despite everything he gave me, I soon realized, however, that his line of research did not suit my temperament. Even if the methods used are quantitative, the conclusions that can be drawn are largely qualitative. I, on the other hand, was (and still am) more interested in molecular mechanisms.

The overwhelmingly dominant influence on my thinking came from Paul Boyer, who was at the time professor on the agricultural campus of the University of Minnesota. Every Wednesday evening he organized an enzyme discussion group, which attracted not only many graduate students but also faculty from several departments, for example, the later famous microbiologist, the late Saul Spiegelman. The presentation of recent research at these meetings (I gave a seminar on the effect of pressure on the rates of enzymic reactions) demonstrated that the problems of enzyme chemistry were gradually coming down to the molecular level. I made a resolve to become an enzyme chemist.

In consultation with Paul, I decided to apply to Brit Chance, at that time a young, rising, enzyme chemist, for postdoctoral work. Brit accepted me but said that I should first learn some protein chemistry, and that the best place for that was Tiselius's laboratory in Uppsala. I saw both Brit and Arne Tiselius in the spring of 1951, in New York, where there was an ACS meeting followed by a IUPAC meeting; Tiselius was President of IUPAC at the time. Tiselius promised to support me, should I choose to come to Uppsala.

My move to Uppsala was encouraged by a Swedish geneticist, Jan Arvid Böök, who was visiting professor in the zoology department in Minnesota. He expected to get a chair in human genetics in the Medical School in Uppsala, where he wanted to build up a unit for biochemical genetics. For this purpose, he wanted to enlist my assistance and, as I will describe later, his intentions materialized to some extent.

My minor in my Ph.D. studies at Minnesota was physical chemistry, and, apart from Boyer, the excellent physical chemists active there at the time provided the most profound influence on my intellectual development. Bill Lipscomb, who later became a good friend, taught me structural chemistry and chemical bonding. Robert Livingston, a prominent photochemist, who told me he had visited Chalmers in Göteborg to see the Swedish photochemist, Hans Bäckström, taught an excellent course in chemical kinetics. Bryce Crawford, primarily a spectroscopist, provided a course in advanced thermodynamics, which supported my keen interest in this subject, fostered already by Rinde.

The most influential event during my Minnesota years was, of course, that I married Betty Hallberg in July of 1951. Betty was trained as a medical technologist, getting her B.S. Degree at the same time as I obtained my Ph.D. After graduating she worked in a maternity hospital, in which our own daughter was born 3 years later. In Sweden, she became a biochemist, however.

Enzyme research in Uppsala

Metal-ion activation of enolase

In December 1951, Betty and I took the Swedish–American line Christmas boat for Göteborg, and I started work in the Institute of Biochemistry in Uppsala, between Christmas and New Year. I had a clear idea of what I wanted to do, and Tiselius accepted this. In my reading I had discovered that physical chemists, for example, Irving Klotz at Northwestern University, had developed methods to measure the binding of metals to proteins. They largely studied proteins, like serum albumin, however, which are easy to purify, but in which the metal binding has no physiological function. From a review by Lehninger (1950), I had learnt that Otto Warburg had shown that the activity of the glycolytic enzyme, enolase, is a function of the concentration of Mg^{2+} in solution, and he suggested that an enolase-Mg^{2+} complex is the active enzyme. An obvious alternative is, of course, that a Mg^{2+}-substrate complex is the true substrate.

My simple idea was to try to distinguish between these two possibilities for the mechanism of Mg^{2+} activation of enolase by measuring metal binding to both enzyme and substrate, and then compare the results with the kinetics of activation. To do this, I first had to purify the enzyme, and the published method (Warburg and Christian, 1942) was described so accurately that I managed to obtain crystalline enolase on my first attempt. Mg^{2+} is bound too weakly, so instead, I measured the binding of another activating ion, Mn^{2+}, by equilibrium dialysis. The activation kinetics was measured by a spectrophotometric method (Warburg and Christian, 1942), taking advantage of the UV absorption of the product of the reaction, phosphoenol pyruvic acid. My results were first presented at a Faraday Society meeting in Cambridge, in the summer of 1952, and then published in *Nature* in February of 1953 (Malmström, 1953a). A fuller account, including measurements of the binding of Zn^{2+}, was published later, together with a detailed analysis of the purity of the enzyme (Malmström, 1953b).

Uppsala friends

Tiselius ran a very stimulating laboratory with many talented young scientists. I liked most of them but became particularly good friends with a few. One was Hans Boman, who investigated acid phosphatase. He later became professor of microbiology, first in Umeå and then in Stockholm, and attracted world renown on the basis of his work on insect defense mechanisms. Another friend was the microbiologist Bengt von Hofsten, who like myself had a keen interest in music. He left academic work for the Swedish Food Administration and unfortunately died a few years ago. One friend had a particularly important influence on my work, as well as that of Betty. This was Jerker Porath, who had just returned from a period with C.H. Li in Berkeley, working on peptide hormones. He invented preparative column electrophoresis, and later, gel filtration, and by introducing me to those methods, he allowed me to keep one step ahead of other investigators in my area.

Thanks to Jan Böök, Betty had a job at the Uppsala Academic Hospital, working with a group studying renal clearance of dextran, which had just been introduced as a blood substitute. At a dinner given by Tiselius, she met Jerker, however, and he persuaded her to join his group. Her first task was to set up the method for amino acid analysis just developed by Moore and Stein (1949), which she did successfully.

Instructor in Minnesota

Teaching and research

Betty and I had agreed to return to the US in September of 1953. The original plan was to go now to the lab of Brit Chance. When I approached him, he still had not received funds for my support. He informed me a few weeks later that he had obtained the necessary grant, but by this time I had already accepted an offer from the University of Minnesota to become instructor in physiological chemistry. One reason for my impatience was the fact that Betty was now pregnant, and I worried about our economy.

My second period in Minnesota was one of disappointment. I had a lot of teaching, which I actually enjoyed, but I was not allowed to continue my independent research. Instead I found myself back working in histochemistry with Glick. We published a paper together in *Journal of Biological Chemistry* on the determination of nanogram quantities of coenzyme A and its distribution in the rat adrenal (Malmström and Glick, 1954). Because of my discontent, I wrote to Tiselius, and he promised me a position, if I wanted to return to Uppsala. I accepted this offer, at which time I was unexpectedly drafted into the American army for 2 years.

I must admit that my year in Minnesota also had some positive sides to it. Rufus Lumry had joined the chemistry department, and we had lunch together at least once a week. During these luncheons he taught me a lot of physical biochemistry, particularly protein chemistry. He told me about his 'rack' concept (Lumry and Eyring, 1954), which 10 years later I was to apply to blue copper proteins (see Malmström, 1994). I also renewed my acquaintance with Bill Lipscomb and may have been a small influence on his later orientation towards protein crystallography.

A short army career

With the aid of the university administration, the Swedish consul and a lawyer, I tried to be exempt from my call into the army, but all to no avail. A few days before Christmas 1953, I was shipped to Fort Riley, KS, for basic training. There, the personnel officer thought that a mistake had been made, and he immediately sent me on leave over Christmas, hoping to have me free on my return. A so-called administrative discharge turned out to be a complicated problem, and the weeks went by. During those weeks I was allowed to work in the lab of the camp hospital.

I made almost daily visits to the personnel officer to learn how my case had progressed, if at all. He became increasingly irritated, until one day he had a brilliant idea. He said that medical discharges could be handled locally, and he immediately picked up the phone and called a doctor at the hospital, saying: 'I have a man here. Find something wrong with him!' The doctor that examined me, Frank Clark, has been a close friend ever since, and he and his wife have visited us in Sweden a couple of times, and we have been guests at their home in Pennsylvania very frequently.

'Docent' in Uppsala

The embryo of an enzyme group

Less than a year after my return to Sweden in 1955, I had to serve in my second army, the Swedish one, which I had escaped during my period in the US. This was very pleasant, however, since my assigned task in an army research unit was to purify serum choline esterase, to be used in an assay for nerve gases, and my superiors judged that this was best done in my lab in Uppsala. I did this, with the aid of some of the experienced protein purifiers in Tiselius's institute, and the work resulted in a publication (Malmström et al., 1956). In my spare time, I worked on my thesis on enolase, which I defended in May 1956, a few days after leaving the army.

In 1955, I started collaborating with Andreas Rosenberg on measuring the infrared spectra of some salts of the enolase substrate, 2-phosphoglyceric acid (Rosenberg and Malmström, 1955). He decided to walk in my footsteps by investigating the metal-ion activation of a peptidase, carnosinase. The impetus for this came partly from a visit by Emil Smith to Uppsala, in the late summer of 1955. Emil had been studying the metal-ion activation of leucine amino peptidase and arrived at the hypothesis that the metal ion forms a bridge between enzyme and substrate (Smith, 1951). He was one of the first investigators to take notice of my enolase studies, and I was very flattered when he wrote to me about a possible visit. We interacted well from the very beginning and have had very close contacts right up to the present. As will be described shortly, in 1958 I spent several month in his lab in Salt Lake City.

The same summer I also had a visit from Bert Vallee. We had been brought together by Glick already, in my Minnesota days, when Bert was analyzing tissues for zinc. By 1955, we had, however, both become interested in enzymes. As will be apparent later, we did not always agree on matters of zinc enzymes, but we have had stimulating and fruitful contacts over the years.

Many things happened in the year 1955–1956. A polymer chemist from Brooklyn Polytech, Ed Westhead, arrived to do postdoctoral work with Svedberg's successor, Stig Claesson. They did not get along well, however, and Ed asked Tiselius if he could move to the biochemistry department. Tiselius approved and recommended that Ed should work with me. We published two papers on the effect of solvent on enzyme kinetics in 1957 (Westhead and Malmström, 1957a; Westhead and Malmström, 1957b), and these were still being quoted decades later. On a more personal side, Ed was baby sitter, when our second child was born.

When Ed returned to the US, I got him a position with Paul Boyer at Minnesota. Scientifically, this was very satisfactory, but Ed was not quite happy with the climate. In his first letter to me from Minneapolis, he wrote something like this: 'As a final test in an army survival program, I can see some point to it, but since people have gone to the trouble of setting up a university and founding a symphony orchestra, it seems they are planning to stay here.' I saw Ed in Minneapolis on my return from Emil's lab in 1958.

Two Italian postdocs, Giorgio Semenza and Giovanni Toschi, worked in the Uppsala lab during the same year (1955–1956). We never published together, but both have been very close friends since that time. Bibo, as Toschi is called, is a physiology professor at the University of Rome, and we had particularly close contact during my 'Italian period' (1968–1971), borrowing his and Nora's apartment in August of 1970. Giorgio married a Swede, Berit, and we see each other frequently both in Sweden and elsewhere. Our professional contacts have been particularly intense after he succeeded Prakash Datta as Editor of *FEBS Letters*.

Tiselius's department during the 1950s was an ideal place in which to work. He provided resources (I did not have a grant of my own before 1958) and a lot of freedom, but he could also come up with extremely useful suggestions in one's research, as I will soon illustrate. He also introduced a form of democracy long before this was the official dogma. Thus, the 'docents' of the department met with him every Saturday morning to discuss scientific as well as organizational matters. I had become a docent in May 1956 and participated actively in these meetings.

An expanding research program

Tiselius approached me one day in 1956, outside the institute, as I was on my way to bicycle home. He said he had just read about a new technique, electron spin resonance (ESR, also known as EPR), by which one could study metal ions and free radicals. He gave me some literature to read and asked me to consider if this method could become useful in my own work. I returned the following morning with a positive reply, and Tiselius responded that he would contact the physics professor, Kai Siegbahn (Nobel Prize 1981), and together they would ask for money for an ESR (and an NMR) spectrometer from the Rockefeller Foundation.

Their application was successful, and Siegbahn engaged Tore Vänngård to start an ESR research program. Tore and I started collaborating in 1957. Our first project concerned one of my old problems, the binding of Mn^{2+} to enolase. I had engaged a new student for this project, Märtha Larsson (later Larsson-Raznikiewicz), and our first results were published in 1958 (Malmström et al., 1958). Later, Märtha worked on the metal-ion activation of another glycolytic enzyme, phosphoglycerate kinase. She could demonstrate that in this case the function of the activating ion is to form a complex with ATP, one of the substrates (Larsson-Raznikiewicz and Malmström, 1961).

In 1957, I also engaged two other new students to work on carbonic anhydrase, my interest from the histochemical days. Sven Lindskog (now professor of biochemistry in Umeå) was asked to purify and characterize the bovine enzyme, whereas Per Olof Nyman (now lecturer in Lund) worked with human carbonic anhydrase in Böök's Department of Medical Genetics. Lindskog and I showed that we could dissociate zinc reversibly from the bovine enzyme (Lindskog and Malmström, 1960), and this laid the foundation for a detailed study of metal binding and catalytic activity (Lindskog and Malmström, 1961).

An interlude in Utah

Emil Smith and I met again at a symposium on protein
structure arranged by Albert Neuberger at the Sorbonne in
Paris, in 1957. Science was still small enough that one could
get everybody who was anybody in the field, plus everybody
who was nobody (like myself at the time), into one lecture
hall. At this meeting were, among others, Linus Pauling, Kaj
Linderstrøm-Lang, Charles Tanford, Arne Tiselius, John
Kendrew, Max Perutz, Eraldo Antonini, Vernon Ingram, Pierre Desnuelle, Hans Neurath, Brian Hartley, Emil Smith,
Stanford More, William Stein, Chris Anfinsen, Heinz
Fraenkel-Conrat, Rosalind Franklin, Rod Porter, Choh Hao
Li and Ieuan Harris. Hans Boman and I were also speakers,
but there were, in addition, several young investigators just
attending, for example, Bob Bray, who was going to work
with me the following year, and Edith Heilbronn, my 'boss'
in the Swedish army. Emil stayed in a small hotel on the
Seine, together with Chris Anfinsen, whom I met for the
first time, and we three spent much of the free time together. Fifteen years later I had the pleasure to present
Chris in the Stockholm Concert Hall, when he was awarded
the Nobel Prize in 1972, for his pioneering work on protein
folding.

Emil and I agreed that I should spend the spring of 1958
in his lab in Salt Lake City, doing some protein chemistry on
enolase, and we published a paper, together with the late
Joe Kimmel, on its amino acid composition and amino-
terminal sequence in *Journal of Biological Chemistry* in
1959 (Malmström et al., 1959). Tiselius had helped me fi-
nance the visit with a Rockefeller Foundation fellowship. On
this trip, I also visited Joe Neilands in Berkeley and Hans
Neurath in Seattle to discuss his carboxypeptidase work. He,
together with Vallee and Williams, maintained that a cysteine sulfhydryl group is one of the ligands to Zn (see Vallee
et al., 1960), but we had evidence that the coordination is

the same as in carbonic anhydrase, the bovine form of which lacks a sulfhydryl group (Malmström et al., 1963). Later, of course, the crystal structures (Liljas et al., 1972; Rees et al., 1983) showed that Lindskog and I had been right: a sulfhydryl group is a ligand neither in carboxypeptidase nor in carbonic anhydrase.

Betty and the children stayed with Betty's parents in Minneapolis during my work with Emil, and I went there on our way back to Sweden. This gave me an opportunity to see Dave Glick, Paul Boyer and, as already mentioned, Ed Westhead again. Dave recommended me to apply for an NIH grant, which I did successfully on my return.

The beginnings of oxidase research

In 1957, a microbiologist, Gösta Fåhraeus, visited Tiselius asking for help with purifying a fungal laccase, and Tiselius sent him to me. I engaged a student, Rolf Mosbach, who purified the enzyme, and we had our first blue oxidase EPR paper published in *Nature* in January of 1959 (Malmström et al., 1959). In the autumn of 1958, Bob Bray from Chester Beatty in London, whom I had met at the protein conference in Paris the previous year, joined the group. Ironically, he wanted to get away from xanthine oxidase, having worked on it too long, so I put him on an enolase project. I also asked him, however, to have some xanthine oxidase shipped to us to put in the EPR machine. We found that the oxidized enzyme was EPR silent, but Mo and flavin radical signals developed on addition of substrate (Bray et al., 1959). We also found a signal with a g-value below 2, which we could not identify. At a conference in 1960, where Tore presented our results, it was called 'the mystery signal', but the participants agreed that it could not come from Fe; we now, of course, know that it represents an FeS center. Bob found our results so interesting that he has worked with xanthine oxidase ever since.

In 1960, Vänngård and I published a paper in *Journal of Molecular Biology*, in which we showed that blue copper proteins have uniquely narrow hyperfine coupling constants compared to other copper proteins and to small copper complexes (Malmström and Vänngård, 1960). I presented this work at a meeting on 'Metal binding in biology' at Penn State (Malmström, 1961), to which I had been invited by Art Martell. After my talk, there was an amusing exchange with Bert Vallee, who thought that I had said that non-blue proteins give no EPR signal, rather than just a wider hyperfine splitting. I tried in vain to get Bert to understand, and finally Mel Calvin intervened, explaining that there was only a question of different parameters. All this is recorded in the printed discussion, which is, however, more polite than what was actually said. In the same discussion, Bob Williams, whom I met personally for the first time at this conference, stated that we had agreed that all blue proteins contain cuprous copper; in fact, I had never agreed to this.

At this conference I also met Paul Saltman, who was on his way to Hans Ussing in Copenhagen. I suggested that he should come to Uppsala to study transferrin by EPR, which he did in 1961 and 1962 (Aasa et al., 1963). Another new acquaintance was Phil Aisen, who studied ceruloplasmin at the time, and later (1966–67), came to our lab on a sabbatical, also to investigate transferrin (Aisen et al., 1967). Both he and Paul became lifelong friends.

The meeting at Penn State was followed by a bioinorganic symposium organized by Gunther Eichhorn at the ACS meeting in New York. There I met Frank Gurd for the first time. He had been one of the first, besides Emil Smith, to give considerable attention to my enolase work, in his review on metal binding to proteins in *Advances in Protein Chemistry* in 1956 (Gurd and Wilcox, 1956). We had dinner together, at which we discussed not only science but also music. Apparently, somebody in the restaurant overheard us, because as we were leaving, a man approached me, asking if

I would like to join their professional old-music group as recorder player. Frank is an excellent horn player, and we often played together later, whenever we met. Frank spent a couple of months with me and Tore in Göteborg in 1966 (Gurd et al., 1967).

At the first session of the bioinorganic symposium, I had a somewhat embarrassing experience. One of the speakers was Gary Felsenfeld, who worked on ceruloplasmin at this time. He reported an attempt to use valence-specific reagents to determine the valencies of the copper ions in the protein. Al Cotton, who had heard my talk at Penn State, was chairman, and when Felsenfeld had finished, he said something like the following: 'We have just heard from Dr. Felsenfeld how not to determine the valence of a metal in a protein. But I see that Dr. Malmström is in the audience, so could he please come up and tell us how to do it'.

In 1961, I got a psychiatrist, Lars Broman, as a student. He had started working on ceruloplasmin, at a time when it was thought to be involved in schizophrenia. He had gotten so interested in this protein that he decided to continue working with it, even after this idea had been abandoned. He showed with Vänngård, Roland Aasa, now a student of Tore, and I, that the same copper in ceruloplasmin and laccase as gives the narrow hyperfine splitting, also is responsible for strong blue color (Broman et al., 1962). This was a necessary experiment, since we had found, both by EPR and susceptibility, the latter in collaboration with Anders Ehrenberg (Ehrenberg et al., 1962), that only half of the copper is paramagnetic, so Williams could have been right. At that time, Tiselius came to me with a paper that he said described a blue copper protein, azurin, which contained a single copper ion. He asked if this was not something for Tore and me! I wrote to the authors, who sent me a sample, with which we showed that the single Cu^{2+} ion gave an EPR spectrum with a narrow hyperfine splitting (Broman et al., 1963).

Our oxidase work brought new visitors to Uppsala. Among them were Helmut Beinert, Howard Mason, Tom Singer and Tsoo King.

The end of the Uppsala period

A very important development in Uppsala in the early 1960s was the establishment of Swedish protein crystallography. This came about because of an interaction between a leading Swedish inorganic chemist, Ingvar Lindqvist, and me. We had first met when he visited Bill Lipscomb during my Minnesota years, but when I moved to Uppsala, we started to see each other regularly. In the beginning, there was mainly the music of Mozart, rather than science, that brought us together, but, in between listening to operas, I tried to persuade him that biochemistry offered a fruitful field for crystallographers. He was eventually convinced and sent two of his students, Bror Strandberg and Carl-Ivar Brändén, to John Kendrew in Cambridge for training in protein crystallography. Strandberg started to collaborate with the carbonic anhydrase group, which, as already mentioned, resulted in a structure of this enzyme in 1972 (Liljas et al., 1972). Lindqvist later became professor at the Agricultural University in Uppsala, where he tried to get me to take an associate professorship. He also served as scientific secretary of the Swedish Natural Science Research Council and as President of the Academy of Sciences. Unfortunately, he died suddenly in 1991, at the age of 69.

The Swedish carbonic anhydrase investigators had close contacts with John Edsall of Harvard, who visited us in Uppsala, and later in Göteborg (the crystallographers stayed in Uppsala), very frequently between 1962 and 1967. We had both first announced publicly our interest in this enzyme at a symposium organized by Hans Neurath at the International Congress of Biochemistry, held in Vienna in 1958. Edsall became foreign member of the Royal Swedish Academy of Science in 1972 and an Honorary Doctor of Science at Göteborg University in the same year.

The Seventh International Conference on Coordination Chemistry (ICCC) was held in Stockholm in 1962, organized by Lars-Gunnar Sillén, an internationally well-known inorganic chemist, with Hugo Theorell as Honorary President. This was the first ICCC with a bioinorganic section, which I chaired. Frank Gurd was plenary lecturer. Other invited speakers in this section included Howard Mason, Paul Saltman, Hans Freeman, Joe Coleman (substitute for Bert Vallee, who had been taken ill), Gunther Eichhorn, Peter Hemmerich, Winslow Caughey, Mel Calvin and Joe Neilands.

In Uppsala, around this time, I became friendly with Peter Reichard, who was professor there for a few years before moving back to the Karolinska Institute. We have had a lot of interactions throughout the years, not only on scientific matters but also on questions of science policy, for example, during the recombinant DNA debate in Sweden in the 1970s. He was member of the Nobel Committee for Physiology or Medicine in the same period that I served on the chemistry committee, and we had many discussions of potential candidates in biochemistry and other common problems in the prize work. As he mentioned in his prefatory chapter in *Annual Review of Biochemistry* (Reichard, 1995), he often came to our group in Göteborg to get advice on the nature of iron in ribonucleotide reductase. Another Uppsala friend was Torvard Laurent, with whom I later worked closely together, when he was secretary and I chairman of the Swedish Biochemical Society (now the Swedish Society for Biochemistry and Molecular Biology), in 1967–70.

Visiting professor in California

A bioinorganic course

In 1962, Paul Saltman offered me a permanent position as professor of biochemistry and biophysics at the University of Southern California (USC). I replied that I could not accept without first seeing how I would like the university, its faculty and facilities, and I was then invited as visiting professor during the spring semester of 1963. I offered a course on 'Role of metal ions in enzymes', with four hours of lectures each week. One of the students in the course was Jim Fee, who was converted from bioorganic chemistry, Tom Fife being his adviser, to bioinorganic chemistry, by my lectures. He asked if he could do postdoctoral work with me after finishing his degree, to which I, of course, agreed.

During this spring I lived with Betty and our two children, Barbro (8) and Jan (6) in South Pasadena, not too far from the Pasadena home of the Saltmans. Little did I know at that time how important Pasadena was to be for both Betty and me 15 years later and for many years to come. The USC Biochemistry Department was divided between the main campus, just southwest of downtown LA, and the Medical School, northeast of the city. The chairman, John Mehl, decided that my lectures should be held on the main campus but that I should have my office in the Medical School, so I had to learn to drive the freeways in our newly acquired Chevy II station wagon. My lectures were between 8 and 10 on Tuesdays and Thursdays, and I often took off by plane around noon on Thursdays and returned on Monday evenings. This gave me an opportunity to visit colleagues and give seminars at other universities. Among those that I visited were Howard Mason in Portland, Tsoo King at Corvallis, Helmut Beinert in Madison, Emil Smith in Salt Lake and Frank Gurd, then in Indianapolis. At the beginning of my stay, I contributed to an ACS symposium in honor of

Mildred Cohn in Cincinatti. From this meeting, Frank drove me to Indianapolis together with Charles Tanford, whom I first got to know well on this occasion.

During my term at USC, I attended the Federation Meetings in Atlantic City, where I gave a short talk on carbonic anhydrase. On my way there, I paid a visit to Duke University, where I saw Bob Hill, who worked in Salt Lake at the same time as I had, Phil Handler, who worked on xanthine oxidase at the time, and Charles Tanford, with whom I shared an interest in physical protein chemistry. Leaving Duke, I went to Johns Hopkins, where Winslow Caughey was at the time. With him, I also made my first trip to Washington to see some beautiful ivory recorders in the Library of Congress.

The transition from Uppsala to Göteborg

By the end of the spring I had been informed by students and friends that if I returned to Sweden, I could probably choose from two permanent positions: a professorship in biochemistry in Göteborg and an associate professorship in chemistry at the College of Agriculture in Uppsala, the professor being my old friend, Ingvar Lindqvist. The job in Uppsala was already in the bag, but that in Göteborg was not formally given to me yet. The faculty had suggested me, but the University Chancellor and the King had to have their say. This was considered a mere formality, but, as I will tell shortly, this turned out not to be entirely true. Anyhow, Betty and I decided to return to Sweden. We and the children took a month crossing the US in our Chevy II and then sailed for Göteborg on the Kungsholm on 17 July.

Professor at Göteborg University

A glimpse of the Swedish academic system of the 1960s

In the autumn of 1962, before going to USC, I had applied for the two professorships mentioned. The one in Göteborg had attracted no less than 15 applicants, including many prominent scientists that soon obtained other chairs: Per-Åke Albertsson, Herrick Baltscheffsky, Hans Boman (who became professor in microbiology, not biochemistry, first in Umeå, then in Stockholm), Anders Ehrenberg (professor of biophysics in Stockholm), Lars Ernster and Jerker Porath. The experts that wrote evaluations for the faculty, one of them being Tiselius, were not unanimous. Two suggested me, whereas one preferred Lars Ernster. Everything is very open in Sweden, and all the applicants were sent the minutes of the meeting at which the members of the faculty voted on the issue. The faculty was very small at the time, and included some professors from Chalmers University of Technology, the total number of members being only nine. Six of them voted for me and three for Ernster. Thus, the faculty decision went to the University Chancellor, and my appointment was now considered a mere formality. What had been overlooked, however, was the fact that one of the applicants had filed a protest, which can be done within the Swedish system, and this had first to be ground through the bureaucratic mills, which could not be done during the summer.

When I arrived in Göteborg by the Kungsholm around 1 August, the Academy Secretary (the title of the chief university administrator at the time) invited me to see him and the Rector, professor of Sanskrit, Hjalmar Frisk. They considered the situation ridiculous and said that it would be only a matter of time before I was appointed; in the meantime, they would make me acting professor from 1 September, the beginning of the autumn term. Tiselius, whom I visited at his summer home close to Göteborg, was furious; he had already given away my space in his department in Uppsala, and now he had to give it back! One November day in my home in Uppsala, where I was having lunch with my family, Tiselius called me and said that I should turn on the radio and listen to the 12 o'clock news. In this it was announced that the King had appointed me professor of biochemistry at Göteborg University from 1 December 1963.

The building of a new department

Five new professorships in chemistry had been created in Göteborg at the same time, so I had four new colleagues in the Section of Chemistry. I already knew all of them, except Gunnar Aniansson, the professor of physical chemistry, who was a second-generation student of Svedberg, being himself a student of Ole Lamm. Two of them, David Dyrssen in analytical chemistry and Georg Lundgren in inorganic chemistry, were students of Sillén, and we had met in various connections, not least in arranging the 1962 ICCC. Lars Melander, the professor of organic chemistry and a well-known authority on kinetic isotope effects, came from the Nobel Institute for Chemistry in Stockholm, where I had contacted him in the summer of 1963, since he and I were supposed to share a floor of the new chemistry building.

The building that was going to house the new chemistry departments (there were as many departments as professors), as well as the Chalmers departments already in existence, was not expected to be finished until the summer of 1964, even if the teaching wing were to be completed by the turn of the year. We five professors wrote a joint petition to the University Chancellor, asking that we be allowed to postpone starting undergraduate teaching until the academic year 1964/65. Actually, this created some sensation in the Chancellor's office, since he claimed that this was the first time in his experience that five professors had agreed on anything. A gratifying experience was how well we worked as a team rather than competing with each other. Largely, this was a result of the wisdom and diplomacy of Georg Lundgren, who became the first Dean of Chemistry and later was to become Rector of the whole university.

Because of the building situation, I decided to stay in Uppsala until the end of 1963 and only go to Göteborg for faculty meetings. I also went down a couple of days each week for matters concerning the organization of the new department, for example, interviewing applicants to a number of positions that were attached to the professorship. I was fortunate to be able to 'steal' a chemical engineer, Lars Strid, from the biochemistry department of the Medical School. He was a specialist in protein chemistry, and was to play an important role in the determination of the primary structure of carbonic anhydrase. From the same department, I also recruited a graduate student, Hedvig Csopák, who was to do outstanding work on another zinc enzyme, alkaline phosphatase.

Märtha Larsson-Raznikiewicz, Sven Lindskog and Per Olof (Pelle) Nyman, as well as my technician, Sven-Olovf Falkbring, had decided to move with me to Göteborg, whereas Lars Broman could not for family reasons. We decided to let the group continue in Uppsala until the summer of 1964, however, and Tiselius agreed to this. He was also very generous in letting me move all equipment that I had bought on my grants, which is not required by the Swedish rules, and this was a welcome addition to the equipment grant I was given as a new professor. Thus, the department got off to a rapid start, and both teaching and research were running smoothly from the autumn semester of 1964.

The professorship entailed a lot of administrative duties, such as serving on local and national committees and research councils. I did a fair share of such work but am not going to discuss it here, where I will instead limit myself to my research activities. I will make one exception, however, namely my 17 years on the Nobel Committee for Chemistry, 12 of them as chairman, for two reasons: what I have to say about this is probably of some general interest and, secondly, the fact that I served the committee so long has turned out to have become important in my present situation, as I will describe towards the end of this essay. I may just mention briefly, that two of my first administrative academic tasks were to evaluate applicants for professorships in biochemistry in Umeå and Stockholm. Somewhat ironically, I played an important role in these positions going to two of my former 'competitors', the one in Umeå to Albertsson (Ernster had not applied, probably because Umeå is too cold for his Hungarian blood) and the one in Stockholm to Ernster.

A couple of years after the start of the department, in 1966, I hired a new secretary, Britt Pehrsson (now Björling), and a new technician, Ann-Cathrine Carlsson (now Smiderot), and they both stayed with me until my retirement in 1993, i.e., for no less than 27 years. Britt, for many years, sacrificed a good portion of her summer vacation to help me in my Nobel work. When she was on leave for child birth, Tore's secretary, Sieglind Billing (now Salo), who started in 1974, took her place, and Sieglind has given me a lot of help over the years. Now we work close to each other on the same floor, whereas Britt has moved with Jan to the floor above. The reason for mentioning these three individuals is not only their long and faithful service, but also the fact that they are by this time well-known to most investigators in my field all across the world.

One difficulty with moving to Göteborg was the fact that, unless I was going to change entirely my direction of research, I had to find a way to move the biophysics group. Tore and Roland had informed me already, in December of 1962, half a year before it was certain that I would get the professorship, that they were willing to move to Göteborg, if it could be arranged. I turned to one of the biochemistry pro-

fessors in the Medical School, Einar Stenhagen, whom I
knew from our common days in Uppsala. He was now a
member of the Natural Science Research Council and a key
person concerning our possible support. He apparently had a
high esteem for our work and promised to help all he could.
Waiting for things to be arranged, Tore spent a year with
Helmut Beinert in Madison. For Roland, I managed initially
to get money from the faculty, a position in geography being
vacant because of the lack of a competent candidate. Eventually there were temporary solutions for everybody, including Karl-Erik Karlsson (now Falk), a new student of
Tore's, who wanted to move with him and Roland. The biophysics group moved to Göteborg in the summer of 1966,
when we also got our new EPR spectrometer with the aid of
a grant from the Wallenberg Foundation. In 1974, Tore obtained a permanent professorship in biophysics.

When I visited Einar Stenhagen in the Medical School, I
also met another professor of medical chemistry, Ulf Lagerkvist, and he was to become one of my best friends in the
science circles of Göteborg (I also have friends in the musical
circles). He had worked at Stanford with Paul Berg and had
established himself as a leading investigator of tRNA syn-
thases. At the time he ran an evening discussion group in
molecular biology and, similarly to Boyer's enzyme club in
Minnesota, it attracted students and faculty from many
branches of science and medicine. The meeting place circu-
lated among the homes of the participants, and it was a tradition that the session started with a glass of sherry. I
learned a good deal from these discussions.

Ulf and I see each other quite regularly, both in and out of
the lab. For a number of years, I had the pleasure of being
invited to sail with him on his boat. We were usually gone
for several days, visiting fishing communities along the coast
north of Göteborg. For many years we have also played
badminton together each week. It was during those games
that we conceived the idea of a small biomedical center, and

in 1994 this became reality. In February of that year, our Department of Biochemistry and Biophysics moved into a new building, adjoining Ulf's department, and housing also Departments of Microbiology, Molecular Biology and Genetics. He became member of the Royal Swedish Academy of Science in 1983, on my initiative, and he then helped me very much in the evaluation of candidates in the DNA-RNA areas.

A number of visitors

The new department was fortunate in receiving several visitors in the early phase of its existence. First of all, we started to have postdoctoral fellows. Philip Whitney came in 1965 from Charles Tanford, with whom he had written a thesis on the hydrophobic effect. Phil worked with Pelle Nyman and me on affinity labeling of carbonic anhydrase with chemically modified sulfonamides (Whitney et al., 1967). Another postdoc was Jim Butzow, who came from Gunther Eichhorn. I put him on an impossible problem, namely, to determine the subunit composition of fungal laccase. Of course, we know now that there are no subunits, but it was current dogma at the time that a protein containing several metal ions must consist of subunits.

Two visitors in 1966, Phil Aisen and Frank Gurd, have already been mentioned. Another stimulating person, who spent a period with us that year, was Hans Freeman. With him, we used spin resonance methods to compare the crystal and solution structure of a Cu(II)-peptide complex (Falk et al., 1967). A few years later (1969), Hans invited me to Sydney as a visiting professor for a month, following the ICCC meeting, at which I gave one of the plenary lectures.

In 1967, when Phil Aisen was still with us, came Brian Hartley from the MRC Laboratory in Cambridge. We had both been speakers in 1964 at a one-day symposium on enzyme chemistry arranged by the Netherlands Biochemical

Society, at which time I started trying to persuade him to come to Göteborg to help Pelle Nyman and Lars Strid in setting up modern methods of protein sequencing. His assistance was a major factor for their successful determination of the primary structure of carbonic anhydrase (Henderson et al., 1973). Brian turned out also to be a major player in the football game at the yearly department picnic.

My research 1964–93

Laccase and other blue copper proteins

Before the biophysics group moved to Göteborg in 1966, I found it difficult to do much work on blue copper proteins. Tore came down from Uppsala occasionally, however, taking EPR samples back, and we did manage to publish a small paper on the high-pH reduction of fungal laccase (Malmström et al., 1965). In 1964, I went to four consecutive meetings in the USA, one of them being the first ISOX (International Symposium on Oxidases and Related Redox Systems), where I presented the concept of rack-induced bonding in blue copper proteins (Malmström, 1965). According to this, the tertiary structure of the protein creates a preformed chelating site with very little flexibility. Almost 20 years later, Harry Gray put this concept on a quantitative basis (Gray and Malmström, 1983), estimating that the rack energy is as large as 70 kJ·mol^{-1}.

Bengt Reinhammar joined the group when we moved to Göteborg. He established type 2 Cu^{2+} as a component of blue oxidases (Malmström et al., 1968). At that time, Jim Fee and Dick Malkin, the latter coming from Jesse Rabinowitz in Berkeley, started their two years of postdoctoral work. They were responsible for no less than eight publications out of our lab, but the main result of all their contributions was undoubtedly the discovery of type 3 copper (Fee et al., 1969). Together with Dick, I wrote a review (Malkin and Malm-

ström, 1970), which has become a citation classic (Malmström, 1983).
The main discovery with laccase during the 1970s, was that of an intermediate in the reaction of the reduced enzyme with dioxygen. This was first observed as an optical intermediate (Andréasson et al., 1973) but was later demonstrated to be a true oxygen intermediate with unusual EPR properties (Aasa et al., 1976). Many investigators contributed to this work: Lars-Erik Andréasson, Rolf Brändén and Joke Deinum as well as Bengt, Roland and Tore.

Three sabbaticals

When I started out as professor, the rules allowed me to take one term off for research abroad, every fifth year. Thus, my first 'sabbatical' (the term is not quite appropriate, since it refers to every seventh year) was due in 1968, and I had decided to spend this with Eraldo Antonini in Rome. We had met throughout the years at various conferences, for example, the Protein Gordon Conferences, and we got to know each other well at a hemocyanin conference, organized by the Ghirettis, in Naples in 1967. At that time we agreed that I should come to Rome the following year. Eraldo was a specialist on stopped-flow kinetics, which he had learnt as a postdoc with Quentin Gibson, then in Sheffield. Having worked mainly with oxygen-binding proteins, he knew how to operate the system anaerobically or with varying oxygen tensions, the same techniques that were required to study an oxidase. Thus, my choice of laboratory was natural for an investigation of laccase kinetics.

An EMBO short-term fellowship made my visit possible. Since I started work in August, Eraldo got the family a house on the beach in Santa Marinella, a small town about 60 km north of Rome. He himself had a summer house just a few hundred meters north of ours, and on the weekends he and his son, Giovanni, often came and picked us up in Eraldo's catamaran. It was actually on one of those sailing excursions that we finally figured out the meaning of the complicated kinetics we had observed in the lab.

Our first result confirmed the presence of type 3 copper, as demonstrated by Jim and Dick in Göteborg, since we observed that anaerobically the enzyme accepts 3 electrons rapidly from $Fe(CN)_6^{4-}$ (type 2 Cu^{2+} is not reduced on the stopped-flow time scale). Furthermore, we found that 1-electron reduced laccase does not react rapidly with O_2, so we suggested that the binuclear type 3 site must have received 2 electrons for O_2 to react (Malmström et al., 1969). We speculated that this may be a general principle for oxidases that reduce O_2 to $2H_2O$ and were anxious to try the idea with cytochrome oxidase. Having no experience working with membrane proteins, we soon abandoned such plans, however.

The following year, Eraldo and I met at the First Harden Conference, where we discussed our idea with Quentin Gibson. As he had no objection, we became even more anxious to put it to an experimental test. Eraldo said that Colin Greenwood, who was also at the conference, had done all the cytochrome oxidase work in Quentin's lab, so we decided to invite him to come to Rome, in 1970, to help us with cytochrome oxidase experiments. After several months of intense experimentation, we had results which clearly confirmed our hypothesis (Antonini et al., 1970). By this time we had acquired a stopped-flow apparatus in Göteborg, and I invited Colin to assist us in starting cytochrome oxidase investigations in our home lab (see next section).

Two months (August–September) of 1971 I also spent in Rome. An important event during this period was the initiative from Bruno Mondovì that we should, together, organize a meeting on copper proteins. We had the first meeting in La Cura, north of Rome, in September of 1971. Bruno has continued to organize copper conferences on a regular basis, and there has been no less than eight such meetings in the period 1971–1995. After the first one, they were moved to Manziana, and the last one (1995) was held in Santa Severa on the Mediterranean coast. Bruno and his wife, Anna, have also become very good friends of Betty and I.

My second sabbatical, in 1973, I spent as a Miller Professor at the University of California, Berkeley, on an invitation from Dan Koshland. My proposal was that Dan and I, on the basis of the crystal structure published the year before (Liljas et al., 1972), should try to estimate if known catalytic effects, such as proximity and nucleophilic attack, were sufficient to explain the catalytic efficiency of carbonic anhydrase. Somehow, Dan could never get started on the project, however, and I ended up spending my time writing a review article (Malmström, 1973), in which I, among other things, analyzed redox interactions in cytochrome oxidase.

At a symposium in honor of Bill Slater in Amsterdam in 1977, I was asked by Hans Vliegenthart from Utrecht, if I was willing to become a visiting professor ('Capita Selecta') in Utrecht the following year. I accepted and spent the autumn of 1978 giving, once more, a course on the role of metals in enzymes. This time it was, of course, more complicated than in 1963, the literature having expanded tremendously, but I was pleased to see that people came, not only from Utrecht, but also from Amsterdam and Leiden, to hear my lectures. Hans worked in two main areas, carbohydrate chemistry and lipoxygenase, and I suggested that we should enlist Roland Aasa to help us with EPR investigations of lipoxygenase. As a result, two of Hans's students, Steven Slappendel and Martin Feiters, did much of their thesis work in Göteborg, defending their theses in Utrecht in 1982 and 1984, respectively. The University of Utrecht made me an honorary doctor in 1986.

Early cytochrome oxidase investigations

Our first cytochrome oxidase paper from Göteborg dealt with

a re-investigation by stopped-flow of the anaerobic and aerobic reaction of the enzyme with cytochrome c (Andréasson et al., 1972). My first student working entirely with cytochrome oxidase was Steffen Rosén, who obtained EPR evidence for 'pulsed' oxidase (Rosén et al., 1977). Steffen also interacted with Steve Lippard, who spent a number of months in our lab in 1972 (see Lang et al., 1974). With two of my students, Boel Lanne and Bo Karlsson, I had already started to investigate oxidase reconstituted into phospholipid vesicles in 1977 (Karlsson et al., 1977), the same year that its proton pumping activity was discovered (Wikström, 1977).

In 1979, Marius Clore came to Göteborg from London to help us set up a low-temperature technique to study intermediates in the reaction of the enzyme with O_2, by EPR and optical spectroscopy (Clore et al., 1980). Our results suggested that one of the oxygen intermediates should be paramagnetic, and we later demonstrated an unusual EPR signal from Cu_B in this species (Hansson et al., 1982). A few years earlier we had shown that it is possible to generate a rhombic, but still more normal, EPR signal from Cu_B (Reinhammar et al., 1980), similar to that seen with half-met hemocyanin (Himmelwright et al., 1978). Mike Wilson, whom I had first met in Rome in 1971, came to Göteborg to work with my student, Peter Jensen, in the winter of 1981 (Wilson et al., 1982), and this collaboration has continued to the present day (Brzezinski et al., 1995), in recent years with support from EU.

My collaboration with Harry Gray

The most profound intellectual influence on my research in the past two decades has undoubtedly stemmed from my interaction with Harry Gray. On a more personal side, the friendship with him and his wife, Shirley, has been a source of immeasurable pleasure both to me and Betty. Our experimental collaboration started in 1978, when Lars-Erik and I provided fungal laccase for a part of his investigations of the electronic structure of blue copper proteins (Dooley et al., 1979). Harry and Sunney Chan invited me to spend 6 months at Caltech, in 1980, as a Fairchild Scholar, and during this period I worked with both of them on the reactions of nitric oxide with tree and fungal laccase (Martin et al., 1981). We also initiated resonance Raman studies of blue copper proteins (Blair et al., 1985).

In 1988, I spent a second 6-month period at Caltech. By this time Harry and his group were deeply involved in the study of long-distance electron transfer in ruthenated proteins, and I had learnt enough about electron transfer to write a review with him (Gray and Malmström, 1989). In addition to my two long stays, I have made visits to Caltech, with durations of between one week and one month, no less than 10 times in the period 1982–95. These trips, like most of my recent travel, were generously paid by the Nobel Institute for Chemistry after applications to the Nobel Committee for Chemistry (Tiselius, when he was chairman of the Nobel Foundation, had changed the Nobel Institutes into what he called immaterial institutes, and the funds liberated are used to support visiting scientists, workshops and travel of those involved in work with the Nobel Prizes, as I will further illustrate later). During 1988, the Beckman Institute was being constructed, and in recent years I have had the privilege to work in this magnificent building. I have also had the pleasure to stay right next to it, in the Tolman House, sometimes together with other visitors: Carl Ballhausen (1990 and 1994) and Ivano Bertini (1995).

My work at Caltech has so far resulted in no less than 14 publications (two of which are still in press). In the late 1980s, work on site-directed mutagenesis was initiated both in Göteborg (Lennart Lundberg) and at Caltech (Jack Richards), and Harry and I soon initiated collaboration in this area as well (Di Bilio et al., 1992). In recent times, our work together has concerned internal electron transfer in cyto-

chrome oxidase (Winkler et al., 1995) and, in particular, the
properties of a component of the oxidase, Cu_A, as will be discussed in a later section. Harry received an honorary doctorate from Göteborg University in 1992, when it had its 100th
anniversary.

Electron transfer and proton pumping

The first student that I assigned a project specifically aimed
at illuminating the proton-pumping properties of cytochrome oxidase, was Per-Eric Thörnström. He started by
characterizing the pH dependence of the steady-state kinetics with enzyme in phospholipid vesicles (Thörnström et al.,
1984). About the same time, Peter Brzezinski began to construct the laser setup used for our measurements of the kinetics of internal electron transfer (Brzezinski and Malmström, 1987). Peter also concluded, from simulations of the
kinetics, that the non-hyperbolic steady-state kinetics dis-
played by cytochrome oxidase, is the result of a redox-
induced conformational change (Brzezinski and Malmström,
1986).
    Direct evidence for a redox-linked conformational change
was provided by a visitor from Albany, Charles Scholes
(Scholes and Malmström, 1986), who returned to our department for a few months this year. Other visitors, Marian
Fabian from Kosice, and Brigitte Maison-Peteri from Paris,
performed experiments indicating the presence of a proton
channel (Fabian and Malmström, 1989) and intrinsic uncoupling at low pH (Maison-Petri and Malmström, 1989). Important work on internal electron transfer, establishing a
rapid redox equilibrium between cytochrome a and cytochrome a_3–Cu_B, was carried out by Mikael Oliveberg
(Oliveberg and Malmström, 1991). Later, Stefan Hallén
could show that this reaction is coupled to the protonation of
a group close to the cytochrome a_3–Cu_B site (Hallén et al.,
1994).

The Cu$_A$ site

Several years ago, Kroneck et al. (1988) suggested that Cu$_A$ is a mixed-valence binuclear site. Sunney Chan and I first questioned this (Li et al., 1989), but later analytical work, in collaboration with physicists in Göteborg (Öblad et al., 1989), established that good preparations of cytochrome oxidase contain 3Cu/2hemes. That Cu$_A$ is really binuclear was clearly established when Matti Saraste and co-workers managed to express a soluble Cu$_A$-binding domain from *Paracoccus* cytochrome oxidase (Lappalainen et al., 1993). This also allowed, for the first time, the recording of the visible spectrum of Cu$_A$, which is completely masked by the heme absorptions in the whole oxidase. The three-dimensional structure of the binuclear Cu$_A$ site has recently been determined (Iwata et al., 1995; Wilmanns et al., 1995).

The Nobel Committee for Chemistry 1972–1988

The composition of the committee

The only one of my various administrative duties that I really enjoyed, was my work on the Nobel Committee for Chemistry. Tiselius had nominated me for the Academy of Sciences with the idea that I should succeed him on the committee, because even if he had not died prematurely in 1971, at the age of 69, he would according to the rules have had to leave the committee on turning 70. I entered the committee in January of 1972, as adjunct member, 45 years old. The five regular members were all of mature age, four of them being born in the period 1900–1903: Arne Fredga, organic chemistry and new chairman after Tiselius, Gunnar Hägg, inorganic chemistry, Arne Ölander, physical chemistry and Karl Myrbäck, biochemistry. The 'youth' among them was the Göteborg medical chemist, Einar Stenhagen (see earlier), who was only 61. A young organic chemist,

Göran Bergson, born in 1934 and favorite pupil of Fredga's, became adjunct member at the same time as I, so that we were a total of 7 chemists on the committee.

Due to the high age of the members, the composition of the committee changed rapidly. Fredga was replaced by Bengt Lindberg, Hägg by Ingvar Lindqvist, Ölander by Stig Claesson, and Myrbäck by Lars Ernster. In addition, Einar Stenhagen died but was not replaced by another biochemist. A prominent physical chemist with a strong biochemical bend, Sture Forsén, soon joined the committee as adjunct member and succeeded Stig Claesson as regular member, when he resigned for health reasons. After 5 years, I became chairman of the committee in 1977. Secretary of the committee was the Stockholm inorganic chemist, Arne Magnéli, whom I had known already as docent in Uppsala. His experience and skill was one of the main reasons that the work on the committee went smoothly and was enjoyable. Arne was succeeded by Peder Kirkegaard in 1987, and I enjoyed, equally well, the short period that I worked with him; sadly, he died suddenly on 29 January 1996, at the age of 67.

Some small reforms

The Nobel work is surrounded by strict rules of secrecy, and I am consequently unable to discuss confidential matters. It was, however, impossible, just looking at the prizes given in recent years, to escape the conclusion that there was an unwritten agreement among the members to circulate among the main branches of chemistry. Thus, it was perhaps not surprising that I managed to get the committee to suggest a biochemical prize, that to Anfinsen, Moore and Stein, in my first year on the committee. Be that as it may, one of my efforts during my term was to select the best candidates regardless of the field. Even so, the committee did not choose biochemically oriented prizes noticeably frequently: in addition to the 1972 prize, Peter Mitchell in 1978, Paul Berg,

Walter Gilbert and Frederick Sanger (his second chemistry prize) in 1980, Aaron Klug in 1982, Bruce Merrifield in 1984 and Johann Deisenhofer, Robert Huber and Hartmut Michel in 1988. I presented all of these at the award ceremony, except Mitchell and Merrifield, who were introduced by Ernster and Lindberg, respectively. I left the committee after 1988, but my successor as biochemist also managed to get a biochemical prize in his first year (1989), namely, that to Sidney Altman and Thomas Cech.

It should be noted that the committee has always been very careful in the formulation of the justification for the prizes. For example, Moore and Stein were not given the prize for their method of amino acid analysis, as many seem to think, but for contributions to our understanding of the structure-function relationship in ribonuclease. When Mitchell got the prize, Ernster and I were well aware of the fact that he was wrong, both concerning the formation of the electrochemical gradient and on its utilization for ATP formation, and our formulation reflects this. Since he was right only on the phenomenological, and not on the mechanistic, level, I suppose that strictly speaking he should have had a physiology prize, but we knew that it was unlikely that the medical committee would award him. Thus, we took a pragmatic view.

During my years on the committee, I managed to introduced a few other reforms. One was joint meetings with the medical committee once a year. I deemed this necessary, because many scientists, biochemists, in particular, were nominated in both committees. Thus, it was theoretically possible that both prizes would be given to the same person in a given year, and, in fact, this had almost happened at one time. Another initiative concerned the role of the chemists in the academy who are not on the committee. When I started, their vote on the committee's proposal was a mere formality. I thought it silly that a number of Sweden's leading chemists should travel to Stockholm to say nothing. It was impractical to have the chemists try to interfere with the current prize, but I introduced the practice that they should discuss and give views on future prizes.

The work on the committee was a pleasure also from a personal point of view. Certainly, we members often had strongly divergent opinions, but the discussions were always civil and cordial. I particularly enjoyed interacting with Lars Ernster, Sture Forsén and Ingvar Lindqvist, the latter two knowing a lot of biochemistry without being biochemists. Sture and I both came to the meetings from outside Stockholm and often went to concerts or the theater together. After the meetings we invariably went shopping for records or, later, CDs, both having a keen interest in music. Sture has spent time both in Rome and at Caltech, partly through my mediation.

The happy life of a professor emeritus

Retirement

My last day as active professor at Göteborg University was 30 June 1993. My successor, Jan Rydström, was already appointed, as the faculty had arranged, so that we would have some overlap. I was very happy that the whole department, faculty, graduate students and other staff, gave me a big farewell, even if they knew I would not leave my office and lab. They were apparently in a quandary what to give me as a present, but Peter Brzezinski's wife, Anna, had suggested wine-making equipment. I received this with pleasure, even if I attempt modesty in its use. I had bought myself a city bike (a hybrid between a mountain and an ordinary bike) on the inspiration of a visit to our house, about a month earlier, from several students on such vehicles. Therefore, I was also given a bicycle computer, which Jim Fee, who was visiting with his wife Shirley, installed in the evening, together with my son, while I was preparing dinner. I had promised Tore,

when I got the presents, not to try to compete with myself.
An even greater celebration was arranged by Tore and Britt, my secretary until a few days earlier, at the Hasselbacken Restaurant in Stockholm, on 7 July. This was during the 22nd FEBS Meeting, the organizing committee of which I chaired. No less than 45 of my friends attended this dinner. There were most of my recent and many of my old students, including those from the Uppsala days. Among my foreign friends were Prakash Datta, Alessandro Finazzi Agrò, with whom I worked on laccase in Rome in 1968, Jim Fee, Oleg Jardetzky, a fellow graduate student at Minnesota, Israel Pecht and Giorgio Semenza. Among the Swedes, I want, in particular, to mention Bertil Andersson, Lars Ernster, Tord Ganelius, former colleague in mathematics in Göteborg and Permanent Secretary of the Academy during my Nobel years, Ulf Lagerkvist, Arne Magnéli and Peter Reichard. Tore, Britt and Ann-Cathrine were, of course, present. I was also very pleased that Jan Rydström was at the dinner and gave one of the many speeches during the evening. I was presented with two magnificent volumes containing my collected works.

At first, I did not understand that retirement was the best thing that had happened to me in 30 years. Not that it is so nice to get old, but at last I had again the opportunity to devote all my time and energy to that which interests me most, namely, research. Jan was generous to provide me with an office and lab space in our new building, the Lundberg Laboratory, and I will end this essay by describing how I have availed myself of this privilege.

Mutants of bacterial cytochrome oxidases

In the early 1990s, Sheilagh Ferguson-Miller, Bob Gennis and their collaborators had developed systems for making site-directed mutants of bacterial cytochrome aa_3 and cytochrome bo (Hosler et al., 1993). Thomas Nilsson, in our de-

partment, and his student, Margareta Svensson (now Svens-
son-Ek), had collaborated with Bob on the quinol oxidase
since 1992, and in January 1993, Bob offered me all his mutants of cytochrome c oxidase for Peter and I to study by
flash photolysis. I thought this opportunity was much too
good to give up, but at the same time I realized that I could
not handle this project alone. Thus, I had to try to recruit a
new student, and so started, what I call, my spring of chas-
ing women, because all the potential students were female.

I thought my problem was solved, when a gifted Chalmers
graduate, Pernilla Wittung, whom Tore had recruited to the
department, came to me and said she was going to do her
thesis project with Peter. We persuaded her to work on our
joint investigations of oxidase mutants. When I went to
Tore, who was department chairman, to ask if we could take
her on, he said it was impossible, since we had no money to
support her (the Swedish Natural Science Research Council
has a policy not to give large grants to emeriti), should she
not get a fellowship. We would probably have lost her anyhow, because her first choice had been physical chemistry,
and when the opportunity presented itself at the end of
January, she moved to the Chalmers Department of Physical
Chemistry to work on spectroscopy of nucleic acids and their
analogues, with Professor Bengt Nordén. As it later turned,
however, we were actually going to carry out several projects
together.

At this time, I discovered that another very good student,
Pia Ädelroth, was doing an undergraduate project in photosynthesis with Lars-Erik. It is perhaps not very ethical to
try to steal students from colleagues, but in my situation I
could not help myself. I went to Pia and asked if she wanted
to work with Peter and me on bacterial cytochrome oxidase.
She said she would like to have time to think about it. While
Pia was thinking, I got a telephone call from a graduate of
the Royal Institute of Technology in Stockholm, Annika
Persson. She said that she wanted to do graduate work in

biochemistry and claimed that through a literature search she had arrived at our projects being the most interesting in Sweden. Peter and I invited her down for a seminar in the beginning of April and then offered her to work with us. She had a hard time, however, deciding if she wanted to leave Stockholm or not, so I continued working on Pia. One morning in early May, she told me she would accept, so my chase was over.

When I now went to Tore again, he said something like the following: 'I have learnt my lesson. I will not make the same mistake again'. It helped, of course, that I still had enough money to promise to support Pia for a year. She could not start until September, however, when she would finish her project and thus get her degree. I was afraid that Bob would get impatient if I did not produce results sooner, so I decided to start preparing oxidase myself, together with Ann-Cathrine, from the strain of *Rhodobacter sphaeroides* that Bob's group had provided. With the money I had left, I also engaged an undergraduate student, Ann Kjellström (now Sörensen), to help us. We three prepared the bacterial membranes by gradient centrifugation and purified the enzyme with FPLC. Peter was so impressed that he photographed me in action, even if my younger co-workers were shocked by my laboratory technique: pipetting with the mouth and not wearing rubber gloves. By the end of the summer, we had prepared enough of the wild-type oxidase to last for Pia's work during the entire autumn of 1993.

Pia and Peter did a thorough study of the internal electron transfer in the wild-type enzyme, showing that it essentially behaved as the bovine oxidase (Ädelroth et al., 1995). The slowest phase, which is limited by proton-transfer reactions, has different pK_a values of the groups involved, however. The distance between cytochrome a and cytochrome a_3-Cu_B, calculated from the electron-transfer rate, agrees well with that found in the crystal structure. In the summer of 1994, one of Bob's students, David Mitchell, brought several mutants, which he and Pia characterized kinetically and by EPR together with Peter and Roland. Among other things, we could demonstrate that a glutamic acid supposed to provide a bridging ligand in the cytochrome a_3-Cu_B unit, is not necessary to give the EPR signal responsible for the bridged species (Mitchell et al., 1995) and that a ligand-exchange mechanism of proton pumping involving a particular tyrosine is excluded (Mitchell et al., 1996).

The Nobel Committee has generously given us money to organize workshops with our international collaborators in 1994 and 1995. In the first one, apart from the Göteborg group, Sheilagh, Bob, Matti and Mårten Wikström participated. To the second one, Jerry Babcock, Sasha Konstantinov and Hartmut Michel were also invited, and Joel Morgan took Mårten's place. Our discussions were extremely useful, particularly, since Hartmut's structure allows us to interpret and plan mutagenesis experiments in a rational way, and we hope to continue next year. The grants were sufficient for us to eat and drink well at the end of each day's hard work. The farewell parties were given at my home, where Margareta, who is an excellent cellist, and I, performed Telemann, together with Betty, in 1994, and with Sasha, a skillful violinist, in 1995.

Cu_A and PNA

During the spring of 1993, I ran into Pernilla several times, when she visited our library or used our cold room, which were both next to my office. I tried, together with Peter and Stefan, to persuade her to return to our department, but without success. She asked, however, if I could not suggest a collaborative project. Pernilla had done her undergraduate project at Imperial College in London, overexpressing and purifying a lysyl-tRNA synthase (Onesti et al., 1994). On her return, she graduated in chemical engineering from Chalmers University of Technology with top grades, and Chalmers, in fact, awarded her a medal in 1994 for being the best chemistry student in her graduating class. To her question about a possible collaboration, I first answered that it was impossible, since she now worked on nucleic acids, whereas my interest was still cytochrome oxidase. When she persisted, I told her that if she was satisfied with a methodological common denominator, she had several techniques available in her new department which I would like to apply to the oxidase.

In the summer of 1993, Pernilla and I tried to detect redox-induced conformational changes in cytochrome oxidase by spectroscopic techniques. We first tried fluorescence without success but then went over to CD. Since Pernilla did this in addition to her regular thesis work, which concerns peptide nucleic acid (PNA; Wittung et al., 1994d), and RecA-DNA interactions (Wittung et al., 1994c), our work was mostly done on weekends and progress was slow. By January 1994, we had enough results, however, to prepare a poster for a Gordon Conference on Protons and Membrane Reactions in California, to which I had promised to take Pernilla as reward for her diligent efforts.

In connection with this trip, we visited Harry Gray at Caltech. Harry arranged a group meeting, at which Jack Richards was also present. He told the group that Jim Fee wanted him to express a soluble Cu_A domain from the aa_3-type oxidase of *Thermus thermophilus* but that he had nobody to put on this project. At this point Pernilla interjected: 'I can do that'. Her rationale was that nowadays a student should know genetic engineering techniques, and that she could regard a couple of months spent on this project as part of her graduate education. As a result she spent July–September 1994 at Caltech, working in Jack's lab. First, she tried expressing a fancy synthetic gene without success, but at the suggestion of another student, Claire Slutter, she finally used the natural gene. In her final week she worked day and night, but fortunately she had expression before it was over.

Despite the fact that I already had two of Matti's Cu_A domains available, it was a big advantage to have the *Thermus* domain. I came over to Caltech in September to do some electron transfer experiments with the *Paracoccus* domain and discovered that it lost the Cu ion in the reduced state. The *Thermus* domain, on the other hand, is very stable in a wide range of temperature and pH, as well as to changes in redox state. When Pernilla left, Claire took over the project, and we have now two papers in press describing a thorough characterization of this protein (Slutter et al., 1996a,b).

Before Pernilla went to Caltech, we used CD to demonstrate that the soluble Cu_A and CyoA domains have the cupredoxin fold in common with blue copper proteins, such as plastocyanin (Wittung et al., 1994a). During this spring, Pernilla also started to guide an undergraduate, Johan Kajanus, whose task it was to use liposomes as model systems for studying the cell penetration properties of PNA. Because of my experience with phospholipid vesicles, I was asked to assist in this project. We found that, just like DNA, PNA has a very low efflux rate from intermediate-sized liposomes (Wittung et al., 1995). In connection with this project, we also analyzed, with theoretical assistance from Mikael Kubista, the phenomenon of absorption flattening in the optical spectra of liposome-entrapped substances (Wittung et al., 1994b). In addition, Pernilla collaborated in a theoretical analysis of the optical spectra of Cu_A domains, together with Sven Larsson in physical chemistry and Bruno Källebring in our department (Larsson et al., 1995). Unlike blue proteins, these have two bands of almost equal intensity around 500 nm. These arise because of an interaction between the two Cu ions in the binuclear site, and the corresponding CD bands are expected to have opposite signs, as Pernilla and I demonstrated experimentally.

It gives me an extreme pleasure in old age to work with two gifted students, Pia and Pernilla. Pia continues her investigations of oxidase mutants. With Pernilla I am attempting to develop methods for delivery of PNA into cells. It is very gratifying to have her teach me a new field of research at this late stage in life.

Acknowledgements

Many of the people mentioned in this essay have checked that my memory is accurate, for which I want to thank them. My recent work mentioned has been supported by the Swedish Natural Science Research Council, the European Union, the Carl Tygger Foundation and the Nobel Institute for Chemistry.

References

Aasa, R., B.G. Malmström, P. Saltman and T. Vänngård, 1963, Biochim. Biophys. Acta 75, 203–222.

Aasa, R., R. Brändén, J. Deinum, B.G. Malmström, B. Reinhammar and T. Vänngård, 1976, FEBS Lett. 61, 115–119.

Ädelroth, P., P. Brzezinski and B.G. Malmström, 1995, Biochemistry 34, 2844–2849.

Aisen, P., R. Aasa, B.G. Malmström and T. Vänngård, 1967, J. Biol. Chem. 242, 2484–2490.

Andréasson, L.-E., B.G. Malmström, C. Strömberg and T. Vänngård, 1972, FEBS Lett. 28, 297–301.

Andréasson, L.-E., R. Brändén, B.G. Malmström and T. Vänngård, 1973, FEBS Lett. 32, 187–189.

Antonini, E., M. Brunori, C. Greenwood and B.G. Malmström, 1970, Nature 228, 936–937.

Blair, D.F., G.W. Campbell, J.R. Schoonover, S.I. Chan, H.B. Gray, B.G. Malmström, I. Pecht, B.I. Swanson, W.H. Woodruff, W.K. Cho, A.M. English, H.A. Fry, V. Lum and K.A. Norton, 1986, J. Am. Chem. Soc. 107, 5755–5766.

Bray, R.C., B.G. Malmström and T. Vänngård, 1959, Biochem. J. 73, 193–197.

Broman, L., B.G. Malmström, R. Aasa and T. Vänngård, 1962, J. Mol. Biol. 5, 301–310.

Broman, L., B.G. Malmström, R. Aasa and T. Vänngård, 1963, Biochim. Biophys. Acta 75, 365–376.

Brzezinski, P. and B.G. Malmström, 1986, Proc. Natl. Acad. Sci. USA 83, 4282–4286.

Brzezinski, P. and B.G. Malmström, 1987, Biochim. Biophys. Acta 894, 29–38.

Brzezinski, P., M. Sundahl, P. Ädelroth, M.T. Wilson, B. El-Agez, P. Wittung and B.G. Malmström, 1995, Biophys. Chem. 54, 191–197.

Carrel, A., 1935, Man, the Unknown (Harper, New York).

Clore, G.M., L.-E. Andréasson, B. Karlsson, R. Aasa and B.G. Malmström, 1980, Biochem. J. 185, 155–167.

Di Bilio, A.J., T.K. Chang, B.G. Malmström, H.B. Gray, B.G. Karlsson, M. Nordling, T. Pascher and L. Lundberg, 1992, Inorg. Chim. Acta 198–200, 145–148.

Dooley, D.M., J. Rawlings, J.H. Dawson, P.J. Stephens, L.-E. Andréasson, B.G. Malmström and H.B. Gray, 1979, J. Am. Chem. Soc. 101, 5038–5046.

Ehrenberg, A., B.G. Malmström, L. Broman and R. Mosbach, 1962, J. Mol. Biol. 5, 450–452.

Engström, A. and B.G. Malmström, 1952, Acta Physiol. Scand. 27, 91–96.

Fabian, M. and B.G. Malmström, 1989, Biochim. Biophys. Acta 973, 414–419.

Falk, K.-E., H.C. Freeman, T. Jansson, B.G. Malmström and T. Vänngård, 1967, J. Am. Chem. Soc. 89, 6071–6077.

Fee, J.A., R. Malkin, B.G. Malmström and T. Vänngård, 1969, J. Biol. Chem. 244, 4200–4207.

Glick, D., A. Engström and B.G. Malmström, 1951, Science 114, 253–258.

Gray, H.B. and B.G. Malmström, 1983, Comments Inorg. Chem. 2, 203–209.

Gray, H.B. and B.G. Malmström, 1989, Biochemistry 28, 7499–7505.

Gurd, F.R.N. and P.E. Wilcox, 1956, Adv. Protein Chem. 11, 311–427.

Gurd, F.R.N., K.-E. Falk, B.G. Malmström and T. Vänngård, 1967, J. Biol. Chem. 242, 5724–5730.

Hallén, S., P. Brzezinski and B.G. Malmström, 1994, Biochemistry 33, 1467–1472.

Hansson, Ö., B. Karlsson, R. Aasa, T. Vänngård and B.G. Malmström, 1982, EMBO J. 1, 1295–1297.

Henderson, L.E., D. Henriksson and P.O. Nyman, 1973, Biochem. Biophys. Res. Commun. 52, 1388–1394.

Himmelwright, R.S., N.C. Eickman and E.I. Solomon, 1978, Biochem. Biophys. Res. Commun. 81, 243–247.

Hosler, J.P., S. Ferguson-Miller, M.W. Calhoun, J.W. Thomas, J. Hill, L. Lemieux, J. Ma, C. Georgiou, J. Fetter, J. Shapleigh, M.M.J. Tecklenburg, G.T. Babcock and R.B. Gennis, 1993, J. Bioenerg. Biomembr. 25, 121–136.

Iwata, S., C. Ostermeier, B. Ludwig and H. Michel, 1995, Nature 376, 660–669.

Karlsson, B., B. Lanne, B.G. Malmström, G. Berg and R. Ekholm, 1977, FEBS Lett. 84, 291–295.

Kroneck, P.M.H., W.A. Antholine, J. Riester and W.G. Zumft, 1988, FEBS Lett. 242, 70–74.

Lang, G., S.J. Lippard and S. Rosén, 1974, Biochim. Biophys. Acta 336, 6–14.

Lappalainen, P., R. Aasa, B.G. Malmström and M. Saraste, 1993, J. Biol. Chem. 268, 26416–26421.

Larsson-Raznikiewicz, M. and B.G. Malmström, 1961, Arch. Biochem. Biophys. 92, 94–99.

Larsson, S., B. Källebring, P. Wittung and B.G. Malmström, 1995, Proc. Natl. Acad. Sci. USA 92, 7167–7171.

Lehninger, A.L., 1950, Physiol. Rev. 30, 393–429.

Lewis, G.N. and M. Randall, 1923, Thermodynamics (McGraw-Hill, New York).

Li, P.M., B.G. Malmström and S.I Chan, 1989, FEBS Lett. 248, 210–211.

Liljas, A., K.K. Kannan, P.-C- Bergstén, I. Waara, K. Fridborg, B. Strandberg, U. Carlbom, L. Järup, S. Lövgren and M. Petef, 1972, Nature New Biol. 235, 131–137.

Lindskog, S. and B.G. Malmström, 1960, Biochem. Biophys. Res. Commun. 2, 213–217.

Lindskog, S. and B.G. Malmström, 1961, J. Biol. Chem. 237, 1129–1137.

Lippard, S.J. and J.M. Berg, 1994, Principles of Bioinorganic Chemistry (University Science Books, Mill Valley).

Lumry, R. and H. Eyring, 1954, J. Phys. Chem. 58, 110–120.

Maison-Peteri, B. and B.G. Malmström, 1989, Biochemistry 28, 3156–3160.

Malkin, R. and B.G. Malmström, 1970, Adv. Enzymol. 33, 177–243.

Malmström, B.G., 1953a, Nature 171, 392–393.

Malmström, B.G., 1953b, Arch. Biochem. Biophys. 46, 345–363.

Malmström, B.G., 1961, Fed. Proc. 20, 60–69.

Malmström, B.G., 1965, in T.E. King, H.S. Mason and M. Morrison (Eds.), Oxidases and Related Redox Systems (Wiley, New York), pp. 207–216.

Malmström, B.G., 1973, Q. Rev. Biophys. 6, 398–431.

Malmström, B.G., 1983, Curr. Contents 26, 18.

Malmström, B.G., 1994, Eur. J. Biochem. 223, 711–718.

Malmström, B.G. and D. Glick, 1951, Anal. Chem. 23, 1699–1703.

Malmström, B.G. and D. Glick, 1952, Exp. Cell Res. 3, 121–125.

Malmström, B.G. and D. Glick, 1954, J. Biol. Chem. 211, 677–686.

Malmström, B.G. and J.B. Neilands, Annu. Rev. Biochem. 33, 331–354.

Malmström, B.G. and T. Vänngård, 1960, J. Biol. Chem. 2, 118–124.

Malmström, B.G., R. Aasa and T. Vänngård, 1965, Biochim. Biophys. Acta 110, 431–434.

Malmström, B.G., A. Finazzi Agrò and E. Antonini, 1969, Eur. J. Biochem. 9, 383–391.

Malmström, B.G., J.R. Kimmel and E.L. Smith, 1959, J. Biol. Chem. 238, 1108–1111.

Malmström, B.G., Ö. Levin and H.G. Boman, 1956, Acta Chem. Scand. 10, 1077–1082.

Malmström, B.G., R. Mosbach and T. Vänngård, 1959, Nature 183, 321–322.

Malmström, B.G., B. Reinhammar and T. Vänngård, 1968, Biochim. Biophys. Acta 156, 67–76.

Malmström, B.G., A. Rosenberg and S. Lindskog, 1963, in P.A.E. Desnuelle (Ed.), Molecular Basis of Enzyme Action and Inhibition (Pergamon Press, Oxford), pp. 172–181.

Malmström, B.G., T. Vänngård and M. Larsson, 1958, Biochim. Biophys. Acta 30, 1–5.

Martin, C.T., R.H. Morse, R.M. Kanne, H.B. Gray, B.G. Malmström and S.I. Chan, 1981, Biochemistry 20, 5147–5155.

Meselson, M. and F.W. Stahl, 1958, Proc. Natl. Acad. Sci. USA 44, 671–682.

Mitchell, D.M., R. Aasa, P. Ädelroth, P. Brzezinski, R.B. Gennis and B.G. Malmström, 1995, FEBS Lett. 374, 371–374.

Mitchell, D.M., P. Ädelroth, J.P. Hosler, J.R. Fetter, P. Brzezinski, M.A. Pressler, R. Aasa, B.G. Malmström, J.O. Alben, G.T. Babcock, R.B. Gennis and S. Ferguson-Miller, 1996, Biochemistry 35, 824–828.

Moore, S. and W.H. Stein, 1949, J. Biol. Chem. 178, 53–77.

Öblad, M., E. Selin, B. Malmström, L. Strid, R. Aasa and B.G. Malmström, 1989, Biochim. Biophys. Acta 975, 267–270.

Oliveberg, M. and B.G. Malmström, 1991, Biochemistry 30, 7053–7057.

Onesti, S., M.-E. Theoclitou, E.P.L. Wittung, A.D. Miller, P. Plateau, S. Blanquet and P. Brick, 1994, J. Mol. Biol. 243, 123–125.

Rees, D.C., M. Lewis and W.N. Lipscomb, 1983, J. Mol. Biol. 168, 367–387.

Reichard, P., 1995, Annu. Rev. Biochem. 64, 1–28.

Reinhammar, B., R. Malkin, P. Jensen, B. Karlsson, L.-E. Andréasson, R. Aasa, T. Vänngård and B.G. Malmström, 1980, J. Biol. Chem. 255, 5000–5003.

Rosén, S., R. Brändén, T. Vänngård and B.G. Malmström, 1977, FEBS Lett. 74, 25–30.

Rosenberg, A. and B.G. Malmström, 1955, Acta Chem. Scand. 9, 1546–1548.

Scholes, C.P. and B.G. Malmström, 1986, FEBS Lett. 198, 125–129.

Slutter, C.E., D. Sanders, P. Wittung, B.G. Malmström, R. Aasa, J.H. Richards, H.B. Gray and J.A. Fee, 1996a, Biochemistry 35, 3387–3395.

Slutter, C.E., R. Langen, D. Sanders, S.M. Lawrence, P. Wittung, A.J. Di Bilio, M.A. Hill, J.A. Fee, J.H. Richards, J.R. Winkler and B.G. Malmström, 1996b, Inorg. Chim. Acta 243, 141–145.

Smith, E.L., 1951, Adv. Enzymol. 12, 191–206.

Svedberg, T. and H. Rinde, 1924, J. Am. Chem. Soc. 46, 2677–2693.

Thörnström, P-E., B. Soussi, L. Arvidsson and B.G. Malmström, 1984, Chem. Scr. 24, 230–235.

Vallee, B.L., T.L. Coombs and F.L. Hoch, 1960, J. Biol. Chem. 235, PC45–47.

Warburg, O. and W. Christian, 1942, Biochem. Z. 310, 384–421.

Westhead, E.W. and B.G. Malmström, 1957a, J. Biol. Chem. 228, 655–671.

Westhead, E.W. and B.G. Malmström, 1957b, Biochim. Biophys. Acta 26, 202–203.

Whitney, P.L., G. Fölsch, P.O. Nyman and B.G. Malmström, 1967, J. Biol. Chem. 242, 4206–4211.

Wikström, M., 1977, Nature 266, 271–273.

Wilmanns, M., P. Lappalainen, M. Kelly, E. Sauer-Eriksson and M. Saraste, 1995, Proc. Natl. Acad. Sci. USA 92, 11955–11959.

Wilson, M.T., P. Jensen, R. Aasa, B.G. Malmström and T. Vänngård, 1982, Biochem. J. 203, 483–492.

Winkler, J.R., B.G. Malmström and H.B. Gray, 1995, Biophys. Chem. 54, 199–209.

Wittung, P., B. Källebring and B.G. Malmström, 1994a, FEBS Lett. 349, 286–288.

Wittung, P., J. Kajanus, M. Kubista and B.G. Malmström, 1994b, FEBS Lett. 352, 37–40.

Wittung, P., B. Nordén, S.K. Kim and M. Takahashi, 1994c, J. Biol. Chem. 269, 5799–5803.

Wittung, P., P.E. Nielsen, O. Buchardt, M. Egholm and B. Nordén, 1994d, Nature 368, 561–563.

Wittung, P., J. Kajanus, K. Edwards, P. Nielsen, B. Nordén and B.G. Malmström, 1995, FEBS Lett. 365, 27–29.

G. Semenza and R. Jaenicke (Eds.)
Selected Topics in the History of Biochemistry: Personal Recollections, V
(Comprehensive Biochemistry Vol. 40) © 1997 Elsevier Science B.V.

Chapter 7

Harland Goff Wood: An American Biochemist

RIVERS SINGLETON, JR.

Department of Biology, University of Delaware, Newark,
DE 19716–2590, USA

I wonder if the new discoveries will ever carry the same impact, thrill, and amazement as those during the past 51 years.

H. G. Wood, 1985

Introduction

There is a popular legend in American culture, which may be called the 'great American success story'. In this story the hero is born in the American mid-west and struggles against the intense economic and social forces of the Great Depression of 1929–1941. Despite the exigencies imposed by this struggle, it is one that ultimately leads the hero to great personal success and widespread public approbation.

In many respects, the life and scientific career of Harland Goff Wood typifies this legendary, heroic picture. His parents were Midwestern, semirural, and middle-class – a cultural system often ridiculed by some American writers, such as Sinclair Lewis. Yet even Lewis recognized that there was potentially great virtue in this culture, for it frequently placed high value on hard work, respect for the views of others, education, and the pursuit of knowledge. The Wood family instilled these best virtues of the Midwestern cultural

Harland Wood at play in his second home at Case Western Reserve University. The note on the chart paper reads: 'My thanks to the members of the Department of Biochemistry, the Medical School and the University for making my scientific life challenging and exciting. Harland G. Wood.' Photo courtesy of Professor Richard Hanson, Department of Biochemistry, Case Western Reserve University. Original published in The Cleveland Plain Dealer and used with their permission.

perspective in all of their children. All six children (five sons and a daughter) of William C. and Inez Goff Wood completed college degrees during various phases of the American Depression. Many of them continued to finish graduate or professional degrees, and they all enjoyed successful and productive careers in their chosen professions.

Like the legendary American hero, Harland Wood (who lived from 1907 to 1991) received much of his education and began his early scientific career during the Great Depression. In fulfillment of the legendary tradition, Wood achieved numerous scientific and professional successes; successes that were recognized and applauded by both his scientific and professional peers and by society in general. These successes were achieved by his insightful, and often tenacious pursuit of biochemical inquiry. While his research work helped expand and clarify many aspects of biochemical theory, Wood himself was primarily an experimentalist. He found theoretical speculations interesting only if they clearly pointed to experimental testing. Wood was a quintessential laboratory scientist, with an intense and lifelong commitment to laboratory inquiry. Despite the widespread approbation of his work in both the scientific and broader cultural communities, as well as the various administrative duties he assumed, the lab bench remained a major passion in Wood's life.

My two goals in this chapter are to explore Wood's passionate commitment to the laboratory and his contributions to our understanding of the biochemical world. To achieve these goals, I must avoid two major difficulties. First, to describe Wood's scientific work, I cannot escape the legendary, almost mythic, nature of his scientific style; this description may imply a depreciation of the style and contributions of other, similarly legendary, scientists. This problem is inherent in any description of the disciplinary explosion of biochemistry during the middle part of this century, for it was 'a time in which giants strode'. We need only consider the names of some of Wood's contemporaries or near contemporaries – such as H. A. Barker, Britton Chance, Carl Cori, Gerty Cori, Authur Kornberg, Ephraim Racker, or Charles Yanofsky, to name a few – to realize the potential magnitude of this problem. Thus, while 'I come to praise' the work of Harland Wood, I intend no slighting of the many other giants of the biochemical world they shared.

A second difficulty in describing Harland Wood's contributions to biochemistry is the shear magnitude and range of his work. He was personally active in laboratory work for more than half of the 20th century. His career began in Chester Werkman's microbiology laboratory at Iowa State doing fermentation analyses of the propionic acid bacteria. Using rudimentary methods of chemical analysis, Wood's graduate studies led to a revolutionary new metabolic concept: i.e., heterotrophic CO_2 fixation. This concept postulated that CO_2 was fixed by *all* living organisms and was not simply a metabolic function restricted to plants and a few specialized groups of bacteria. For the next 50 years, Wood elucidated the metabolic steps of the fermentation, isolated many of the enzymes responsible for the reactions, and physically characterized many of those proteins. At the end of his career, he was analyzing the structure of and sequencing the genes coding for a central enzyme in the pathway. During the same time, he also described an entirely new pathway of CO_2 fixation in a different group of organisms and was studying the role of pyrophosphate (PP_i) in bioenergetics. Throughout all of this work, 'Wood's Hole' (as it was often referred to by many inhabitants) was always working on the frontiers of biochemical innovation.

Thus, whereas Wood's career almost perfectly reflects the way biochemistry developed in this country, it is difficult to describe that career adequately within the limited confines of an essay of this nature. In encapsulated form, some of his major contributions to modern biochemistry include:
– the notion of heterotrophic CO_2 fixation;

- application of isotopes (especially radioisotopes) to our understanding of metabolic processes;
- elucidation of the propionic acid cycle in the propionic acid bacteria;
- structural and mechanistic analysis of transcarboxylase;
- characterization of the role of pyrophosphate (PP_i) in metabolic processes;
- and elucidation of the acetate biosynthesis pathway in *Clostridium thermoaceticum*.

Relating the various ways these complex studies were brought to fruition is beyond the limit of this chapter. Consequently, I restrict my scope to defining the human being responsible for this broad scientific achievement and will focus on his description of the propionic acid cycle in the bacterial genus *Propionibacterium* to illustrate Wood's practice of science.

Several important dimensions of Wood's personal character were vital for his scientific contributions; these same characteristics describe many truly successful and creative scientists. Wood was, first of all, a model experimental scientist. Virtually all of his work was experimentally driven, and his persona brought experimental creativity and excitement to science. A related dimension was Wood's constant innovation and modification of his experimental approach to new questions. As I noted previously, his career began using organic chemical methods to do fermentation analysis and ended doing gene sequences and X-ray diffraction analyses of protein structure.

Wood, like most truly productive scientists, was rarely fully satisfied with answers to research problems. No matter how elegant or adequate a research question might be resolved, the answer frequently raised new questions; questions that on occasion spawned an entire new avenue of research inquiry, which was only peripherally connected to its predecessor. For example, as understanding of the propionic acid cycle became more complete, the Wood lab discovered the role of PEP carboxytransphosphorylase in the cycle (see below). In this reaction, the phosphoryl group of PEP is transferred to P_i to form PP_i, which has a ΔG approximating that of ATP. This led Wood into consideration of the role of PP_i in bioenergetics and a search for other enzymes that utilized PP_i in a bioenergetic fashion.

Like many of his successful contemporaries, Wood was a model collaborator and was constantly driven to develop unique collaborations with other investigators to answer important questions of interest to both groups. Wood did not see proprietary value in knowledge and willingly shared both information and methods. Despite this collaborative nature of his personality, Wood was very competitive and wanted to be the first to understand a problem. It was a competition always tempered by a sense of fairness and good humor. Furthermore, despite these collaborative and competitive natures, Wood often pursued questions with dogged determination for protracted periods of time; he spent 40 years trying to understand the physiology of the propionic acid bacteria [1] and three decades solving the problem of acetate biosynthesis in *Clostridium thermoaceticum* [2, 3].

Focus on Wood's description of the propionic acid cycle illustrates these personal character traits. This project, from beginning to end, was experimentally driven. One set of experiments answered specific questions but simultaneously opened new avenues of experimental inquiry. Often, these new avenues demanded novel experimental tools for their resolution. Although he was able to develop these new experimental approaches, the work also demanded that Wood establish a variety of collaborative efforts in order to complete the problem. Furthermore, complete elucidation of the complex reaction sequence involved in the propionic acid cycle demanded four decades of determined research inquiry.

Early years and education

Harland Wood, the third son of six children, was born on September 2, 1907 in Delavan, MN. Within a year, his family moved to the town of Mankato, MN, where they spent the remainder of their lives. His father, William C. Wood, operated a prosperous real estate business as well as a small family farm.

Inez Goff Wood, who taught school before and for a while after marriage, set about 'operating the family', which was no small task considering its size and the economic times. It was a task at which she was obviously successful; she received official recognition by the State of Minnesota as the '1950 Minnesota Mother of the Year'. At the award ceremony, the Governor commented that she had been fortunate 'to send all of your children through college'. Mrs. Wood immediately interrupted the Governor and said: 'Let's get this straight Mr. Governor, we are people of modest means. Our children had to *work* to get through college and earn their degrees' [4].

Her comment is interesting for several reasons. First, she clearly states the economic status of the Wood family. Second, she documents the means whereby Harland, and the other Wood children, obtained their education. But the comment is perhaps most important for its directness and refusal to allow a perceived untruth go unchallenged. It was a style of interaction characteristic of many of the Wood children. Anyone who dealt with Harland Wood throughout his professional career was quite familiar with that style of interaction; he was always direct, to the point, and refused to allow falsity to go unchallenged.

Like all of the Wood children, Harland received his early education in the public schools in Mankato. He was 'frail as a child' and spent 2 years in kindergarten and 2 years in first grade [5, 6]. According to his brother's journal, he de-

veloped an interest in science while a student at Mankato
High School [4], however Harland's early career goal was
medicine. It was a goal diverted, in part, by an encounter
with Latin. In an oral biography [6], he commented that
Latin was a requirement for medical school admission. He
took *Cicero* in the 9th grade and was doing poorly in the
work. He talked with the Latin teacher about his difficulty,
and the teacher responded: 'Well, Wood, you ought to take
chemistry. Chemistry is very important in industry. It's
very important training. Why don't you drop your Latin and
take chemistry?' [6]. A good high school chemistry teacher
influenced him to take other science courses, including
physics.

Wood completed his undergraduate studies at Macalester
College, a small Presbyterian liberal arts college in St. Paul,
MN, where he majored in chemistry and mathematics. His
Macalester years perhaps best illustrate his mother's indig-
nation when Governor Youngdahl suggested that she sent
'all of her children through college'. Harland worked both in
the college dormitory kitchen and as a laboratory assis-
tant/paper grader in the Chemistry Department for financial
support. Extra financial support came from work as a
clothing company representative on campus.

Wood's undergraduate study at Macalester was a bit extraordinary for his time; he and Mildred (Millie) Davis were
married in September of 1929 during their junior year. To
place this event into economic perspective, consider that
during the next few months, stocks listed on the New York
Stock Exchange would lose 50% of their value. As Robert
McElvaine noted, 'The Great Depression was under way' [7].
Surely, this was not the most propitious time to begin a
marriage that was to last for over 60 years! To further com-
plicate the relationship, they needed the college President's
permission to remain in school [5].

He was also active in college athletics. In these days before rigorous NCAA regulation, athletics – especially football

– also provided additional income. Wood stated that if he continued to participate in athletics after his marriage, he 'would be paid by the hour if [he] taped up the players before the games. In addition, an alumnus offered $50.00 toward tuition which, at that time, was $87.50 per semester' [5]. As in much of his life, Wood was diversified and competitive in athletics and earned 11 letters in swimming, track, and football during his four undergraduate years.

    With the onset of the economic depression, any remaining considerations of medical school became unrealistic, and Wood's attention turned to a career in chemistry. He decided that a graduate degree was necessary if he was to be successful and began applying for fellowships in chemistry. A Macalester biology professor suggested that he apply to the bacteriology department at Iowa State College (now Iowa State University) in Ames, IA, which was chaired by the prominent and influential American bacteriologist Robert E. Buchanan. His application was reviewed by Chester H. Werkman,[1] who offered Wood a fellowship to work in his laboratory [5]. Wood accepted the offer, and in the Fall of 1931 began graduate work with Werkman. Although Wood was formally trained in bacteriology, his interest in biochemistry began at this time.

Affiliation with C. H. Werkman at Iowa State

Chester Werkman completed his Ph.D. in immunology with Buchanan at Iowa State in 1923 and continued in Buchanan's laboratory for a post-doctoral year. After a year as a

---

1 Apparently, because of his affiliation with the Agricultural Research Station, Werkman was able to offer financially attractive fellowships. He was thus able to attract highly qualified graduate students, with good skills in chemistry, who might have gone elsewhere. For example, Merton Utter began graduate work in Henry Gillman's organic chemistry lab at Iowa State. Werkman offered him a fellowship with a larger stipend and within a year Utter joined Werkman's research group (Mrs. M. F. Utter, personal communication).

faculty member at the University of Massachusetts, he returned to Iowa State in 1925 as an assistant professor, where he joined Buchanan and Max Levine in developing a powerful bacteriology department that was deeply influenced by chemical perspectives. Werkman spent the remainder of his professional career at Iowa State, where he became Department Head in 1945 and was elected to the National Academy of Sciences in 1946. He retired as Department Head in 1957 because of ill health and died in 1962.

Werkman's research program at Iowa State initially focused on a variety of food bacteriology questions, especially problems of food spoilage. To facilitate this research, his laboratory developed an assortment of methods to isolate and quantify bacterial fermentation products. When Wood joined his laboratory in 1931, Werkman had already published several papers that used quantitative analytical methods to understand bacterial fermentations. He had also published an initial study of the classification of the propionic acid bacteria, which were an important group of food microorganisms. An underlining theme of Werkman's program was a desire to understand these microbial processes in chemical terms, and his research began an evolution into fundamental bacterial physiology. During this time, he was profoundly influenced by the Dutch microbiologist and biochemist, A. J. Kluyver.

Apparently at Buchanan's invitation, Kluyver spent the months of May, June, and July of 1932 in Ames as a 'visiting professor of bacteriology and chemistry'. During this time he presented a series of 25 lectures on the 'biochemistry and physiology of microorganisms'. The series opened with a general lecture that focused on a major Kluyver theme: the 'unity and diversity of microorganisms'. The rest of the lectures covered all aspects of microbial metabolism from various fermentations to respiratory processes and included a lecture on 'Specificity in Biocatalysis'.[2] Perhaps the two lectures that were of greatest significance for Harland Wood, as a graduate student, were lectures 13 – Propionic Acid Fermentation' and 24 – Redox Potential and Metabolism'.

The Kluyver lectures apparently played a significant role in the intellectual life of the Iowa State students and faculty. News stories and interviews with Kluyver were published in the local newspapers. Mid-way through the lecture series, the college awarded Kluyver an honorary D.Sc. at the Spring commencement; this was the first foreign recognition of Kluyver's scientific contributions. In his annual report to the graduate faculty, Buchanan (who also served as Dean of the Graduate College) devoted an entire page to the impact of Kluyver's lectures. He noted that the lectures 'made possible and available to our *graduate students* an important European scholar and made real to them the existence and importance of European scholarship'.[3]

Kluyver's lectures were listed as formal graduate courses in the Iowa State Catalogue,[4] and Wood's academic transcript shows that he completed both the Spring and Summer courses, as well as the 'Seminar in Physiology and Biochemistry of Bacteria' Kluyver organized and conducted both terms. Although Wood rarely mentioned Kluyver during his later career, given the general tone and content of these lectures, it is hard to imagine that Kluyver did not play a significant role in shaping Wood's views about biochemistry.

Another important influence on Wood's scientific career was Iowa State's chemistry program, which was very strong. Wood completed a minor in chemistry, and his academic

---

2 Unpublished papers of A. J. Kluyver, Technical University of Delft.
3 'Annual Statement of Dean of Graduate College to the Graduate Faculty, Iowa State College, Sept. 24, 1932.' R. E. Buchanan Papers, Archives, Iowa State University, Ames, IA.
4 Iowa State College of Agriculture and Mechanical Arts Official Publication; 'Catalogue Number Announcements for 1932–1933' 30(39): 130 (Feb. 24, 1932).

transcript shows completion of courses in qualitative and quantitative organic chemical analysis, physiological and nutritional chemistry, and a variety of biochemistry courses, including courses in electro- and colloid chemistry. Werkman (and his students) also had access to the expert organic chemical advice of Henry Gilman, one of the most productive American chemists. (As an indication of the Gilman laboratory productivity, during the decade of 1920–1929 he published almost 100 papers; the following decade he published over 183 papers [8].)

Graduate research

Wood stated that when he arrived in Werkman's laboratory he was handed a copy of C.B. van Niel's thesis on the propionic acid bacteria [9] and essentially told to discover a thesis problem in it. Except for his broad interest in food related microorganisms, there seems to be little in the historical record to suggest why Werkman was interested in van Niel's work or how he obtained a copy of the thesis. Wood said 'Werkman offered no explanation of the purpose of my studies or what he expected me to find out that was new' [5]. In van Niel's thesis, Wood noticed that members of the genus *Propionibacterium* consistently produced succinate during glucose fermentation. This observation seemed to contradict van Niel's proposal that succinate arose from aspartate present in the yeast extract added as a medium growth supplement (see p. 353). Thus, as his thesis problem, Wood set about completing a set of fermentation balances,[5] which enabled him to speculate about succinate formation in glucose metabolism by the propionic acid bacteria.

He began a complex series of fermentation balance experiments with various strains of propionic acid bacteria. The experiments were both terminal, in that all of the glu-

---

5 See Appendix A for a description of the fermentation balance technique.

cose was consumed during the fermentation before analysis, as well as kinetic, i.e., aliquots were removed for analysis at various times during the fermentation. Some fermentations were conducted in the presence of inhibitors, such as bisulfite, that allowed the isolation and characterization of potential intermediates in the fermentation process. For example, using bisulfite as an inhibitor, Wood confirmed van Niel's speculation regarding the role of pyruvate in glucose fermentation by *Propionibacterium arabinosum* [10].

Within a few years, Wood essentially clarified the core of his thesis problem, i.e., glucose fermentation by propionic acid bacteria (see below). He then began, 'without any specific reason', to examine glycerol utilization [5]. The phosphate buffer used in the glucose fermentations was replaced with calcium carbonate. Any CO_2 evolved during the fermentation was precipitated and weighed, and at the reaction conclusion, residual carbonate was determined. Some of the data, summarized in Table 1 [11], contradicted two cardinal requirements of a fermentation balance (see Appendix A): the products failed to account for all of the carbon present in

TABLE 1

Fermentation products of *Propionobacter* cultures growing on glycerol
(data from Wood and Werkman [11])

Culture number	Products per 100 mmol of fermented glycerol (mmol)			CO_2 per 100 mmol of fermented glycerol (mmol)[a]
	Propionate	Succinate	Acetate	
11W	89.3	3.9	2.6	−1.1
15W	78.4	7.8	5.8	−12.3
34W	59.3	34.5	2.0	−43.2
49W	55.8	42.1	2.9	−37.7
52W	78.4	8.7	5.9	−20.0

[a]The minus sign indicates that CO_2 was consumed during the fermentation process.

TABLE 2

Carbon and oxidation-reduction balances of data from Table 1 calculated on the basis of non-assimilation or assimilation of CO_2 (data from Wood and Werkman [11])

Culture number	Carbon balance (%)		"O/R Index"	
	Glycerol only	Glycerol plus CO_2	Glycerol only	Glycerol plus CO_2
11W	96.8	96.5	1.162	1.135
15W	92.6	89.1	1.376	1.047
34W	106.0	93.1	2.270	0.925
49W	114.0	101.2	2.550	1.081
52W	101.0	94.6	1.386	0.918

the substrate and the oxidative processes were not balanced by concomitant and equal reductive processes.

As he was writing his thesis, Wood wanted to incorporate the glycerol experiment data [12], however the anomalous fermentation balances in those experiments were difficult to explain. He commented that he was suddenly struck by the idea that if the fermentation process reduced CO_2, then the resulting product might permit succinate production. This hypothesis would simultaneously explain both the carbon balance and oxidation-reduction ratio (see Table 2 [11]). He said:

> I hurriedly calculated the oxidation reduction balances on the basis that the missing CO_2 had been reduced. The balances were beautiful when the reduced CO_2 was included.... I rushed to the Bacteriology Department to give Professor Werkman the news and to tell him the thesis would have to be rewritten completely [12].

Werkman apparently listened patiently and, according to Wood, replied: 'The thesis is all typed except the bibliography, we don't want to write it again. Let's let the thesis go as it is and we'll take care of this question of CO_2 fixation later' [12].

Wood never fully understood Werkman's reasoning. Much later he noted that after doing the calculations, he was so convinced of the heterotrophic CO_2 fixation hypothesis he '... knew then and there that CO_2 was used by these bacteria'. For Wood, this conviction meant that his Ph.D. thesis – at the time almost completely finished – needed to be rewritten. The CO_2 fixation hypothesis convinced him that succinate originated via fixation of CO_2 with a three carbon compound, rather than via reaction of two acetates, with two carbons each (see Fig. 1). 'Clearly, if CO_2 was utilized, *what I had written about the mechanism was wrong*' [5] (emphasis added). Wood obviously had more confidence in the concept than his mentor, and Werkman argued for pragmatism over innovative – but speculative – science in the thesis.

Recounting this oft told story of the initial discovery of what was subsequently called the Wood–Werkman Reaction [13], illustrates an important dimension of Wood's scientific character. It does not require close examination of the data in Table 1 to see that they vary greatly from one strain of *Propionibacterium* to another. For example, succinate and propionate production were roughly equivalent in strain 49W. In the other strains, however, the ratio of propionate to succinate varied from almost two-fold to over twenty-fold. Furthermore, CO_2 utilization seems significantly high (relative to propionate production) only in strains 34W and 49W.

His reaction to these seemingly anomalous data is noteworthy for it reflects a research style that characterized Wood's entire career. It was a style noted for tenacious pursuit of 'hunches' about the meaning of data. A less determined investigator might easily conclude from the significant variation in carbon recovery for the various cultures that the differences arose from experimental error. Wood, however, apparently was not focused on carbon recovery but rather was concerned with the fact that he could not account for all of the CO_2 originally added as $CaCO_3$. This observation, coupled with the significant production of succinate (which distorted the oxidation/reduction ratio), convinced Wood that there was something unusual about the way *some* of these organisms metabolized glycerol.

Wood's graduate work is also noteworthy for another reason; it was a period of intense productivity. During his time as a graduate student (1931–1935), he produced sufficient data for 13 major publications. This rate of publication is even more remarkable in light of the fact that for several years Wood's appointment was a 9-month contract. He was married and had one child; financial considerations were important.[6] Consequently, Wood spent some summer time during his graduate career working in the family business in Mankato, MN and was not able to work in the laboratory.

Postdoctoral work

Wood spent 1 year, of a 2-year NRC Post-Doctoral Fellowship, at the University of Wisconsin with W.H. Peterson, where he and E.L. Tatum studied propionic acid bacterial growth requirements. He had intended to spend the Fellowship's second year in Germany working with Meyerhoff. For a variety of reasons [14], however, he returned to Ames where he resumed work with Werkman on the problem of CO_2 fixation. Within a relatively short period of time, using methods no more sophisticated than fermentation analysis coupled with various inhibitors, he was able to fairly conclusively demonstrate that the process involved:

$$CO_2 + \text{pyruvate} \Rightarrow \text{succinate} \tag{1}$$

---

6 For example, Wood's widow notes (personal communication) that of the $50 monthly income the family received, they spent almost $34 on rent. To make financial ends meet, Wood tended the furnace of the apartment building in which they lived.

For Wood, as well as for other members of the scientific community, questions and doubts still remained about the reaction. Thus, Wood began to explore other experimental approaches to the problem. At the 1939 International Congress of Microbiology in New York he heard about using ^{11}C, which was produced in the Berkeley cyclotron, as a radioactive tracer of metabolic processes. It was immediately obvious that [11]CO_2 as a tracer in the glycerol fermentation would provide powerful evidence to support the heterotrophic CO_2 fixation hypothesis. The experiment had the further potential to unequivocally identify the product of carbon fixation. The short half-life of ^{11}C posed two experimental difficulties: (i) the fermentation reactions and product isolation would have to be completed very rapidly; and (ii) the experiment would have to be conducted at the isotope source, the Berkeley cyclotron.

Wood returned to Ames and immediately began designing the experiment and running it under sham conditions. He quickly demonstrated that he could readily carry out the fermentation and isolate the appropriate products in sufficient time for an experiment with [11]CO_2 to demonstrate CO_2 fixation. He approached Werkman about traveling to Berkeley to do the experiment and even offered to pay his own expenses. Werkman, for reasons never explained, was unwilling to allow Wood to go [5, 12, 14].

Wood's brother, the American physiologist Earl Wood, was a Ph.D.–M.D. student at the University of Minnesota and was familiar with A.O. Nier's work on developing techniques to isolate and measure the heavy (non-radioactive) carbon isotope, [13]C. On his brother's suggestion, Wood visited Nier and arranged a collaboration between the two laboratories. As he later described it, there 'then began a very useful and exciting collaboration and with Professor Werkman's blessing' [12]. Nier's laboratory provided Wood with [13]CO_2, which was used as a substrate in fermentation reactions with the propionic acid bacteria in Ames. The fermentation products were isolated and then taken to Nier's laboratory where the isotope distribution was determined.

Wood's description of the collaboration as 'useful and exciting' remarkably understates its success. In slightly more than a year of work, the collaborative effort produced nine major papers that confirmed the heterotrophic CO_2 fixation hypothesis and demonstrated that $^{13}CO_2$ was incorporated into the carboxyl group of succinate (see below). The two laboratories also began to extend the role of heterotrophic CO_2 fixation into other metabolic processes, work which ultimately led to clarification of carbon flow in the Krebs citric acid cycle [14].

In addition to its high productivity, the Werkman–Wood and Nier collaboration in some respects illustrates a concept that Clarke and Fujimura [15] describe as 'the right tools for the right job'. The phrase conveys the notion that a scientific investigation may take a direction often contingent on factors outside of the immediate area of inquiry. For example, while collaboration with Nier's laboratory was exceptionally successful, it became clear to Wood that the Ames group would have to master the techniques of working with isotopically labeled compounds. With Nier's concurrence and assistance, Wood and his colleagues in the Werkman laboratory (especially Lester Krampitz) constructed their own gas diffusion column to isolate $^{13}CH_4$ and their own mass spectrometer to measure the isotope. In addition to constructing the equipment, they also became proficient at its operation. Installation of this physical instrumentation into what was primarily a microbiology laboratory created a powerful environment for exploring biochemical processes.

Thus, by the early 1940s, Wood had helped construct one of the finest microbial physiology research facilities in the country in Werkman's laboratory [14]. They blurred the disciplinary boundaries of biochemistry and microbiology by converting a somewhat conventional bacteriology laboratory into a modern biochemistry facility. This achievement did not go unrecognized, for Wood received the Eli Lilly Award from the Society of American Bacteriologists (later known as the American Society for Microbiology) in 1942.

The success of the research facility Wood helped to create in Werkman's laboratory is reflected by the fact that they co-authored more than 40 papers together in less than a decade. It is especially noteworthy that Wood had no solo authored papers during this time, despite the fact that the research community recognized the magnitude of his contributions, as evidenced by the Lilly Award. It is also remarkable that Wood held no regular faculty appointment for most of this time; he held various research positions (e.g. Research Assistant, Research Associate, Research Assistant Professor, etc.) in the Bacteriology Section of the Agricultural Experimental Station. His first faculty appointment was part-time in 1942, when he held a 1/3 time appointment as Assistant Professor in the Bacteriology Department [14].

The Lilly Award seemingly opened a breach in the collaboration between Wood and Werkman [14]. The award carried a cash prize, and the Wood family planned to use the money as a down payment on a house in Ames. When word of the planned real estate purchase reached Werkman, he asked Wood 'Why did you do that, do you think you can stay here forever?' [5]. Werkman's comment elicited a major argument that led to a permanent break in the relationship. Later, Wood stated that the argument involved the lack of independence Werkman allowed. He told Werkman 'I can't order any chemicals. I don't have any graduate students. I have been here for seven years, and in a way I am practically running your research' [6]. Werkman was unmoved and essentially told Wood to find a job elsewhere. Wood called Maurice Visscher at the University of Minnesota who offered him a position as Associate Professor of Physiological Chemistry. Within months the Wood family left Iowa.

The propionic acid cycle[7]

In later years, Wood reflected that the move from Iowa was probably fortunate and commented, 'Werkman kicked me out of a place where I probably couldn't have gotten very far ahead' [6]. Historical assessments of that nature are, of course, difficult to evaluate. Nevertheless, it is true that once Wood left Werkman's laboratory, he rapidly moved 'very far ahead'. During his 3 years at Minnesota, his research focused on using ^{13}C to trace metabolic processes in polio virus infected rats [20–22] as well as to trace the incorporation of CO_2 into glycogen [23–26].

In 1946, Wood was invited to organize and Chair the Department of Biochemistry at Western Reserve University (now Case Western Reserve University) in Cleveland, OH, where he helped create an intellectual atmosphere vastly different from that of Ames. It was a climate in which both laboratory equipment and research ideas were freely shared. A common theme from interviews with former department associates refers to the 'democratic fashion' by which the department functioned. Perhaps because of his relationship with Werkman, unless Wood actually spent 'bench-time' working on a problem, he was hesitant to put his name

---

7 There is frequent confusion in the term *cycle* in biochemical parlance, and the word is used to describe a variety of processes (see for example [16]). Consider the way the term *cycle* is used in the following metabolic activities. In the *urea cycle*, various nitrogenous compounds are degraded in a cyclic fashion to convert ammonia to urea. In the *citric acid cycle*, citrate functions only as a transitory intermediate whereby various carbon compounds are oxidized to CO_2 and a portion of the concomitant $-\Delta G$ released is conserved in various ways useful to the cell (e.g., substrate-level ATP synthesis, proton gradient formation). In the *propionic acid cycle*, various carbon compounds are oxidized to propionic acid (and other acids) and CO_2 for purposes of cellular energy conservation. Finally, as noted elsewhere, Bechtel [17–19] emphasizes the important cyclic aspects of oxidation-reduction and phosphorylation co-factor regeneration, which function in all metabolic processes.

on his student's papers.[8] During this period, the department became one of the leading biochemistry departments in the country, noted both for its original metabolic research programs and its innovative approaches to medical education.

After the move to Cleveland, Wood's personal research program began the explosive expansion and diversification described previously. His research activity, over the next four and a half decades, involved a variety of prokaryotic and eukaryotic organisms. Nevertheless, much of that work involved, as Wood himself described it [1], 'Trailing the Propionic Acid Bacteria'. With the exception of his work on bovine lactose synthesis and acetogenesis in *Clostridium thermoaceticum* (both of which were major research projects), virtually all of Wood's research inquiry had its origin in aspects of propionic acid metabolism in the genus *Propionibacterium*.

Early studies on propionic acid metabolism

Bacteria in the genus *Propionibacterium* grow anaerobically in a mixed acid fermentation. They consume a variety of C_3 to C_6 sugars and alcohols and yield propionate as a major product and lesser amounts of acetate, succinate, and CO_2 [27]. In early metabolic studies to explain the mixed acid fermentation, Kluyver's student, C.B. van Niel concluded that glucose was converted to lactate, which in turn was

---

8 Isadore Bernstein commented about the professional disadvantages of this practice (personal communication). While completing his Ph.D., he published several papers from his thesis research. Because his mentor's name was absent from the papers, potential employers did not realize his association with Wood. Bernstein pointed out this professional difficulty to Wood, who modified his view of authorship. In later years, Wood placed his name on associate's publications, if he made what he considered to be significant intellectual contributions to the work.

both reduced to propionate and split into acetate and CO_2. He believed that succinate arose from reactions involving the yeast extract added to the growth medium [9]. During the late 1920s, van Niel's thesis was the guiding paradigm to explain propionic acid bacteria metabolism.

As I noted previously, Wood's interest in the metabolism of the propionic acid bacteria began as a graduate student. He found several difficulties in van Niel's thesis and was unconvinced by it. One problem was the highly speculative nature of van Niel's proposal, and Wood noted that, 'none of the intermediate compounds [of the mechanism] were isolated and identified' [28]. Wood was especially unconvinced of van Niel's conclusion regarding the origin of succinate. Two factors played a role in this assessment [28]. First, while the carbon recovery in van Niel's proposed pathway was acceptable, the oxidation/reduction ratio was not balanced; the products were more reduced than glucose. van Niel explained this difficulty by claiming that excess reducing potential came from hydrogen donors present in the yeast extract, which led to succinate accumulation in the fermentation products. Second, earlier work by Virtanen and Karström [29] suggested to Wood that succinate served a more central role in the propionate fermentation.

For these reasons, Wood's goal in his graduate research was to investigate the 'mechanism of glucose dissimilation and to derive a scheme to represent this mechanism'. He also noted that:

'To accomplish such a purpose it is essential to have not only accurate knowledge of the quantities of end products of fermentation but it is also necessary to have information concerning the intermediate compounds which are involved in the steps leading to these final conversion products' [28].

In his thesis, Wood drew on the ideas of Virtanen, van Niel, and Neuberg, as well as contemporary notions of Embden, Meyerhoff, and Lohman to devise a metabolic pathway that would account for the massive amount of fermentation data he accumulated. His scheme, which is only summarized in Fig. 1 [28] and does not show full stoichiometry, was fully balanced in terms of oxidations and reductions and recovering all of the glucose elementary composition. It was especially important for Wood to write a pathway that was fully balanced in terms of oxidative and reductive processes. I do not present this pathway for its great biochemical validity; as I noted previously, Wood was ready to discard portions of it even before the thesis was finished. Rather, the scheme is important because it reflects Wood's maturation as a scientist and also illustrates a profound shift in biochemical thought during this time.

Wood's thesis proposal for propionate formation illustrates a major conceptual change in biochemical thinking during his career, for the scheme in Fig. 1 is what William Bechtel refers to as a 'view of metabolism as a linear, nearly decomposable process' [17]. What Bechtel intends by this phrase is that, during the early decades of this century, biochemists did not view metabolic processes as integrated systems, in which reducing equivalents and phosphoryl carriers were mobile elements capable of reacting in multiple processes. Indeed, even the role of phosphorylation was unknown. So the question arises, why did Wood invoke hexose and triose phosphate and phosphoglyceric acid in his scheme? He had not experimentally demonstrated their presence. They certainly are not inconsistent with any aspect of the proposal, and most likely were included because the developing biochemical view was that 'phosphorylation was significant only as a means to 'tailor' the molecule ... for further transformation' [30]. As Bechtel notes this linear view of metabolism underwent rapid change during the 1930s and 1940s as the roles NAD/NADP and ATP were recognized [18, 19].

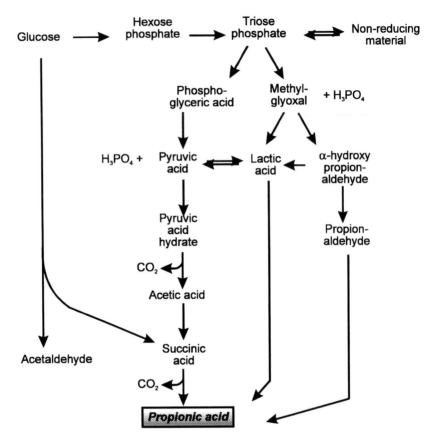

Fig. 1. Fermentation of glucose to propionate as originally proposed by Wood [28].

Expansion of the research project

During the early stages of his career, Wood's research work was accomplished using methods no more complex than fermentation analysis coupled with metabolic inhibitors. This included establishing the fundamental stoichiometry of the

Wood–Werkman Reaction (reaction (1)). His methodological perspective expanded greatly during the collaboration with Nier's laboratory. The $^{13}CO_2$ experiments confirmed his hypothesis that members of the genus *Propionibacterium* fixed CO_2 when grown on glycerol and further supported the stoichiometry of reaction (1). The $^{13}CO_2$ work also established that the bacteria fixed CO_2 when grown on glucose and suggested that propionic acid was derived by decarboxylation of succinate. The later inference arose from the observation that propionate isolated from the fermentation products contained about 50% of the ^{13}C found in the succinate. Wood argued that because succinate is symmetrical, either carboxyl group could be lost in propionate formation, thereby effectively diluting the potential amount of isotope found. Thus, by the end of the 1940s, the following general pathway guided the research [31]:

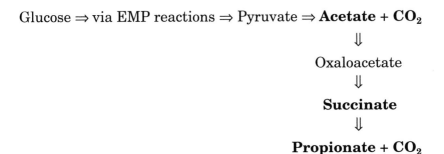

Glucose ⇒ via EMP reactions ⇒ Pyruvate ⇒ **Acetate + CO₂**

⇓

Oxaloacetate

⇓

Succinate

⇓

Propionate + CO₂

Although they are not shown in this schematic, Wood clearly realized the cyclic role of phosphoryl and electron carriers in the formation of propionate. Thus, the sequence is not strictly what Bechtel refers to as a 'linear process' [17]. Nevertheless, from Wood's perspective, carbohydrate metabolism in the propionic acid bacteria was linear, rather than cyclic, in terms of carbon flow; i.e., there was a linear flow of carbon from glucose to propionate, acetate, and CO_2.

358 R. SINGLETON JR.

Indeed, Wood did not realize the cyclic nature of the process until the 1960s.

After the move to Cleveland, Wood resumed work on propionate formation. During this time his laboratory was intensely using ^{14}C-labeled glucose to trace metabolic pathways (see Wood's review [32] for a partial summary of this research activity). In this method, an organism ferments glucose in which specific carbons are labeled with [14]C. The reaction products are isolated and sequentially degraded to CO_2 in order to identify the distribution of [14]C. Based on the distribution pattern of ^{14}C from the growth substrate into the products, it was feasible to reconstruct a theoretical

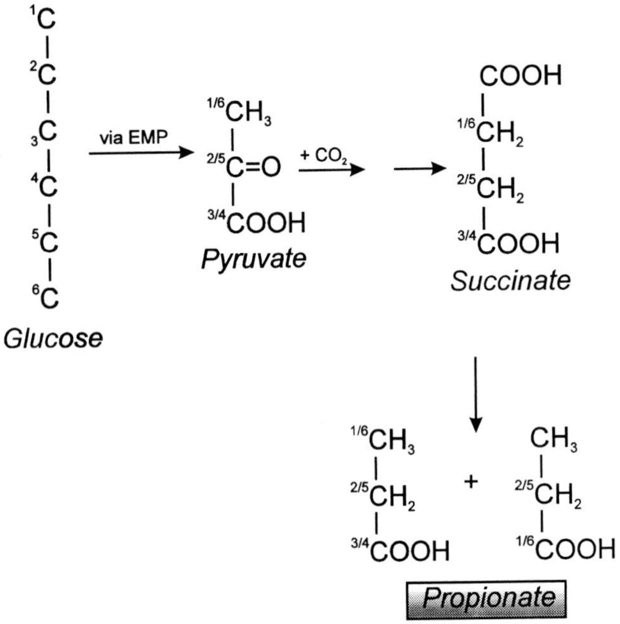

Fig. 2. Predicted labeling patterns of propionate from glucose, assuming direct decarboxylation of succinate (superscript numbers refer to carbon number in original glucose molecule).

TABLE 3

Distribution of ^{14}C in fermentation products of *Propionibacterium arabinosum* strain 34W (data from Tables 1 and 2 [33])

Glucose substrate	Specific activity as a percentage of labeled position in glucose							
	Propionate			Succinate		Acetate		CO_2
	CH_3	CH_2	COOH	CH_2	COOH	CH_3	COOH	
[1-^{14}C]-	14.2	15.7	8.1	15.5	4.5	15.1	5.2	19.7
[2-^{14}C]-	20.2	19.4	3.1	13.5	5.2	13.5	24.1	3.1
[3,4-^{14}C]-	6.6	8.1	55.4	–	41.4	6.5	5.1	48.1
[6-^{14}C]-	11.1	11.9	4.1	10.0	3.3	6.4	3.9	3.3

metabolic pathway connecting substrate and products. This technique was used to clarify metabolic processes in the propionic acid bacteria and demonstrated a major flaw in the hypothesis that guided the research.

The general pathway outlined above predicted a distribution of label, from glucose to propionate, shown in Fig. 2 [33]. The experimental data were quite different from this prediction, however, and are partially summarized in Table 3 [33]. The most striking feature of the labeling data is the generalized appearance of label in all of the product carbons; this feature did not lend itself to facile interpretation. Wood later commented that 'The isotope labeling patterns ... are the most complicated that I have ever seen' [34]. To explain these complex labeling patterns, he postulated the simultaneous operation of two pathways: one pathway involved reactions typical of EMP metabolism, and a 'second type of cleavage which is not yet identified' [33].

Discovery of the transcarboxylation reaction

Two observations were crucial to define this 'second type of cleavage'. One observation was that when $^{14}CO_2$ was fixed during bacterial growth, only the carboxyl group of propionate, succinate, and acetate was labeled. However, [^{14}C]formaldehyde was fixed into every carbon of the resulting acids [31]. Second, when [3-^{14}C]propionate was incubated with intact cells of *P. arabinosum* an exchange reaction occurred that rapidly labeled C_2 carbon of propionate [35]. These observations led to speculation that the metabolic pathway from glucose to propionate involved a 'C_1' compound, and interest began to focus on the potential role of CoA derivatives in the process.

During a 1955 sabbatical in New Zealand, Wood developed a cell-free system from *P. shermanii* capable of carrying out the propionic acid fermentation from glucose [36, 37]; this system was later used by Bob Swick in Wood's lab to characterize the nature of the 'C_1' compound. Swick found that the cell-free system catalyzed a series of exchange reactions involving ^{14}C-labeled CoA compounds and proposed that *P. shermanii* carried out the following set of reactions [38]:[9]

Succ-CoA $\Rightarrow$ MM-CoA   (2)

Enz + MM-CoA $\Rightarrow$ Prop-CoA + Enz-C_1 (3a)

Enz-C_1 + Pyr $\Rightarrow$ OAA + Enz   (3b)

OAA + 4H $\Rightarrow$ Succ   (4)

Succ + Prop-CoA $\Rightarrow$ Succ-CoA + Prop   (5)

Net: Pyr + 4H $\Rightarrow$ Prop   (6)

---

9 Abbreviations used: MM-CoA, methylmalonyl-CoA; OAA, oxaloacetate; ProP, propionate; Prop-CoA, propionyl-CoA; Pyr, pyruvate; Succ, succinate; Succ-CoA, succinyl-CoA.

Swick and Wood [38] referred to reactions (3a) and (3b) as a *transcarboxylation* and called it a 'novel biochemical reaction'. The reaction sequence was inhibited by avidin, which suggested that transcarboxylation was biotin dependent. Because avidin inhibited transcarboxylation and not isomerization of MM-CoA (reaction (2)), reactions (2) and (3a,b) were clearly separate processes. Reactions (2)–(5) were combined with other reactions known to occur in propionic acid bacteria to create the major pathway leading to propionate formation outlined in Fig. 3.

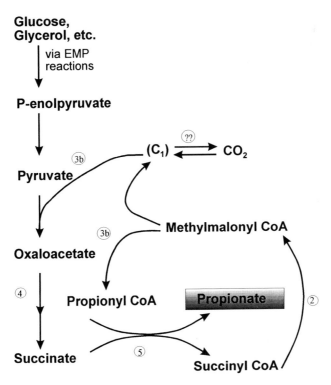

Fig. 3. Propionate formation via transcarboxylation involving CoA intermediates. The numbers refer to Reactions 2–5 in text. Based on [38].

Although minor portions of this pathway were later changed, the Swick and Wood paper was pivotal because: (i) it helped clarify the metabolism of the propionic acid bacteria; and (ii) it played a decisive role in shaping Wood's subsequent career. There were several reasons for its metabolic importance. First, it resolved previous difficulties in understanding propionate synthesis in *P. shermanii* extracts. For example, previous data demonstrated that the turnover of CO_2 measured by $^{14}CO_2$ exchange was inconsistent with the rate of propionate synthesis [31]. The potential for exchange with CO_2 is greatly reduced, however because most of the compounds shown in Fig. 3 are present at only catalytic levels. Of greater importance, however, the mechanism proposed by Swick and Wood shifts the view of propionate formation away from the linear process that dominated Wood's thinking since his Ph.D. thesis to a cyclic mechanism, driven by the continuous reformation of CoA derivatives present only at catalytic levels. Lastly, the Swick and Wood paper directly led to purification and characterization of its central enzyme, transcarboxylase. Wood's major research program, which clarified the physical structure and chemical mechanism of this extremely complex protein, justly earned his reputation as one of the foremost enzymologists of modern biochemistry.

Because members of the genus *Propionibacterium* could catalyze CO_2 fixation, Swick and Wood speculated that carbon from CO_2 could enter the pathway via carboxylation of phosphoenolpyruvate (PEP) to form oxaloacetate (OAA) (not shown in Fig. 3). Several observations supported this speculation. One was the discovery of PEP carboxykinase in Mert Utter's laboratory [39]. A second reason for proposing CO_2 fixation with PEP was previous evidence that CO_2 fixation required either PEP or a phosphorylating mechanism [40, 41]. The possible involvement of PEP in the Wood-Werkman reaction had postulated since 1940, however its role was not clarified until Patrick Siu demonstrated the presence of PEP carboxytransphorylase (CTrP) activity in *P. shermanii* in 1961 [42].[10] This enzyme catalyzes the reversible carboxylation of PEP and is energetically coupled to pyrophosphate (PP_i) (reaction (7)) [43].

$$PEP + CO_2 + P_i \rightleftharpoons OAA + PP_i \qquad (7)$$

Wood's description of CTrP further illustrates his capacity to uncover new research problems in seemingly solved problems. The CTrP reaction provided an elegant molecular resolution for the experimental hypothesis of heterotrophic CO_2 fixation generated three decades previously. However, Wood realized that when the reaction is viewed in the reverse direction, pyrophosphate served as a phosphorylation source in a manner analogous to ATP. This realization led Wood into a rich avenue of research inquiry about the role of pyrophosphate as a metabolic bioenergetic source that occupied his attention for the major part of the following decades [44, 45].

Finally, there is an important historical aside in the discovery of the transcarboxylation process. Despite its importance in Wood's career, his name almost did not appear on the paper describing it.[11] Bob Swick spent a sabbatical year in Wood's lab, and much of his lab work was carried out in Wood's absence. Because of a family illness, Wood was out of the laboratory for a major portion of Swick's work. When it

---

10 Rune Stjernholm and Wood were able to demonstrate that extracts of *P. shermanii* fixed $^{14}CO_2$ into oxaloacetate, but their intensive attempts to isolate the enzyme in ammonium sulfate fractions met with little success. In a matter of days, Patrick Siu, who was a graduate student, demonstrated the enzymatic activity when he dissolved the ammonium sulfate fractions in phosphate buffer, rather than the Tris buffer that Stjernholm and Wood used previously. In so doing, Siu demonstrated that Pi, in addition to PEP, was required for the Wood–Werkman reaction [1].

11 Robert Swick, personal communication.

came time to prepare the work for publication, Wood objected to his name being on the paper because he had not been 'at the bench' while the work was carried out. Swick, with the assistance of Mert Utter, insisted that Wood's intellectual contribution to the work justified putting his name on the paper.

Isolation of the transcarboxylase enzyme

Since the demonstration that enzymes are protein molecules capable of isolation and crystallization in the 1930s, a major goal of biochemists has been to purify enzymes responsible for metabolic processes [46]. This research activity especially accelerated during the 1950s and 1960s. Although his colleague and close friend Mert Utter began working with cell-free systems in Werkman's laboratory, Wood did not attempt to such work until 1955 [36]. This practice changed radically in the early 1960s, when his laboratory began a concentrated effort to isolate and characterize many of the enzymes responsible for propionate formation in *P. shermanii*.

The first of these enzymes, transcarboxylase (methylmalonyl-CoA:pyruvate carboxytransferase, EC 2.1.3.1), played a dominant role in Wood's research program for the rest of his life. The enzyme is both mechanistically and physically complex, and its properties have been extensively reviewed elsewhere [47]. Nevertheless, because of its central importance in Wood's career, I will briefly describe some of its salient properties.

In 1961, Wood and Stjernholm [48] reported the purification, via ammonium sulfate fractionation and DEAE-cellulose chromatography, of a protein that catalyzed the reaction:

$$\text{MM-CoA} + \text{Pyr} \rightleftharpoons \text{Prop-CoA} + \text{OAA} \tag{8}$$

The enzyme was assayed in the forward direction by measuring OAA formation with malic dehydrogenase; the reverse reaction was measured by coupling with lactate dehydrogenase. While the enzyme preparation was not homogeneous (ultracentrifugation demonstrated the presence of contaminating proteins), it was free of malic and lactic dehydrogenases, NADH oxidase, and thiol deacylase. The enzyme was non-discriminating for CoA esters, and acetyl-CoA, butyryl-CoA, and acetoacyl-CoA served as carboxyl acceptors from OAA. Keto acid specificity was more restricted, however, and only pyruvate served as substrate. The enzyme had a K_{eq} of around 2, indicating that the reaction was freely reversible. Avidin inhibited the purified enzyme, confirming Swick and Wood's previous observation that the transcarboxylation process was biotin dependent. Although Wood and Stjernholm did not present data for the purity of their preparation, they speculated that the reaction was catalyzed by a single protein.

In a companion paper, Stjernholm and Wood [49] reported the purification of MM-CoA isomerase, which catalyzed reaction (9):

$$Succ\text{-}CoA \rightleftharpoons MM\text{-}CoA \tag{9}$$

The enzyme was assayed by a coupled reaction sequence. Purified transcarboxylase was used to carboxylate pyruvate to OAA, which was in turn measured by coupling to malate dehydrogenase. The purified isomerase required added 'B$_{12}$ coenzyme' (e.g., dimethylbenzimidazolylcobamide) for activity and was free of lactic and malic dehydrogenases, transcarboxylase, CoA deacylase, and prop-CoA transferase activities.

Completion of the propionic acid cycle

Harland Wood's 'style' of practicing science was typical of most successful scientists. An important attribute of that 'style' is an ability to continuously realize that new approaches of scientific investigation are necessary to make progress on ongoing research problems. Successful scientists either adapt new insights and disciplinary developments from others or develop new approaches themselves in order to advance their own research program [50]. For example, well into the late 1950s, the Wood laboratory rarely worked with cell-free extracts. Indeed, it is curious that although Wood viewed himself as an enzymologist for all of his career,[12] much of his research work prior to the 1960s involved whole cells or tissue slices. By the mid-1960s, however, this practice changed, and his laboratory was purifying and characterizing enzymes in a way that Arthur Kornberg described[13] as 'brilliant'.

Within a few years after describing the transcarboxylase process and its central role in propionate formation, Wood's laboratory isolated, in varying degrees of purity, the major enzymes involved in the propionate cycle. In addition to isolating the proteins, they began to characterize the physical properties and reaction mechanisms of the enzymes. There was a sense of unity in the laboratory's approach. Whenever possible, common procedures were used to purify individual enzymes. For example, bacterial cells were grown and harvested in large quantity. Assay systems for individual enzymes often utilized other enzymes under investigation in the laboratory, as I described earlier for the MM-CoA isomerase. Often, a protein fraction from one stage in the

---

12 Although he did not take formal course work in enzymology offered by the ISC Chemistry Department as a graduate student, Wood refers to his work as that of a 'zymologist,' in his Ph.D. thesis [28].
13 Authur Kornberg, personal communication

isolation of a particular enzyme would be used to isolate another enzyme involved in the cycle.[14]

The work that brought the propionic acid cycle to completion clearly illustrates this communal approach to enzyme purification. In a single paper [51], the Wood laboratory reported purifying phosphotransacetylase, acetyl kinase, malic dehydrogenase, and CoA transferase from *P. shermanii.* Spectrophotometric assays were developed for each enzyme. In addition to developing assay methods and purifying the enzymes, estimates of molecular size (sedimentation coefficients) and substrate binding affinities (K_m) were also determined.

Based on an ability to partially reconstruct the propionic acid cycle by combining the purified enzymes, Allen *et al.* [51] proposed a cyclic pathway for glucose fermentation to acetate and propionate. The cycle is outlined in Fig. 4, and its stoichiometry is summarized in Fig. 5. Each of the enzymes shown in the proposed pathway had either been isolated or demonstrated to be present in crude extracts by the Wood laboratory, or had been demonstrated by other investigators.

The cyclic pathway proposed by Allen *et al.* explains numerous metabolic features of the genus *Propionibacterium.* For example, van Niel's classic work on the bacteria [9] observed that glucose fermentation produced 2 mol of propion-

---

14 I had a particularly unpleasant experience with this communal approach to enzyme purification. My postdoctoral research in Wood's lab involved the reaction mechanism of CTrP. To purify the enzyme, I obtained an ammonium sulfate fraction isolated during the transcarboxylase purification, which contained the major CTrP activity. I had no difficulties with the published purification until a crucial cellulose-Pi column, which yielded an enzymatically inactive protein fraction in approximately the proper position of the gradient. The problem arose, in part, from the communal laboratory effort. All of the exchange celluloses used in the laboratory were recycled in batches by laboratory aides; we used exchangers from this communal supply. For some reason, the cellulose-Pi was left in the protonated form after this recycling. Apparently none of the other enzymes under investigation were affected by this fact, whereas the CTrP was immediately inactivated by the decrease in pH that accompanied binding to the cellulose-exchanger.

368 R. SINGLETON JR.

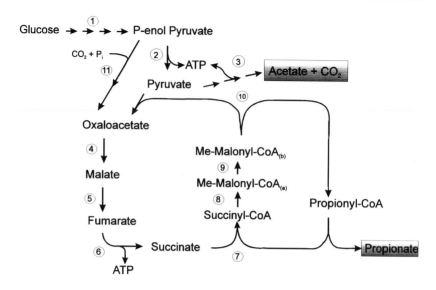

Fig. 4. Contemporary view of the propionic acid cycle. Enzymes (or pathways) indicated are: 1, Embden–Meyerhoff–Parnas pathway; 2, pyruvate kinase; 3, pyruvate DH, acetyl kinase, etc.; 4, malate DH; 5, fumarase; 6, fumarate DH; 7, CoA transferase; 8, succinyl-CoA isomerase; 9, methylmalonyl-CoA racemase; 10, transcarboxylase; 11, carboxytransphosphorylase.

EMBDEN-MEYERHOFF REACTIONS:

$$1.5 \text{ Glucose} + 3 \text{ ADP} + 3 \text{ P}_i + 3 \text{ NAD} \longrightarrow 3 \text{ Pyruvate} + 3 \text{ NADH} + 3 \text{ H}^+ + 3 \text{ ATP} \tag{1}$$

ACETATE FORMATION:

$$\text{Pyruvate} + \text{ADP} + \text{P}_i + \text{NAD} \longrightarrow \text{Acetate} + \text{CO}_2 + 2 \text{ H}_2\text{O} + \text{NADH} + \text{H}^+ + \text{ATP} \tag{2}$$

PROPIONATE FORMATION:

$$2 \text{ Pyruvate} + 4 \text{ NADH} + 4 \text{ H}^+ + 2 \text{ ADP} + 2 \text{ P}_i \longrightarrow 2 \text{ Propionate} + 4 \text{ NAD} + 2 \text{ H}_2\text{O} + 2 \text{ ATP} \tag{3}$$

NET:

$$1.5 \text{ Glucose} + 6 \text{ ADP} + 6 \text{ P}_i \longrightarrow 2 \text{ Propionate} + 1 \text{ Acetate} + \text{CO}_2 + 2 \text{ H}_2\text{O} + 6 \text{ ATP} \tag{4}$$

Fig. 5. Summary of reactions involved in the propionic acid cycle showing stoichiometry of reactants and products. ATP yields are based on both substrate level and oxidative phosphorylation processes. Based on [51].

ate and 1 mol of acetate per glucose. The propionic acid cycle clearly explains the basis of this stoichiometry. However, even van Niel observed variation in the ratio of propionate to acetate, and Wood and Werkman reported a variation that ranged from 2.1 to 14.7 depending on the strain of organism and conditions of fermentation [52]. Wood occasionally noted that the scheme does not explain this variation in fermentation products. Although it most likely arises from strain differences, the variation is an issue that remains unresolved [1, 34].

In one sense, the cycle resembles a classical *fermentation* metabolic pathway (i.e., reducing equivalents generated during an oxidative process are disposed of to an internally produced electron acceptor) rather than a *respiratory* pathway (i.e., an electron transport linked oxidative phosphorlation, in which reducing equivalents are disposed of to an external electron acceptor such as O_2 or sulfate [53]). Such tidy metabolic definitions do not seem suitable for the propionic acid cycle, because NADH generated in glucose and pyruvate oxidation is used to reduce fumarate to succinate. Although fumarate is an internal product of the reaction sequence (and would therefore fit the notion of a *fermentation*), the process is coupled to electron transport linked oxidative phosphorylation (and thus resembles a *respiratory* pathway).

Fumarate reductase action helped explain the previously reported high ATP yield for the process [54]. As Fig. 5 illustrates, the sole function of propionate formation (reaction (3)) appears to be disposal of electrons generated by glucose and pyruvate oxidation (reactions (1) and (2)). If this were the case however, substrate level phosphorylation in reactions (1) and (2) would generate only 4 mol of ATP. Oxidizing NADH via fumarate reductase generates additional ATP via electron transport-coupled phosphorylation [55], thereby increasing the overall ATP yield. The complexity of these bioenergetic mechanisms illustrates some of the difficulties arising in precisely defining metabolic processes [53].

Another curious historical side-light of this story is that the reaction that initially ignited Wood's interest in the propionic acid bacteria, i.e., heterotrophic CO_2 fixation, seemingly plays only a peripheral role in the pathway. The cyclic mechanism, however, explains some of the early experiments that initially led Wood to look for CO_2 fixation. As a graduate student, Wood observed that the propionic acid bacteria seem to fix more CO_2 in fermentations in which significant amounts of succinate accumulated. Fig. 4 shows that if the cycle is disrupted so that succinate accumulates, the amount of MM-CoA necessary to generate OAA will decrease. In this situation, the amount of CO_2 fixed via CTrP – the enzyme responsible for the 'Wood–Werkman Reaction' – (reaction (11) in Fig. 4) will increase. This relationship may, in part, explain Wood's earlier observation of the close connection between CO_2 fixation and succinate formation [11, 41].

The CTrP reaction also serves another important role in this cycle. As Fig. 4 illustrates, carbon flow through the propionate pathway leads to disposal of reducing equivalents and ATP generation. Yet, like TCA cycle intermediates, many intermediates in the propionate cycle serve biosynthetic roles. Thus, CO_2 fixation via CTrP serves an anaplerotic function and generates OAA and other compounds necessary for biosynthetic pathways.

Conclusions

Arthur Kornberg observed that Harland Wood was one member of a group of scientists with 'a lifetime record of stellar achievements' [56]. Elucidation of the propionic acid cycle was only one of Wood's 'stellar achievements'. But propionate metabolism played a central role in Wood's overall research program. His work in clarifying the pathway illustrates several important aspects of Wood's research style, which typify a great many successful scientists.

As I noted earlier, an important aspect of Wood's approach to research was his dogged pursuit of research questions. His interest in the metabolism of the propionic acid bacteria began as a graduate student in Werkman's laboratory during the 1930s. Using the best techniques available to him, isolation of organic compounds from complex mixtures and precise quantitative analysis of those compounds, he produced an initial answer to some of the metabolic questions posed by the organisms (Fig. 1). While the proposed pathway satisfied the various fermentation analysis data, it left many issues unresolved.

When carbon isotopes became available, Wood realized their great potential to clarify and identify metabolic pathways. At Iowa State, he helped construct a facility to isolate and measure ^{13}C [5, 12, 14]. When ^{14}C became available, his laboratory at Western Reserve University rapidly became a leading center for using it as a metabolic tracer. Wood used ^{14}C as both a practical tool to clarify specific pathways (see [57–62] for example) and as a theoretical method for a general understanding of carbon flow in metabolic pathways [63, 64]. When ^{14}C was used as a tracer to sort out propionate metabolism, however, it raised more questions than it answered [1, 34] and forced him to turn to other approaches. It seems reasonable to claim that the inadequacy of radioisotope tracers for solving the problem of propionate formation forced Wood into working with enzymes.

Harland Wood's contributions to enzymology were considerable and would require several chapters to fully chronicle. His laboratory isolated and completed structural and mechanistic characterizations of several propionic acid cycle enzymes. Resolving the complex molecular structure of transcarboxylase was a major achievement of this work. Like his resolution of the propionate cycle, work on transcarboxylase structure occupied Wood's attention for over 25 years. In achieving an understanding of the protein's structure he drew on an immense variety of techniques including classical kinetics, sophisticated electron microscopy coupled with careful analytical ultracentrifuge analysis, and ending with molecular biology approaches such as site-directed mutagenesis.

In addition to pursuing research questions for extended time periods, another aspect of Wood's research style was a seemingly inexhaustible capacity for work. As I noted previously, when he assumed the Biochemistry Department Chair at Western Reserve University in 1946, he rapidly helped build the department into one of the top ten departments in the United States. At the same time, he was a major leader in reshaping the Western Reserve medical curriculum [65].

In 1953 he was elected to the National Academy of Sciences and increasingly played major professional roles at both the national and international levels. For example, he was on numerous grant agency study panels, served various editorial functions for a variety of journals, and was President of the International Union of Biochemistry from 1979 to 1985. In 1968 he was appointed a Presidential Science Advisor. Most people, when faced with such mounting external pressures on their time, leave direct involvement in the laboratory. Wood's love of the laboratory, however, made that route impossible. Somehow, he managed to successfully complete all of these external activities and simultaneously maintain a vigorous and active role in his laboratory. His success in balancing all of these activities with his drive to remain active in the laboratory grew out of what appeared to be an inexhaustible supply of energy.

I experienced Wood's legendary capacity for work first hand many times as a postdoctoral associate. One particular incident is especially memorable. The Wood brothers were very close and all loved to hunt. In the Fall of 1971 the oldest brother organized an Alaskan hunting trip, which was to last for almost three weeks. Shortly after the hunting trip, Harland had an IUB meeting in Varna, Bulgaria. In a period of slightly more than 24 h, he flew from Anchorage, Alaska, to Cleveland and on to Varna.

During his time in Cleveland, he talked, in depth, with everyone in the laboratory about our research. How had the experiments we planned before his departure come out? If things had not worked out as expected, what went wrong? What was next? They were not trivial questions, for he clearly remembered the planned experiments and expected detailed responses. His familiarity with the research was so intimate that it was difficult to believe that this man, who was a year short of the 'age of retirement', had just spent almost 3 weeks in the wilderness of Alaska. After a 'blitzkrieg' visit in the lab, he and Millie rushed to the airport for the trip to Europe. A few weeks after their return, I asked Millie what was Harland like after the extensive hunting in Alaska and crossing more than 10 time zones in a little over a day. Her reply was something like, 'Well, after part of a night's sleep, he went to the IUB meeting just like he was home in Cleveland'.

Coda

Like the American legendary hero, Harland Wood had many achievements throughout his life. He helped expand biochemical knowledge through his numerous and significant scientific contributions. He was a vigorous academic administrator and involved in innovative medical education. He was elected to various honorary memberships and received numerous awards. Like most prestigious scientists, he played important professional roles as journal editor and organizer, as a Presidential Science Advisor, and as a leader in national and international organizations.

Despite all of these achievements, however, Harland was passionately tied to the laboratory all of his life. Despite his obvious love for and commitment to family and friends, despite his great skill as an administrator, despite the great pleasure he took from social engagement and his enjoyment of outdoor recreation (e.g., learning to ski, in his 60s, with Fritz Lynen as instructor), Harland's true 'heart and soul' were at the lab bench. Two memories, separated by decades, illustrate Harland's passion.

During the late 1960s – a period of social unrest, especially on many college campuses – Harland served as Dean of Sciences of Case Western Reserve University. Shortly after my arrival in the laboratory, Harland's secretary from his Dean's office called the lab and asked to speak with Dean Wood. A protesting student group had occupied his office in Adelbert Hall, and the secretary called to tell him that he would not be able to work in the Dean's office for a while. Harland calmly replied, 'Good! Tell the SOBs they can keep the place. I never liked it very much anyway!' Those of us in the lab wondered what it was like to be a student protester who had finally occupied a pinnacle of power on the campus, only to be told by its occupant to 'keep it, he never liked it anyway'. I have always wondered what that historical period in this country might have been like if other administrators had held similar perspectives on their positions.

I last saw Harland a year or so before his death. I had called to say I would be in Cleveland and would like to stop to visit for a bit. He had developed an interest in the physiology of sulfate-reducing bacteria, and, we planned to talk, in part, about their biochemistry. I went up to his lab and as typical, he was sitting at the Gilford, pipette in hand, starting a series of assays. 'Go away', he said, 'I'm right in the middle of an experiment. Come back in about 20 minutes'. So I wandered down the hall to chat with some friends and went back to his lab after around a half hour. He finished up a few things and slipped a timer into his lab coat pocket. We went down to his office where we talked about a wide variety of both scientific and personal topics for the better part of an hour and a half. It was a conversation in which we both took obvious pleasure and enjoyed. Then his timer went off. He said, 'Well, my incubation is finished; I need to get back to the lab'.

Acknowledgments

Many colleagues at the University of Delaware, as well as numerous family members, former students, and associates of Harland Wood assisted me with their discussion and criticism, in preparing this paper; I thank them all for that invaluable help. I also thank Dennis Harrison and Tyler Walters, and their staffs, in the Archives of Case Western Reserve and Iowa State Universities who provided invaluable information and assistance both in person and electronically. Jane Mainenshein and Joseph Fruton also contributed many valuable suggestions by critically reading various manuscript drafts. Finally, I thank the College of Arts and Sciences and the General University Research program of the University of Delaware and the National Science Foundation (Grant Number: SBR 9602023) for their financial support.

Appendix A: Fermentation balances

For many years, fermentation balances were a fundamental way to assess metabolic pathways. The technique involves allowing a fermentation to occur on a known quantity of growth substrate, often in the presence of metabolic inhibitors or trapping agents (e.g., bisulfite to trap carbonyl compounds), followed by an accurate and complete quantitative analysis of all fermentation products. Based on the ratio of various products, it was possible to speculate about possible pathways connecting the substrate with the products. Although fermentation balances are still used in some industrial applications, few modern biochemical readers are familiar with the method. Thus, some of the basic concepts of fermentation chemistry are reviewed in this appendix. Much of this material is based on the classical monograph written by A.C. Neish [66].

Fermentation balances are derived by careful analysis of the growth products produced by an organism grown on a known quantity of growth substrate. A key assumption of the method is that all of the products observed are derived from the growth substrate. Many growth media used in early experiments were poorly defined (i.e., they contained various amounts of non-defined components such as yeast extract), thus the validity of this assumption was highly variable.

Two fundamental principles were essential for an accurate fermentation balance. First, the requirement of conservation of chemical mass mandated that all of the substrate elements had to be recovered in the fermentation products. This was especially true for carbon, since many of the experiments were attempting to understand the utilization of carbon compounds (referred to as a 'carbon balance'). Second, the dictates of oxidation–reduction chemistry required that there could be no net change in the oxidation state of the products when compared to the substrate fermented; the oxidation/reduction (O/R) ratio of the fermentation must equal one.

To calculate the carbon balance, the molar yield of all products was multiplied by the number of carbon atoms in each compound; this resulted in a value for the millimoles of C_1. In a similar fashion the millimoles of C_1 represented by the growth substrate was also calculated. Since the yield of products was conventionally expressed as a function of 100 mmol of substrate, the millimoles of C_1 in substrate was a simple multiplication of 100 by the number of carbon atoms present in the substrate (e.g., C_1 for glucose was 600, C_1 for glycerol was 300, etc.). Carbon recovery was then the ratio:

$$= \frac{\text{millimoles of } C_{1\text{products}}}{\text{millimoles of } C_{1\text{substrate}}} \times 100$$

The ratio of oxidation to reduction was calculated by comparison to carbohydrates (carbon hydrate) such as glucose, with the empirical formula $C_n(H_2O)_n$. Initially, an O/R value can be calculated for any compound by comparing its formula with the empirical formula for carbohydrates. The O/R value is equal to the number of hydrogen atoms that must be added or subtracted to the molecular formula to make the ratio of hydrogen to oxygen equal 2:1. Three examples serve to illustrate the concept. Glycerol has the molecular formula, $C_3H_8O_3$. In order to make the ratio of hydrogen to oxygen (H/O) equal 2:1, two hydrogens must be subtracted from the formula ($H_8O_3 - 2H = H_6O_3$); thus the O/R value for glycerol is –2. The molecular formula for succinic acid is $C_4H_6O_4$ and requires that two hydrogens be added to make the H/O = 2:1 ($H_6O_4 + 2H = H_8O_4$), and the O/R value for succinate is +2. Finally, the molecular formula for carbon dioxide is CO_2, which needs the addition of four hydrogens [$C(H_0)O_2 + 4H = C(H_4)O_2$]. The O/R value for CO_2 thus is +4.

The "O/R Index" for the fermentation is determined by multiplying the millimoles of product by its corresponding O/R value. The sum of the positive values (oxidations) is divided by the sum of the negative values (reductions) to generate the "O/R Index" (or balance) for the process. For fermentations involving neutral substrates (e.g., those having a general molecular formula $C_n(H_2O)_n$), the substrate oxidation state does not need to be considered. For compounds more or less oxidized than glucose, their O/R value must be considered in assessing the O/R balance. For example, when glycerol, which has an O/R value of –2, serves as a fermentation substrate, its utilization must be represented as an oxidation. Because product yields of fermentations are ex-

pressed as millimoles of product per 100 millimoles of substrate, the contribution of the substrate to the *O/R* balance is
simply 100 times its *O/R* value.

REFERENCES

1   Wood, H. 1976. Trailing the propionic acid bacteria, 105–115. In A.
    Kornberg, B. Horecker, L. Cornudella and J. Oro (eds.), Reflections
    on Biochemistry in Honor of Severo Ochoa. Pergamon Press, New
    York.
2   Kamen, M. 1994. Reflections on the first half-century of long-lived
    radioactive carbon (^{14}C). Proc. Am. Philos. Soc. 138: 48–60.
3   Wood, H. and L. Ljungdahl. 1991. Autotrophic character of the ace-
    togenic bacteria, 202–250. In J. Shively and L. Barton (eds.), Varia-
    tions in Autotrophic Life. Academic Press, New York.
4   Wood, C. W. 1976. The Wood-Goff Family Chronicle. Privately
    printed, Leesburg, FL.
5   Wood, H. 1985. Then and now. Ann. Rev. Biochem. 54: 1–41.
6   Bohning, J. 1990. Harland G. Wood. Unpublished interview, the
    Beckman Center for the History of Chemistry.
7   McElvaine, R. 1993. The Great Depression. Times Books, New York.
8   Eaborn, C. 1990. Henry Gilman. Biograph. Mem. Fellows R. Soc. 36:
    153–172.
9   van Niel, C. 1928. The Propionic Acid Bacteria. J. W. Boisevain,
    Haarlem, NY.
10 Wood, H. and C. Werkman. 1934. Pyruvic acid in the dissimulation
    of glucose by the propionic acid bacteria. Biochem. J. 28: 745–747.
11 Wood, H. and C. Werkman. 1936. The utilization of CO_2 in the dis-
    simulation of glycerol by the propionic acid bacteria. Biochem. J. 30:
    48–53.
12 Wood, H. 1972. My life and carbon dioxide fixation, 1–53. In J. F.
    Woessner, Jr. and F. Huijing (eds.), The Molecular Basis of Biologi-
    cal Transport. Miami Winter Symposia, Vol. 3. Academic Press, New
    York.
13 Barron, E. S. 1943. Mechanisms of carbohydrate metabolism. An
    essay on comparative biochemistry, 149–189. In F. Nord and C.
    Werkman (eds.), Advances in Enzymology, Vol. 3. Interscience, New
    York.
14 Singleton, R. 1997. Heterotrophic CO_2-fixation, mentors and stu-
    dents: the Wood–Werkman reactions. J. Hist. Biol. 30: 91–120.

15 Clarke, A. and J. Fujimura (eds). 1992. The Right Tools for the Job in Twentieth Century Life Sciences: Materials, Techniques, Instruments, Models and Work Organization. Princeton University Press, Princeton, NJ.

16 Estabrook, R. and P. Srere (eds). 1981. Biological Cycles. Current Topics in Cellular Regulation, Vol. 18. Academic Press, New York.

17 Bechtel, W. 1986. Biochemistry: a cross-disciplinary endeavor that discovered a distinctive domain, 77–100. In W. Bechtel (ed.), Integrating Scientific Disciplines. Martinus Nijhoff, Dordrecht.

18 Bechtel, W. 1986. Building interlevel theories: the discovery of the Embden–Meyerhoff pathway and the phosphate cycle, 65–97. In P. Weingartner and G. Dorn (eds.), Foundations of Biology. Hölder-Pichler-Tempsky Verlag, Vienna.

19 Bechtel, W. 1988. Fermentation theory: empirical difficulties and guiding assumptions, 163–181. In A. Donovan, L. Laudan and R. Laudan (eds.), Scrutinizing Science: Empirical Studies of Scientific Change. The Johns Hopkins University Press, Baltimore, MD.

20 Wood, H., I. Rusoff and J. Reiner. 1944. Anaerobic glycolysis of the brain in experimental poliomyelitis. J. Exp. Med. 81: 151–159.

21 Utter, M., J. Reiner and H. Wood. 1945. Measurement of anaerobic glycolysis of brain as related to poliomyelitis. J. Exp. Med. 82: 217–226.

22 Utter, M., H. Wood and J. Reiner. 1945. Anaerobic glycolysis in nervous tissue. J. Biol. Chem. 161: 197–217.

23 Wood, H., N. Lifson and V. Lorber. 1945. The position of fixed carbon in glucose from rat liver. J. Biol. Chem. 159: 475–489.

24 Wood, H., B. Vennesland and E. Evans. 1945. The mechanism of carbon dioxide fixation by cell-free extracts of pigeon liver: distribution of labeled carbon dioxide in the products. J. Biol. Chem. 159: 153–158.

25 Lorber, V., N. Lifson and H. Wood. 1945. Incorporation of acetate carbon into rat liver glycogen by pathways other than carbon dioxide fixation. J. Biol. Chem. 161: 411–412.

26 Utter, M. and H. Wood. 1945. Fixation of carbon dioxide in oxalacetate by pigeon liver. J. Biol. Chem. 160: 375–376.

27 Cummins, C. and J. Johnson. 1992. Chapter 37. The genus *Propionibacterium*, 834–849. In A. Balows, H. Trüper, M. Dworkin, W. Harder and K. Schleifer (eds.), The Prokaryotes, Second Edition. Springer-Verlag, New York.

28 Wood, H. 1934. The Physiology of the Propionic Acid Bacteria. Ph. D. Dissertation, Iowa State College, Ames, IA.

380   R. SINGLETON JR.

29   Virtanen, A. I. and A. Karström. 1931. Über die Propion-
     säuregärung, III. Acta. Chem. Fennica. B 7: 17.
30   Korman, E. 1974. The discovery of fructose-1,6-diphosphate (the
     Harden–Young ester) in the molecularization of fermentation and of
     bioenergetics. Mol. Cell. Biochem. 5: 65–68.
31   Wood, H. and F. Leaver. 1953. CO_2 turnover in the fermentation of
     3,4,5 and 6 carbon components by the propionic acid bacteria. Bio-
     chim. Biophys. Acta. 12: 207–222.
32   Wood, H. 1955. Significance of alternate pathways in the metabo-
     lism of glucose. Physiol. Rev. 35: 841–859.
33   Wood, H., R. Stjernholm and F. Leaver. 1955. The metabolism of
     labeled glucose by the propionic acid bacteria. J. Bacteriol. 70: 510–
     520.
34   Wood, H. 1982. The discovery of the fixation of CO_2 by heterotrophic
     organisms and metabolism of the propionic acid bacteria, 173–250.
     In G. Semenza (ed.), Of Oxygen, Fuels, and Living Matter. John
     Wiley & Sons, New York.
35   Wood, H., R. Stjernholm and F. Leaver. 1956. The role of succinate
     as a precursor of propionate in the propionic acid fermentation. J.
     Bacteriol. 72: 142–152.
36   Wood, H., R. Kulka and N. Edson. 1955. Fermentation of glucose-1-
     C^{14} in cell-free extracts of Propionibacteria. Proc. Univ. Otago 33:
     24–25.
37   Wood, H., R. Kulka and N. Edson. 1956. The metabolism of ^{14}C-
     glucose in an enzyme system from Propionibacterium. Biochem. J.
     63: 177–182.
38   Swick, R. and H. Wood. 1960. The role of transcarboxylase in propi-
     onic acid fermentation. Proc. Natl. Acad. Sci. USA. 46: 28–41.
39   Kurahashi, K. 1985. The discovery of phosphoenolpyruvate car-
     boxykinase, 71–97. In G. Semenza (ed.), Comprehensive Biochemis-
     try. Vol 36, Selected Topics in the History of Biochemistry. Personal
     Reflections. II. Elsevier Science Publishers, Amsterdam.
40   Wood, H. and C. Werkman. 1940. The fixation of CO_2 by cell suspen-
     sions of Propionibacterium pentesaceum. Biochem. J. 34: 7–14.
41   Wood, H. and C. Werkman. 1940. The relationship of bacterial utili-
     zation of CO_2 to succinic acid formation. Biochem. J. 34: 129–138.
42   Siu, P., H. Wood, and R. Stjernholm. 1961. Fixation of CO_2 by phos-
     phoenolpyruvic carboxytransphosphorylase. J. Biol. Chem. 236:
     PC21–PC22.
43   Siu, P. and H. Wood. 1962. Phosphoenolpyruvic carboxytransphos-
     phorylase, a CO_2 fixing enzyme from propionic acid bacteria. J. Biol.
     Chem. 237: 3044–3051.

44   Wood, H. 1988. Squiggle phosphate of inorganic pyrophosphate and polyphosphates, 581–602. In H. Kleinkauf, H. von Döhren and L. Jaenicke (eds.), The Roots of Modern Biochemistry. Walter de Gruyter & Co., Berlin.

45   Wood, H. and J. Clark. 1988. Biological aspects of inorganic polyphosphates. Ann. Rev. Biochem. 57: 235–260.

46   Kohler, R. E. 1973. The enzyme theory and the origins of biochemistry. Isis. 64: 181–196.

47   Wood, H. 1979. The anatomy of transcarboxylase and the role of its subunits. CRC Crit. Rev. Biochem. 7: 143–160.

48   Wood, H. and R. Stjernholm. 1961. Transcarboxylase. II. Purification and properties of methylmalonyl-oxalacetic transcarboxylase. Proc. Natl. Acad. Sci. USA. 47: 289–303.

49   Stjernholm, R. and H. Wood. 1961. Methylmalonyl isomerase. II. Purification and properties of the enzyme from Propionibacteria. Proc. Natl. Acad. Sci. USA. 47: 303–313.

50   Hull, D. 1988. Science as a Process. University of Chicago Press, Chicago, IL.

51   Allen, S., R. Kellermeyer, R. Stjernholm and H. Wood. 1964. Purification and properties of enzymes involved in the propionic acid formation. J. Bacteriol. 87: 171–187.

52   Wood, H. and C. Werkman. 1936. Mechanism of glucose dissimilation by the propionic acid bacteria. Biochem. J. 30: 618–623.

53   Singleton, R. 1993. Introduction to the sulfate-reducing bacteria. In J. Odom and R. Singleton (eds.), Sulfate-reducing Bacteria: Contemporary Perspectives. Springer-Verlag, New York.

54   Bauchop, T. and S. Elsden. 1960. The growth of micro-organisms in relation to their energy supply. J. Gen. Microbiol. 23: 457–469.

55   De Vries, W., W. van Wyck-Kapeteyn and A. Stouthamer. 1973. Generation of ATP during cytochrome-linked anaerobic electron transport in propionic acid bacteria. J. Gen. Microbiol. 76: 31–41.

56   Kornberg, A. 1989. For the Love of Enzymes: The Odyssey of a Biochemist. Harvard University Press, Cambridge, MA.

57   Schambye, P., H. Wood and M. Kleiber. 1957. Lactose synthesis. I. The distribution of C^{14} in lactose of milk after intravenous injection of C^{14} compounds. J. Biol. Chem. 226: 1011–1021.

58   Wood, H., P. Schambye and G. Peeters. 1957. Lactose synthesis. II. The distribution of C^{14} in lactose of milk from perfused isolate cow udder. J. Biol. Chem. 226: 1023–1034.

59   Wood, H., P. Siu, and P. Schambye. 1957. Lactose synthesis. III. The distribution of C^{14} in lactose of milk after intra-arterial injection of acetate-l-C^{14}. Arch. Biochem. Biophys. 69: 390–404.

R. SINGLETON JR.

60 Wood, H., S. Joffe, R. Gillespie, R. Hansen and H. Hardenbrook. 1958. Lactose synthesis. IV. The synthesis of milk constituents after unilateral injection of glycerol-1,3-C^{14}. J. Biol. Chem. 233: 1264–1270.

61 Wood, H., R. Gillespie, S. Joffee, R. Hansen and H. Hardenbrook. 1958. Lactose synthesis. V. C^{14} in lactose, glycerol and serine as indicators of the triose phosphate isomerase reaction and pentose cycle. J. Biol. Chem. 233: 1271–1278.

62 Hansen, R., H. Wood, G. Peeters, B. Jacobson and J. Wilken. 1962. Lactose synthesis. VI. Labeling of lactose precursors by glycerol-1,3-C^{14} and glucose-2-C^{14}. J. Biol. Chem. 237: 1034–1039.

63 Wood, H. and J. Katz. 1958. The distribution of C^{14} in the hexose phosphates and the effect of recycling in the pentose cycle. J. Biol. Chem. 233: 2719–1282.

64 Wood, H., J. Katz and B. Landau. 1963. Estimation of pathways of carbohydrate metabolism. Biochem. Z. 338: 809–847.

65 Williams, G. 1980. Western Reserve's Experiment in Medical Education and Its Outcome. Oxford University Press, New York.

66 Neish, A. 1952. Analytical Methods for Bacterial Fermentations, Second Edition (Report No. 46-8-3). National Research Council of Canada, Saskatoon.

G. Semenza and R. Jaenicke (Eds.)
Selected Topics in the History of Biochemistry: Personal Recollections, V
(Comprehensive Biochemistry Vol. 40) © 1997 Elsevier Science B.V.

Chapter 8

Fate has Smiled Kindly

S.V. PERRY

Department of Physiology, School of Medicine, University of
Birmingham, UK

Early years

My interest in chemistry was stimulated at school in my
early teens. Up to that time, the applications of chemistry to
biology had not been brought to my attention for during the
1930s there was not much emphasis on biology as a discipline for secondary education in Britain. At my school in
Southport, Lancashire, biology was not taught at the higher
level and the preparation for science and medical courses at
university was invariably through the Higher Certificates in
mathematics, physics and chemistry. Whilst still at school,
my imagination was stimulated by several books from an
American author, Paul de Kruif, describing major discoveries in the medical sciences. His absorbing account of the
isolation of insulin, by Banting and Best, in particular, fired
me with an intense interest in the relatively new discipline,
biological chemistry, with its application of chemistry to
medical and biological problems.

Although neither of my parents had academic back-
grounds, I was strongly encouraged by my mother to aim at
attending university. At that time, a university education
was not widely considered as an essential background to all
careers, as is the case today. Certainly, it was the route to

the established professions such as teaching, medicine, dentistry and science generally. On the other hand, if you were aiming for a career in business, it was considered that you were probably wise to start straight from school. There were financial problems if I was to carry on with my education. My father, a commercial traveller for a mail order firm, suffered from a heart condition resulting from rheumatic fever contracted during service in the trenches in France during World War I and died at an early age when I was 13. Unlike the current situation in Britain, grants based on a means test were not then available to all students who had been admitted to university. Nevertheless, a small scholarship, supplemented by my mother at considerable personal financial sacrifice, enabled me to enter the University of Liverpool in the autumn of 1936. Liverpool was about 20 miles away and I travelled daily to the university from home, as did most of my contemporaries.

It so happened that Liverpool was one of the three British universities, at that time, providing undergraduate courses in biochemistry or physiological chemistry. The total number of persons graduating in Britain with first degrees in biochemistry, in the late 1930s, was small, about 15 to 20 students per year. At Liverpool, aspiring biochemists took the first 2 years of the honours course in chemistry and transferred to biochemistry for the third and final year. As a first year student allocated alphabetically to space in the practical laboratory of the chemistry department, I found myself sharing a bench with Rodney Porter. From this contact started a close friendship which lasted until his death in 1985. Later to become a Nobel Laureate for his work on the structure of the immunoglobins, Rod was within a few weeks of retiring from the Whitley Professorship of Biochemistry at the University of Oxford when he was tragically killed in a road accident.

Rod and I transferred from chemistry to the biochemistry department in 1938, to join the final year class of six students. We both graduated with first class honours degrees in the summer of 1939, full of hopes and ambitions, but events were moving rapidly in Europe. By midsummer, war seemed inevitable. Despite the uncertainties as to the future, we were keen to stay on for research and were accepted in the biochemistry department at Liverpool as post-graduate students. In the meantime, the British government decided that, as a precaution, young men of my age group were to be called up for national service in the Militia, as it was then called. At that time it was not certain how soon the call would come or whether there was any possibility of deferment for students. On the declaration of war on 3 September, it was obvious that I could not expect to be able to defer call-up for long. The head of the department of biochemistry, Professor H.J. Channon, also made it very clear to Rod and I, that he saw little chance of either of us being able to complete a Ph.D. in view of the current state of emergency. A recruiting office had opened in the university so I enlisted in the army and joined the University Officer Training Corps on 18 September. No doubt because of my degree in biochemistry, I was offered the choice of joining a chemical warfare unit or the Royal Artillery. I had no hesitation in selecting the latter.

Although now officially a member of the armed forces, it became apparent that it would be some time before I was called up for training. Mobilization of existing reserves had left little capacity for the training of raw recruits and some attempt was being made to permit students to finish at least part of their courses. Despite the uncertainty, it seemed sensible to carry on in the laboratory and gain some experience for as long as I could. I had a feeling that Channon did not have a clear idea of how, under the circumstances, I could be best employed but he gave me a variety of tasks, not all of which could really be described as research. One of these was to repeat a recently published preparation of glutamine from beetroot. This experience I remember well from the large volumes of brightly coloured extract that I processed, but the reasons for carrying it out were not clear to me. Most of my time was spent attempting to investigate the distribution and metabolism of iron in rat tissues. This was not an ongoing research interest of the department but was designed to take advantage of a sporadic supply of radioactive iron and other radionuclides that had become available to the department. Radio-active isotopes were not, at that time, generally available in the United Kingdom but samples were passed on to Channon by Sir James Chadwick, head of the adjoining physics department, who was awarded the Nobel Prize for the discovery of the neutron. Chadwick was receiving samples of these newly produced isotopes from Lawrence in California, with whom he had strong research links. After the physicists had studied the new isotopes, they were passed on to the biochemists to use in metabolic studies. Counters were not commercially available and those we used in the physics department were very simple by modern standards. They were made by the physicists themselves and used in primitive holders with loose wires issuing from them. Likewise, precautions to prevent contamination were virtually non-existent, such was the current ignorance of the dangers in handling these substances. At the same time, Rod was using ^{32}P to study the metabolism of phospholipids, a subject closer to the main research interest of the department. These experiments were probably some of the earliest of their kind employing radioactive isotopes of biological importance carried out in the United Kingdom.

For the first few months after the outbreak of war, life at Liverpool did not change very much. By late spring of 1940, World War II had started in earnest, the British and the French were in retreat in France and the indications were that very soon we would be called to our units. In the meantime, Rod and I spent some time each day drilling in the university quadrangle, just in front of the biochemistry department, with the Officer Training Corps. In June, when the invasion of Britain was beginning to look to be a very serious possibility, we decided with a few student friends who were also waiting for call-up, to contribute directly to the war effort by assisting the farmers. For the next few weeks we worked as farm labourers at the village of Blockley, in the Cotswolds. Most of our time was spent removing, by hand, wild mustard that was threatening to smother a large field of kale. When we started, the whole field was one great expanse of yellow and each weekend we would walk across the valley and look back to see how far the yellow line had receded. One by one notifications to join units arrived and by August we were all in the forces.

War service

August to November was spent training as a field gunner at an isolated camp on the moors near Harrogate, in Yorkshire. After further training in field and anti-tank gunnery at the School of Artillery on Salisbury Plain in Wiltshire, I was commissioned as a 2nd lieutenant, early in the summer of 1941. At that time, newly commissioned officers were given the option of choosing the war theatre to which they wished to be posted. I opted for the Middle East, unaware then that this was to be the last time during service in the forces when I was ever given any choice. As a young man seeking a little adventure, this seemed an interesting part of the world and if you had to fight a war, the desert was a sensible place to do so. It turned out to be a momentous decision and one which meant that I was in due course able to return to biochemistry.

Early one morning in July 1941, at Liverpool, I boarded the *Empress of Australia*, an ancient passenger vessel confiscated from the Germans after World War I, as part of the reparations and now employed as a troopship. After the first day at sea, the convoy finally assembled in the North Channel between Ireland and Scotland. It was an impressive sight consisting of four large passenger liners containing troops, some twenty to twenty five cargo ships loaded with supplies and vehicles, many of which could be seen lashed to the decks. At this time, Britain was alone in the war and losing hundreds of thousands of tons of shipping per month in the North Atlantic with the German submarine campaign at its height. Twice a day we assembled on deck for lifeboat drill wearing our old-fashioned kapok lifejackets which we were told to keep close by at all times. I must confess that on these occasions I often thought of the chaos that would ensue if a torpedo was to hit our crowded troopship. It was comforting to see that we had a very strong naval escort. A screen of destroyers and cruisers surrounded the convoy with the accompanying battleship, the *Repulse*, ploughing through the seas at the centre. From this display, and the fact that an aircraft carrier remained with us until we reached the South Atlantic, it was apparent that the convoy was an important one. We subsequently learnt, when we had been at sea for a short time, that it contained the elements of an armoured division and reinforcements for a major offensive to be launched in Libya later in the year. After about 6 weeks, much of which was spent zig-zagging across the Atlantic to avoid the enemy submarines and was broken by stops at Freetown and Durban, we landed at Port Tewfig, Egypt.

After a brief period of training in desert navigation and warfare, I was sent forward as an officer replacement when the November offensive to relieve Tobruk started. I contracted severe dysentery and was sent back to hospital in Cairo. On recovery, I was sent forward once again to the battle areas which had moved a good deal further west. Tobruk had been relieved and Rommel had retreated to behind the salt marshes at El Agheila. By the time I reached my unit, an artillery regiment armed with 25 pounder guns, Rommel had counter-attacked earlier than expected, catching our forces off balance. My regiment was part of a small column acting as a rearguard covering the retreat along the coast road from Benghazi towards Derna. After several days in contact with the German forces, we were finally overrun on 2 February 1942, in the open desert south west of Derna. We were pinned down by intense machine gun, mortar and artillery fire and were unable to withdraw. I was captured trying to do what I could for one of the gunners with a shrapnel wound in his chest. At the time, I was unarmed for my revolver had been stolen from me some weeks before in Cairo. In retrospect, this was perhaps fortunate for if I had had a weapon I would have most probably used it. In that case these recollections would not have been written. A revolver is not a very effective weapon to take on heavily armed German infantry advancing in the open desert and I would have soon been silenced.

Captivity

Two days after capture, my German captors handed me over to the Italians and I was transported in stages to a large prisoner of war (POW) camp at Tarhuna some 60 km SW of Tripoli. Subsequently, I was taken to Tripoli and loaded into the hold of a Italian cargo boat, one of three transporting Allied prisoners. On 23 March, the small convoy escorted by four destroyers of the Italian navy left for Italy. Our apprehensions about this dangerous trip across the Mediterranean were confirmed for we were shadowed by British submarines and the convoy split up. One of the cargo boats containing prisoners, including men from my battery, was torpedoed but fortunately that in which I was travelling reached Naples 3 days later with two of the four destroyers of our original escort.

Most of the time in Italy was spent at Padula, a small village in the mountains about 60 miles SE of Salerno. Here POWs were housed in an old Carthusian monastery, La Certosa di San Lorenzo, surrounded by a barbed wire fence. My constant concern as POW, both in Italy and later in Germany, was to get enough to eat. According to the Geneva Convention governing prisoners of war that was observed by combatants in the west European theatre, officer prisoners, unlike other ranks, were not allowed to work. This meant that our rations were those of the non-working civilian, a very rare species in Europe during the war years. I calculated these rations provided 1000–1100 calories per day. On such an intake I steadily lost weight and if I had had to depend solely on it for the whole three and one quarter years I spent as a prisoner, I very much doubt whether I would have survived. Fortunately, we were able to supplement these basic rations by sporadic supplies of food parcels sent from Britain and Canada and other Dominions, to Switzerland, via neutral countries and distributed by the Red Cross.

After the Allies invaded Sicily, we were moved north to a camp outside Bologna. Whilst here, the Italian armistice was announced on the evening of 8 September 1943, and we were all ready to break out and disappear into the countryside. Despite our strong protestations, the Italian camp commandant ignored the conditions of the armistice and refused to remove his guards from the camp perimeter. In the middle of the night, a few hours after the armistice announcement, a German infantry unit burst in, took over the camp and replaced the Italian guards. In the first few minutes of confusion, I and about 50 others managed to break out through a gate at the back of the camp. It was clear that the German takeover had been carefully planned in advance with the whole camp complex carefully surrounded to prevent any escape for most of us were recaptured. I was caught trying to get through the German cordon by crawling through a maize field about 400 m south of the camp. A few days later we were loaded on to cattle trucks at Modena station, bound for Germany via the Brenner Pass.

The thought of spending the rest of the war in a POW camp in Germany filled me with despair and I resolved to get off the train if I possibly could before we reached the Brenner. Whilst our train was stationary in a siding close to Mantua station, I managed to get out of the bolted truck by pleading with the German soldier on guard that I needed to obey a very urgent call of nature. In the crowded cattle trucks there were no toilet facilities of any kind. When his attention was momentarily distracted, I quickly sneaked away and hid in an empty passenger train until it was dark and safer to move around. During the evening, I spent some hours attempting to find my bearings in Mantua, a city with which I was totally unfamiliar. The streets were strangely deserted for, as I realized later, the Germans had imposed a curfew. After some time wandering around the streets, I came upon an Italian who was making his way to his father's farm, north of Mantua. I joined him for I intended to travel in that direction to make my way to neutral Switzerland. This was not to be for I was recaptured by the German guard on a bridge I was attempting to cross, on the road out of the city. He allowed my companion across for he had a pass, but recognized me as the Englander his section had guarded when I was recaptured 5 days earlier, outside the camp near Bologna. By some amazing coincidence, after loading us on a train at Modena, his unit had moved 60 miles NW to Mantua to protect the lines of communication to the Brenner and he was now on guard on this particular bridge. I was quite taken aback by this turn of events and felt that fate was not on my side, but the Germans treated me very well. I was escorted to a villa in Mantua which the German unit had commandeered as their headquarters. There I was taken to room where about eight German officers were sitting round a table drinking wine. Among them was the lieutenant who, after my recapture at Bologna, had driven me back to the POW compound a few days previously. He inquired how I had managed to escape again and introduced me to the other officers. I was given a seat at the table and offered wine which I gladly accepted.

We discussed the war and later I was locked in the bathroom for the night. The following day, I was returned to Mantua station and, with three other recaptured British officers, sent to Germany locked in a cattle truck once again.

The prisoners rushed out of Italy after the armistice put a strain on POW accommodation in Germany. I was held in camps at Moosberg near Munich, Fort Bismark near Strasbourg, and Weinsberg near Kahlsruhe, before reaching a more permanent camp at Mahrish Trübau in the Sudeten Gau, Silesia. In May 1944, as the Russians advanced in the east, we were moved further west to Brunswick. During the journey to Brunswick, a hole was cut in the end wall of the cattle truck and with a friend I jumped off the train as it was moving through Bohemia. Once again the exhilaration of freedom was short-lived for we became separated and I was picked up by a German patrol in a region where we did not expect to find one. My companion was also recaptured soon afterwards and landed up in the Gestapo jail in Prague. As I learnt after the war, this was a very dangerous time to escape in this part of Czechoslovakia. A few weeks earlier, there had been a massive escape via a tunnel from a nearby Royal Air Force officers' POW camp in Sagan, Silesia. Hitler was so incensed by this that he gave instructions that all escapees from this camp should be rounded up and shot. Of the 76 who escaped, 50 were shot. Fortunately I did not get caught up in this exercise.

According to the Geneva Convention, prisoners should not be punished for escaping, nevertheless I was charged with damaging the property of the Deutsches Reichbahn (Fig. 1). In due course I appeared before a military court in Hildesheim. My trial was conducted in an air-raid shelter during a heavy raid by the American airforce. As expected, I was found guilty and sentenced to one month's solitary confinement. On return to Brunswick, we discovered that the city too had suffered a heavy raid and of more personal concern, the camp had been hit. Our camp was located close to an

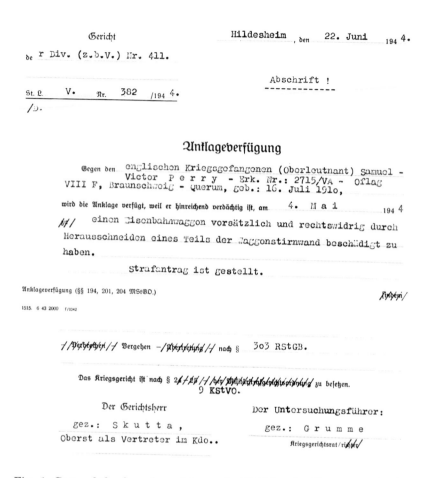

Fig. 1. *Copy of the form issued to me in 1944 by the German authorities indicating the offence for which, in due course, I was to be court-martialed.*

aircraft engine works on the perimeter of Brunswick airport, obviously to discourage the bombing of these sensitive targets. Nevertheless the engine factory was destroyed in the raid but about 200 bombs of assorted sizes landed in the compound. There were a number of casualties, dead and wounded, among the prisoners, but by some miracle, less than might have been expected. That day was the only occasion upon which I had been outside a POW camp, except during transportation between camps, during the whole of my stay in Germany.

My sentence was carried out in a military jail, in which I was carefully segregated from the other occupants, disaffected German soldiers. The only contact with them was when, under supervision of a guard, I was allowed a daily 30 min exercise period walking round a small courtyard. In the walls of the courtyard were cell windows, at each of which, was the face of the occupant. Despite the efforts of the guard to prevent any contact with the other prisoners, each time I passed a window a few words were exchanged. It was not the most satisfactory way to carry out a conversation but nevertheless, when the news reached the jail that the Ardennes offensive had started in the winter of 1944, the wave of despair amongst the occupants was all too obvious. For most of them, the only hope of release was an early end to the war. One of the most worrying features of this period was the constant air raids, during which I was left locked in my cell. I would lie under the bed and worry about what I should do if an incendiary bomb, which were widely scattered during the Allied raids, were to arrive in the cell. After this experience, I was quite pleased to be discharged into a normal POW compound when I had served my term. We were liberated at Brunswick on 12 April 1945, by the advancing Americans, and flown home to England a few days later.

Looking back, it is difficult to remember how one managed to endure the passage of time as a prisoner, 1165 days in my case. When books were available, one read a lot and during more settled periods many classes were organized. I gave courses in biochemistry and agricultural chemistry. As there were no suitable books, I asked in one of the infrequent postcards we were allowed to send home, if some biochemical literature could be sent to me in Padula. Channon arranged for me to be made a member of the Biochemical Society. As probably the only member of the society ever to be elected in a POW camp, journals were sent to me through the Red Cross. In fact, few arrived and if they did, they were in a very tattered state. I was, however, delighted to receive the *Annual Review of Biochemistry* for 1942, Volume 11.

Just prior to the arrival of this volume at Padula, there had been an escape and the Italians suspected that maps and money was being smuggled into the camp concealed in the hard backs of books. The books already in the camp were identified by an official stamp and it was decreed that all new books coming into the camp would have their hard backs removed. I learnt through British contacts who worked in the camp post room, that the annual review had arrived and was waiting to have its back removed. As it was the only book I possessed, I was very anxious that it should be preserved intact. I therefore arranged for it to be smuggled out of the post room and marked with the appropriate stamp, forged in the camp, so that it would pass future inspections. As time passed by, I became very attached to this volume. On the three occasions on which I escaped, it was left behind, usually in cattle trucks, but on recapture I managed subsequently to retrieve it. The most remarkable recovery was when I left it in the truck at Mantua station. Sometime later at a camp in Bavaria, I caught up with fellow prisoners who were in the truck from which I had escaped. They had held on to my few possessions, including Volume 11. The book remained in my possession for the rest of my travels (Fig. 2). I became so attached to it and its provenance that it was one of the few items that I took with me on repatriation to Britain after liberation. In some ways, it is a monument to my incompetence as an escapee. If I had escaped successfully, this volume would not have survived the war. It is now in the archives of the Biochemical Society.

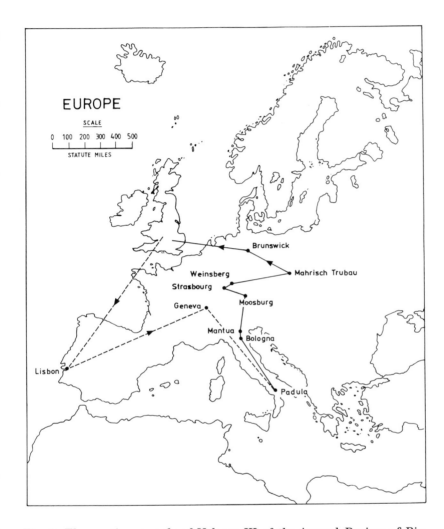

Fig. 2. The wartime travels of Volume XI of the Annual Review of Biochemistry. ------, Under auspices of the International Red Cross; ———, with
Lieutenant Perry.

Halcyon days

After repatriation and a couple of months leave, I was sent
to north Yorkshire for a refresher course in gunnery and put
on the list for active service overseas. A Far East posting
seemed imminent. Fortunately, the dropping of the atomic
bomb on Japan changed all that and I started planning my
return to biochemistry. With this in mind, I went to see Professor R.A. Morton, now the head of my old department at
Liverpool, who had built up an internationally distinguished
group working on vitamin A and the carotenes. Although
Morton was prepared to offer me a place as research student, as a POW, I had had a lot of time to think about my
future on demobilization and the idea of working at Cambridge attracted me very much. Soon after the Japanese ca-
pitulation, Trinity College, Cambridge, advertised a Rouse
Ball Research Studentship for which I applied and was fortunate enough to be awarded. On the strength of this studentship, Professor A.C. Chibnall offered me a place in the
department of biochemistry at Cambridge and I decided not
to return to Liverpool. I knew that Rod Porter, who was then
in the forces in Italy, was also anxious to return to biochemistry, so I wrote and suggested that he join me. On his next
home leave Rod went to see Chibnall, with the result that we
both started research early in January 1946, having as students been granted accelerated demobilization.

To be returning to biochemistry in Cambridge, immediately after the war, was a tremendous experience. Due to the
wartime achievements of the atomic physicists, science had
a high profile and there was a great optimism in the air.
Many of the students returning to their studies had been
seasoned by wartime experiences and a high proportion of
them were very able and hard working. They were very
anxious to make up for the time lost to academic expression
by the demands of national service. The price that had been
paid was only apparent from the occasional burnt or scarred

face and disfiguring wounds sometimes revealed in swimming pools and changing rooms. In my whole academic career, now spanning well over 50 years, there was never a period quite to compare with that of the immediately post war
years in Cambridge. I shall never forget the pleasure of
walking for the first time through the Backs of Trinity College and seeing the crocuses breaking through in February
1946. Spring-time during the previous 4 years had been
spent pacing round the interior of barbed wire compounds in
north Africa and various parts of Europe.

The biochemistry department at Cambridge, in 1946,
seemed to me to be almost top heavy with senior scientists
possessing impressive international reputations. These included Malcolm Dixon, Robin Hill, Joseph and Dorothy
Needham, Charles Chibnall, Charles Haines, Marjory Stephenson, Earnest Gale and Kenneth Bailey. All were either
current fellows of the Royal Society or were elected whilst I
was in Cambridge. The former head of department, Sir
Gowland Hopkins, Nobel Laureate, now retired and half
blind, could often to be seen slowly making his way up the
stairs to the laboratory that he still occupied. Although we
were not to appreciate it at the time, a group of young men,
still in their twenties, were laying the foundations of careers
to be of enormous importance for the development of biochemistry. Peter Mitchell was in the department, as was
Fred Sanger, now joined by Rod Porter as his research student, all later to become Nobel Laureates. At one time or another, between 1946 and 1949, they all worked in the same
small research laboratory on the first floor of the building,
which I was fortunate to share with them. It is doubtful
whether there is any other laboratory room in the world that
can claim that its occupants, over such a short period, were
in the course of time to be awarded four Nobel Prizes.

Kenneth Bailey had just discovered tropomyosin and was
naturally much interested in the new muscle protein, actin.
News of the isolation of actin by Straub in Albert Szent-

Gyorgyi's laboratory in Szeged, Hungary, was beginning to percolate through to Cambridge after the end of hostilities in Europe. The original papers describing the work of the Szeged group had been privately published in 1941–1943, by Szent-Gyorgyi, in three volumes of the *Studies from the Institute of Medical Chemistry, University Szeged* [1]. Due to their limited circulation and the state of war in Europe, these journals were not generally available. Fortunately, when the European war ended, copies had been sent by Szent-Gyorgyi to his old friend W.T. Astbury, at Leeds. These papers are now in my possession. Astbury pioneered the X-ray structure of fibrous proteins and had shown that a number, including the muscle proteins myosin and Bailey's tropomyosin, all possessed a characteristic spacing along the fibre axis of 5.1 Å. This periodicity he ascribed to what he called the α-fold and which we now know to arise from the α-helix. As Astbury was keen to see if the structure of the fibrous form of the new protein actin was similar to the other muscle proteins, he sent the papers to Bailey, with whom he had many common interests and had in the past collaborated on protein structural problems. The papers were made available to me, and as Bailey's new research student in January 1946, I was given the task of preparing actin and thus started my career in muscle biochemistry.

Samples of F-actin, probably the first preparations to be made in the UK, were dried down on glass plates and taken up to Leeds for X-ray analysis. The actin diffraction pattern was new and unlike that of the α-proteins, but it enabled the contribution of actin to the X-ray pattern of whole muscle, earlier reported by Bear [2], to be identified [3]. It was concluded that in F-actin, the corpuscular units 'are not strung together in an arbitrary fashion but always in the same way with atomic precision'; a view which subsequent work has confirmed.

An aim of my thesis work was to try and understand the nature of the interaction between actin and myosin and the remarkable effects of ATP on it. At high ionic strength when both proteins were in solution, low concentrations of the nucleotide dissociated the viscous complex, whereas at low ionic strength when the complex existed as a gel it contracted, a model motile system. One of our hypotheses at this time, was that actin bound to myosin because the actin contained a prosthetic group composed of ATP or some similar compound containing a pyrophosphate group that acted as a pseudosubstrate for the enzyme. If the affinity of myosin for the pseudosubstrate was much less than for ATP, the dissociating effect of the nucleotide on actomyosin in solution could be explained. Sensitive methods for ATP determination were not then available but its presence was indicated by the estimation of acid labile phosphate.

A significant amount of labile phosphate was detected, even in well dialyzed preparations of actin [4], but the amounts were much less than would be expected from the minimum levels of ATP required to dissociate the system. For this reason we rejected the hypothesis. Our conclusions about the presence of this nucleotide were correct, but we should have explored the acid labile phosphate content further for we were detecting the bound ATP of actin. This was shown some 3 years later by Straub and Feuer [5] to be present in all G-actin preparations and to be converted to ADP on polymerization to F-actin. The study was also concerned with the role of sulphydryl groups on the interaction. As such, it was the one of the earliest to demonstrate their importance both for the ATPase activity and the interaction with myosin.

For much of this period I worked in a small research laboratory in the basement of the biochemistry building. This contained four research bays, with Porter and I at either end of the room, and our supervisors, Sanger and Bailey in the middle bays adjacent to their supervisees. I suspect the two rather elderly students, fresh from the army, were somewhat boisterous and at times expressions of concern at our behaviour could be detected in our supervisors' faces. Nevertheless, it was a very happy laboratory and the period spent there was one of the most enjoyable of my working life [6]. There was a succession of foreign visitors, most of them very distinguished, anxious to meet our two supervisors. Many came to see Sanger for his method for end group assay of peptides and proteins was an important approach to the determination of protein structure, then still the holy grail for many biochemists.

I remember the visit of Albert Szent-Gyorgyi to Cambridge, in 1946, very clearly. His lecture was given in the Molteno Institute, where he had worked in the 1930s, and David Keilin, now the director of the Institute, was in the chair. This was the first time Szent Gyorgyi had been to Cambridge since before the war and I was particularly anxious to hear first hand details of the Szeged work which had so changed our conceptions of muscle. His development of the actomyosin thread, which responded to the addition of ATP by contraction, provided the bridge between biochemical and mechanical events in muscle for the first time. The lecture was given with characteristic flamboyance and towards the end he produced his model. It consisted of a series of child's coloured wooden building bricks, on one side of each was inserted a staple. Through each staple was passed an elastic band under tension. He held, outstretched, the elastic with the attached bricks and then suddenly let go. With the tension released the model shortened to a coiled structure with the bricks distributed radially. 'Muscle contraction' he exclaimed triumphantly [7]. I, and I am sure most of the audience, failed to see the connection with the actomyosin system but no one was prepared to question him on it. It is more than likely that even he did not believe that he was demonstrating the actual mechanism but was making a point in a dramatic way. He was, however, taken to task over one incident. During the course of the lecture he was eulogizing on the properties of the myosin B, the name

given to what was known to be crude preparation of actomyosin and from which he could make his contractile models.
He happened to mention that it was a succinic dehydrogenase. One could almost see Keilin bristle. At the end of
the lecture, Keilin proceeded to demolish this loose statement. By asking a series of precise questions that required
yes or no answers, Szent-Gyorgyi was made to slowly withdraw his claim and agree, as we all had supposed, that succinic dehydrogenase was simply a contaminant in his preparation.

Extramural activities

As a research student I was about 7 years older than most of
my contemporaries who had progressed to the post graduate
level by the normal route straight from school. I worked very
hard in an effort to make up the lost time but it would be
wrong to give the impression that every minute of my time
was spent in the laboratory. At school and university I had
been very keen on rugby football and felt that I was getting
into my prime as a player in the first winter of the war before I went into the army. Thence onwards, with active
service and subsequent capture there was no opportunity to
pursue this interest. When I returned to Britain after liberation, aged 26, run down in health and about 40 lbs under-
weight, I considered that my prospects of playing the game
at any reasonable level were over. Nevertheless, after feed-
ing up over the summer and training to get fit, I found myself playing again and enjoying the game. At Cambridge I
was fortunate to get into the university team and subsequently into the English international XV for 1947 and 1948.
These activities made considerable demands on my time and
I consider myself fortunate in having a supervisor, such as
Kenneth Bailey, who was prepared to tolerate my frequent
absences from the laboratory. During the rugby season, as a
member of the university team, I was expected to train most

afternoons, whilst for international matches, I would disappear on Thursday morning for a game at the weekend. I am rather ashamed to confess that in later years I did not exhibit such forbearance to absences from the laboratory by my own research students.

My associations with rugby football were responsible for early contacts with muscle scientists in Japan. In the early 1950s, the Japanese Government felt dominated by the American presence and the influence of its culture. MacArthur was in power and the Korean war was in progress with many American troops in Japan. As part of a policy of increasing contacts with Britain, approaches were made to the rugby and rowing clubs of Oxford and Cambridge universities, inviting them to send teams to tour the country. In response, the Cambridge University rugby union football club toured Japan in the late summer of 1953. As by that time I had given up playing myself, I was invited to accompany the team as the manager. This opportunity appealed to me as I had never travelled to the Far East before, but I was concerned as to whether, as a university staff member, I could justify a 6 week absence from the department for a reason that could hardly be considered to be biochemical. I approached the head of the department, explaining that although my reasons for going may appear to be rather unacademic, the trip was valuable in promoting the university's image abroad. Permission was given as my absence would not interfere with my teaching duties, but I gained the impression that he felt that I could spend my time more profitably at the bench.

It would be fair to say that in the first instance I had some apprehension about the trip, conditioned by the current anti-Japanese sentiment that was strong in certain quarters in Cambridge at the time. This was a reflection of the fact that a Cambridge regiment had been captured at the fall of Singapore and many of its members had died in Japanese POW camps. I need have had no fears for we were treated as VIPs with great friendliness and hospitality. Indeed as Setsuro Ebashi once said to me, many years later, 'You were treated like princes when life was very difficult for us'. At that time rugby was not a major sport in Japan but was played mainly by universities. It had a good standing and support in high places for the Emperor's late brother, Prince Chichibu, had been fond of the game. Indeed, the national ground was named after him and his widow, Princess Chichibu, attended most of our major receptions. I was delighted that we won all our games, three against All Japan.

As we played in most of the major cities in Japan, I made a point of visiting university biochemistry departments when possible for very few British biochemists had visited Japan since the war. In Tokyo, I gave a lecture on some aspects of my research on the actomyosin system and met, for the first time, Setsuro Ebashi, Kosak Maruyama and Hiroshi Kumagai, colleagues whose friendship I have continued to enjoy over the years. I think our hosts were surprised that a rugby team manager could manage to give a lecture on somewhat specialized biochemistry. One of my most vivid memories of that trip was provided by Reiji Natori. After introduction, he took me into the laboratory to show me his skinned muscle fibre preparation, then unknown in the west. He dissected out a single frog muscle fibre, immersed it in a drop of oil and sharpened a match-stick. This he held in his hand and ran it carefully along the surface of the fibre, peeling off the membrane as he went. Much impressed with the elegance of this preparation and its potential as a research tool, I did what I could to publicise it when I returned to Britain [8] for the technique was not then described in any readily available literature.

Post doctoral work in the United States

The evenings and early hours of the autumn of 1947 were spent writing a dissertation on my research, to be submitted for the Trinity College Prize Fellowship competition. To my surprise, and great pleasure, I was successful in this endeavour. The Trinity Fellowship, in addition to providing a small salary and allowing me to take my meals at the High Table in college, also permitted me to walk across the college lawns, a concession restricted to fellows. To be even a very junior member of the Fellowship of Trinity was a great privilege and experience. Trevelyan, the historian, was master at the time; Bertram Russell was in residence and it was not unusual for me to find myself occasionally sitting next to him at lunch. There were also many other very distinguished fellows including Frisch, the atomic scientist, Broad, the philosopher, Hardy, the mathematician, and the outstanding physiologists Adrian, Hodgkin and A.F. Huxley. In the summer of 1948, I was married and with my wife Maureen, sailed to the United States where I was to hold a Commonwealth (now Harkness) Fund Fellowship at the University of Rochester, New York.

I had elected to spend my Commonwealth Fund Fellowship at Rochester, New York, where Wallace O. Fenn, the distinguished muscle scientist, was head of the physiology department . My approach to muscle had, up to that time, been very much that of the biochemist. As such, muscle was a tissue from which you made extracts after mincing or converted to an acetone powder before you carried out any experiments. One thing that was clear about muscle was that its characteristic structure was very important for its function. For this reason, one could not hope to understand the mechanisms involved without paying much more attention to the structural aspects of muscle than was the habit of biochemists at that time. As a background to this approach, I wanted to get some experience in handling the intact tissue. Fenn's department seemed at the time the place to do this. I was grateful to Fenn who, even though he was quite busy with various national committees, found time to show me personally, how to dissect and generally manipulate the isolated frog sartorius muscle preparation. I was much impressed with the ease with which this muscle could be handled and its durability. He suggested a number of lines of research that I might follow but did not press them when I did not appear too enthusiastic, as they were rather more physiological than I felt capable of undertaking with any confidence.

It so happened my interest had been aroused by an observation that I had made with frog sartorius muscles soon after my arrival in Rochester. After dissection, they could be stored for a few days just above 0°C and still responded normally to stimulation. If they were frozen, however, on thawing they immediately shortened irreversibly to 15–30% of the original length. Much struck by this phenomenon, which was completely new to me, I discovered that there had been brief reference to it in the literature but little attempt made to explain the mechanism. By the use of iodoacetate I was able to show that if the ATP stores of the muscle were depleted, 'thaw rigor', as I called the phenomenon, did not occur [9]. The shortening was a result of the uncontrolled effect of ATP on the myofibrils in the muscle cell disorganized by low speed freezing. This disrupted the sarcoplasmic reticulum, allowing the stored calcium to leak out on thawing and trigger off the actomyosin ATPase. Today all this seems obvious but at that time, the properties of actomyosin had just been discovered and nothing was known about the intracellular regulatory systems of muscle. Another striking feature of thaw rigor was a loss of about 35% of the muscle weight due to the extrusion of fluid that accompanied the shortening process. A process similar to the synaeresis that occurs when actomyosin suspensions are treated with ATP. I was to become more conscious of the consequences of this phenomenon much later in life when I became associated with research in the meat industry, where thaw rigor is often a problem with meat frozen soon after slaughter. The liquid issuing from such meat on thawing is known in the meat trade as 'drip' and packaging is often designed to accommodate this problem.

The experience with the frozen sartorius muscles increased my interests in the structure of the muscle cell. This had been stimulated earlier by the meticulous work of the 19th century histologists in defining the organelles of the muscle cell and their distribution. Much of this work was collected in a splendid article entitled 'Die Contraktile Substanz' by Heidenhain [10]. By a lucky chance, I had discovered in a Cambridge secondhand bookshop, a copy of the volume of his book, *Plasma und Zelle*, containing this chapter. The muscle cell contained unique structures and it seemed the time was now ripe to attempt to isolate them for their biochemical study. It so happened that at the Federation meeting in Detroit, in the spring of 1949, I heard a short communication [11] describing a small scale method for isolating myofibrils using trypsin on thin frozen sections of skeletal muscle. Immediately on return to Rochester I attempted to prepare myofibrils by this method. In determining the ATPase activity as an index of the quality of the preparation, I noted that not all the enzymic activity centrifuged down under conditions when I would expect myosin or actomyosin to be insoluble. This effect was somewhat variable and depended on the concentration of trypsin used, suggesting that there might be partial degradation of the enzyme. By now my time at Rochester was running out, but although the method showed promise it required some work to produce satisfactory myofibrils in quantities that could be used for biochemical studies. Further work had to be deferred until I returned to Britain.

One of the attractive features of the Commonwealth Fund Fellowships was that fellows were expected to make a tour of the United States visiting institutions and individuals carrying on work of interest to them. To enable them to do this, fellows were provided with a generous travel grant. Maureen and I spent the last 2 months of our stay in a grand tour which took us down to Mexico, to California, thence up the Pacific coast to Vancouver, and back through the northern states to Rochester. In the course of this trip we drove about 20 000 miles and I visited most of the laboratories carrying on research of interest to me. Thanks to letters of introduction from Chibnall, we were entertained by the Coris in St Louis and in his flat in Bethesda I had a talk with Otto Meyerhof, then approaching the end of his life. With Wayne Kielley he had just reported the isolation of a new ATPase from muscle, the so-called Kielley–Meyerhof ATPase, an enzyme in which I was later to become interested. I also visited Ernst Fisher at Richmond and admired his splendid art collection which included paintings by Paul Klee. In addition, I met John Gergely at Bethesda, Wilfred Mommaerts at Duke and many others with whom I have continued to keep in contact over the years.

I made a special visit to see the great man himself, Albert Szent-Gyorgyi, then established at Woods Hole. After a stimulating talk on the beach close to his house, he suggested that we should go for a swim. This we did and at one time I found myself struggling somewhat to get back to the beach. My host, although considerably older than I, swam apparently effortlessly, with no problems at all. This impressed me but later I learnt that Szent-Gyorgyi was an expert on the local currents and how to take advantage of them when swimming; information he did not always impart to casual guests when they swam with him.

In September 1949, we left Rochester very conscious of how hospitably we had been treated by Wallace Fenn, his family and his colleagues in the department. My lasting memory is of Fenn sitting in his sailing boat on Lake Canadaigua, hand on the tiller, with his strong New England profile silhouetted against the sky. Many in Rochester expressed surprise that we were returning to the UK which was still suffering from the effects of war, with food and many other items still rationed. The call of Cambridge was still strong, however.

Return to Cambridge

On return to Cambridge in September 1949, I was very anxious to get back to the myofibril work, preferably in the department of biochemistry. Whilst I was in the USA, Chibnall had given up the chairmanship of the department and had been replaced by F.G. Young. In consequence, there had been a considerable change in research emphasis and I was uncertain of my position as I was not on the tenured staff. Fortunately, I was given research facilities on the strength of my college fellowship. As was normal at that time, with people in my position, I had no personal research funds and relied on the department to provide me with materials and equipment for research. Young did, however, encourage me to apply to the Medical Research Council for a small personal grant for research expenses, of £100 per annum. This, subsequently slightly increased, was the only external support for research expenses I received the whole time I was in Cambridge. The stipend associated with the college fellowship was small and certainly not adequate enough to support my wife. I was grateful to Astbury who provided a supplement which brought my income close to a level appropriate for a post doctoral fellow. The only obligation I had for this support was to supply him with muscle proteins when he required them for his X-ray work. It was possible also, to make a little extra income by undertaking tutorial work for the colleges, the standard practice for postdoctoral fellows. In 1951, I was appointed to the tenured post of lecturer in the department of biochemistry, a position that provided me with some financial security. Although it gave me the opportunity to develop my research interests, it also meant that more of my time had to be devoted to teaching duties.

Muscle organelles

It soon became clear that tryptic digestion was not a satisfactory procedure for the preparation of myofibrils, as it was
degrading the myosin in situ to a soluble form [12]. John
Gergely independently reported a similar observation on
isolated myosin. These observations led, later, to the preparation of heavy meromyosin by Andrew Szent-Gyorgyi [13].
In the light of the experience with trypsin, I turned to the
use of collagenase to break down the muscle cells. This enzyme was not generally available but had been isolated as
one of the toxins of C welchii by a group set up in Britain
early in the war, to investigate this organism. It had infected
wounds producing gas gangrene during World War I with
devastating effects and during World War II, the Ministry of
Defence wished to be ready to cope with this problem if it
became serious again. Elizabeth Bidwell, who had been engaged in this research and was then working at Burroughs
Welcome Research Laboratories, kindly sent me supplies of
this, then unique, enzyme.

Using collagenase, I was able to made excellent myofibrils
on the larger scale, the only problem being to avoid the thaw
rigor occurring in the frozen sections of fresh muscle when
they were transferred to buffer solutions for digestion [12]. A
somewhat ironical turn of events in view of my earlier work.
By cutting thin sections and transferring them immediately
into large volumes of ice-cold buffer, this could be mini-
mised. The amounts of collagenase available to me were
strictly limited and not adequate for the preparation of myofibrils on the scale I needed. After some trials, it became apparent that acceptable myofibrils could be made simply by
homogenization in a suitable buffer [14]. Unless special precautions were taken to remove the calcium by EDTA [15],
these were partially contracted by the action of the endogenous ATP when fresh muscle was used. Uncontracted myofibrils could be prepared from muscle strips which had been

held at rest length before allowing them to go into rigor. About this time, Sir John Randall and Jean Hanson visited me at Cambridge to inquire about myofibril preparations, and I described to them how satisfactory uncontracted myofibrils could be prepared without the use of enzymes. This technique was used later by Jean Hanson in her work with Hugh Huxley at MIT in the USA.

In the course of preparing myofibrils by a non-enzymic method, I noticed that the supernatant left when the myofibrils had been removed was slightly turbid and possessed ATPase activity that was clearly not of the myosin type. High speed centrifugation removed the turbidity and yielded a preparation that was granular in appearance in the microscope and was rich in lipoprotein [16]. I concluded that this consisted of the muscle microsome fraction and that the ATPase was identical with the so-called Kielley–Meyerhof enzyme that the authors had isolated from muscle using the standard protein preparative methods rather than cell fractionation procedures. They concluded that their enzyme was probably of microsomal origin because of the association of lipoprotein with their preparation. As I was to learn later, I was in fact isolating crude preparations of the sarcoplasmic reticulum for the lipoprotein granule fraction, as I called it at the time, which was later shown by Kumagi et al. [17] to have relaxing factor activity. Indeed, I later used the preparation for study of its inhibitory activity on the myofibrillar ATPase [18].

With the first report of the isolation of intact mitochondria from liver, by Hogeboom and Schneider in 1946, there was much interest in the biochemical activities of these organelles. Virtually all the work since that time had been carried out on mitochondria from liver and kidney because of the relative ease of preparation from these tissues where they were the major components of the cytoplasm. Cardiac muscle had long been used for the preparation of oxidative enzyme systems but in general muscle mitochondria had not been studied as isolated organelles. Indeed it was not clear how similar they were to liver mitochondria for Heidenhain [10] had distinguished two types of sarcosomes in skeletal muscle, namely A and I granules, depending on their localization along the sarcomere. Against this background, Brian Chappell, my first research student, set about studying the properties of muscle mitochondria. It is much more difficult to separate mitochondria from muscle for, unlike the liver, the major cytoplasmic organelle with similar sedimentation properties is the myofibril. Although fast skeletal muscle of the rabbit was the tissue of choice for myofibril preparations, it contains very few mitochondria. Preparations were, therefore, made from the highly oxidative pigeon breast muscle which is rich in mitochondria. We could find no evidence for two types of muscle mitochondria whose properties were very similar to those of liver mitochondria [19]. Nevertheless, when prepared under comparable conditions, they lacked the features associated with the fresh intact liver preparations, i.e., pronounced latent ATPase activity and high P:O ratios when oxidizing substrates. In oxidative skeletal muscle, the mitochondria are usually ramified in structure and wrapped round and between the myofibrils, the major site of utilization of the ATP they produce. Partly due to the mechanical disruption involved in their preparation, they undergo morphological change and swelling which is responsible for their impaired properties. The inclusion of ATP in the preparation medium much improved the properties of the isolated mitochondria by preventing and reversing the uptake of water [20]. The effect of ATP was so dramatic that it seemed, at the time, that we were dealing with a contractile system rather than with an osmotic effect. Chappell was an outstanding student who later went on to develop an international reputation for his work on ion transport in mitochondria. He retired in 1995 as head of the biochemistry department, University of Bristol.

Myofibrillar ATPase and its regulation

By the late 1940s, it was clear that the contractile system of muscle consisted of actin and myosin and that the hydrolysis of ATP by the myosin enzyme accompanied contraction. Myosin ATPase was known to be activated by calcium and surprisingly inhibited by magnesium [21]. At the time, it was not certain which cation activated the myofibrillar ATPase in situ. Much of the earlier work had been focused on the calcium activation of myosin ATPase. The ATPase of extracted actomyosin was known to be magnesium, as well as calcium, activated but the former activation is so salt sensitive that at ionic strengths corresponding to that existing in the muscle cytoplasm, it could not account for the rate of ATP hydrolysis in contracting muscle. Intact myofibrils, however, were shown to possess an active MgATPase at higher ionic strengths, indicating that MgATP could act as the substrate for muscle contraction in vivo [12].

At this time, interest was beginning to focus on the mechanisms for relaxation in muscle. It was clear that contractile activity in model actomyosin systems could be regulated by controlling the ATPase. This was done by artificial means but it was not possible to demonstrate the cycle of contraction followed by relaxation as occurred in muscle with the actomyosin models. I was much struck by a report from Bozler [22] that he could produce relaxation in glycerated fibres by EDTA. This implied that EDTA might be inhibiting the MgATPase and prompted me to test the chelator on the enzymic properties of isolated myofibrils. Inhibition was obtained but, strikingly, at concentrations of the chelator which were very much lower than those of the magnesium in the system. Virtually complete inhibition was obtained when the chelator concentration was less than 5% of that of the magnesium in the assay system. It was concluded that some cation other than magnesium, with a high affinity for EDTA, was essential for the MgATPase of myofibrils

[23]. In following my other research interest in the swelling of mitochondria, I came across a paper by Raaflaub [24], in which he had used the chelator, then known as 'glycolcomplexon', to show that the ability of this compound to inhibit swelling by liver mitochondria was due to it specific chelating effect for calcium. This compound which is now known as EGTA (ethylene glycol bis[β-aminoethyl ether] N, N'-tetraacetate) had recently been synthesized by the Schwarzenbach group in Switzerland. It was not generally available but was described in the thesis of one of Schwarzenbach's students [25]. I wrote to Schwarzenbach who kindly sent me a sample of 'glycolcomplexon' which I tested on the activity of the MgATPase of myofibrils. Complete inhibition was obtained with the concentration of chelator at less than 5% of that of the magnesium, indicating that a trace of calcium was required for the MgATPase of the myofibrils [26]. This was the first use of EGTA in a muscle system, although it is not widely recognized as such, for most workers are not aware that 'glycolcomplexon' is in fact the old name for EGTA. Weber [27] confirmed later, by the careful use of calcium EGTA buffers, that 10^{-6} to 10^{-5} M calcium was required for activation of the myofibrillar MgATPase.

Protein components of the myofibril

It was clear that there were proteins in the myofibril, in addition to actin myosin and tropomyosin. With Andrea Corsi and Marius Zydowo, who joined me from Italy and Poland respectively, I attempted to characterize these additional components. The aim was to devise selective extraction procedures which would leave the bulk of the actomyosin behind. Very low ionic strength buffers, particularly at pH 8, selectively extracted the I band [28] to give a viscous solution which contained actin, tropomyosin and a third fraction, named component C [15], which I now know was troponin. Indeed, after Ebashi [29] reported his discovery of 'natural tropomyosin' (tropomyosin-troponin complex), I used this procedure to make crude troponin preparations. Characterizing complex protein fractions, at this time, was not easy for gel electrophoresis of proteins had not been developed and boundary electrophoresis in the Tiselius apparatus was the standard procedure used. The resolution was poor and with the apparatus then available, at least 5–10 mg of protein was required for a run.

At that time, we had shown that the MgATPase activity of a crude preparation of actomyosin, which we euphemistically called 'natural actomyosin', was calcium sensitive whereas that of 'synthetic actomyosin', made from separately purified actin and myosin, was not [26]. In retrospect, it should have been obvious to us that the additional proteins present in the myofibrils that we were isolating were in some way responsible for conferring calcium sensitivity on the MgATPase. We did try a few simple reconstitution experiments, without success, and directed our attention in other directions. It was left to Ebashi [29], some 7 years later, to report the isolation of the system that conferred calcium sensitivity on the actomyosin MgATPase.

The Cambridge contribution to protein structure

In the 1950s, the foundations of our current knowledge of protein and nucleic acid structure were being laid in Cambridge. Sanger was spending much of his time eluting fractions from strips of chromatography paper and drying them down on strips of plastic film to identify the amino acid residues in the small peptides that he had obtained from specific cleavage fragments of the A and B chains of insulin. I have particularly strong memories of the departmental seminar he gave describing his determination of the sequence of the insulin B chain. It was, without doubt, the most significant seminar that I have ever attended. The name of each amino acid was written on a card which was mounted vertically on a wooden block. As the lecture proceeded, the cards were placed in the order indicated by his selective cleavage studies. At the end, laid out before us on the lecturer's bench, were cards arranged to represent, for the first time, the complete amino acid sequence of the B chain of insulin, 60% of the molecule. One felt that history was being made. Soon afterwards, the sequence of the A chain was completed and in 1958, we were drinking champagne in the entrance hall of the department to celebrate Sanger's first Nobel Prize. Soon after his return from the medal presentation ceremony in Sweden, a small group from the department went out to dinner with him at a restaurant in Kings Parade in Cambridge. The evening went well and everyone was in high spirits. As we emerged from the restaurant someone called out 'Let's have a look at the medal, Fred'. Sanger pulled the case from his pocket, removed the medal and rolled it along the pavement. I am glad to say that as it rolled towards the drain in the gutter, I was able to get my foot in the way. There cannot have been many Nobel medals in such danger.

The X-ray crystallographers in the Cavendish laboratory had turned their attention to proteins. Perutz made an attempt on haemoglobin but the early results were somewhat disappointing and attention was directed to myoglobin. An approach was made to Chibnall and I remember Karl Schmidt, who was a postdoctoral fellow from Switzerland, preparing large crystals of whale myoglobin which were of interest to Kendrew. Crick would attend the departmental seminars particularly if they were concerned with protein structural topics. In the ensuing discussions, he usually had something perceptive to say, although his remarks were rarely presented with any touch of humility.

Kenneth Bailey

It was my good fortune to be allocated to Kenneth Bailey as a postgraduate student when I started research in Cambridge. Bailey and Sanger were the two members of staff with whom Rod Porter and I were placed when we joined the department. I have often wondered on what basis Chibnall made the decision to allocate Rod to Sanger and me to Bailey and what would have happened to both our careers if the arrangement had been the other way round. Bailey taught me how to handle proteins and stimulated an interest in muscle that stayed with me for the rest of my scientific life. He had great technical skill in the laboratory and his advice was widely sought by colleagues. He was of sensitive nature, enjoying the arts. Self-taught, he played the piano competently, and was an able sculptor and painter.

Bailey's knowledge and experience in classical protein chemistry was enormously helpful to me, attempting to re-establish myself in biochemistry after almost 6 years in the army. After my Ph.D. and the year in the USA, I worked independently of him, as is often the case with post-doctorals wishing to establish themselves outside the shadow of an internationally distinguished supervisor. Nevertheless, we regularly discussed our work and remained good friends until his tragic death in 1963.

In 1955, his mental heath deteriorated and he went into an acute depressive state, unable to work. Knowing him well, I became very concerned for his behaviour was such to suggest that he might be suicidal. For a few days, Maureen and I looked after him at home so that we could keep him under observation until Frank Young and I arranged for him to be admitted to the National Hospital for Nervous Diseases, London. At the National Hospital he was given electrical shock therapy and I was amazed at the change in his mental condition when I visited him on the following day. Apart from a little retrograde amnesia, it was possible again to carry on a normal conversation, both about science and more general matters. This breakdown in health kept him from effective work for about 5 months but when he felt sufficiently recovered he spent a year's con-

valescence leave at the Stazione Zoologica, Naples. He was
much attached to this laboratory, knew many of the workers, and had frequently spent summers working there.
During this convalescent period, he brilliantly isolated and
crystallized tropomyosin A. He returned to his post in Cambridge, where the last 7 years of his life were marred by
bouts of depressive illness which were treated both in London and in Cambridge. During the periods of normality he
continued excellent research, particularly on the biochemistry of fibrin formation and just before his death he described
with T. Weis-Fogh, the isolation of yet another new protein,
resilin.

   After the autumn of 1959, when I moved to Birmingham, I
lost regular contact with him and was shocked to learn in
1962 that he had left the country and returned to Naples.
Apparently the police had arrested him in his car after he
had been drinking in a public house. There was a young man
with him and he was charged with driving under the influence of alcohol and with indecent assault. This was a time
when homosexuality was an indictable offence in Britain and
when individuals whom the police suspected were often pursued assiduously. I had long suspected that Bailey was homosexual, both from some of his acquaintances and his attitude to me on one occasion. I made it very clear then what
my preferences were but this never affected our friendship.
At the time, Bailey was in one of his depressive phases and,
as he often did on these occasions, resorted to alcohol as a
form of self therapy. He was so shocked by the charge and
the consequences of it, in the current climate, to him as a
Fellow of the Royal Society and of a Cambridge college that
he felt that he could not stay in the country. By absconding
in this way and not presenting himself at the local Newcastle-under-Lyme court when his case was to appear before the
magistrates, he was liable for immediate arrest if he returned to the UK. I wrote to him suggesting that he should
return, as we would do what we could with the authorities to

facilitate this. He could not stay abroad indefinitely for he would soon run out of cash and be unable to support himself.

Whilst I was on holiday in Wales during the summer of 1962, I received an urgent phone call asking me to get in touch with one of his acquaintances at the Naples laboratory, Andrew Packard. From the latter I learnt that Bailey was causing some embarrassment at the Stazione Zoologica for he was clearly a sick man. Through Packard's efforts, Bailey was persuaded to go into a private mental clinic where he was treated for alcoholism, but he refused EST. All his many friends thought he should return to the UK where he could be looked after properly. At the same time, I was receiving letters from him which clearly indicated that his mental state was far from normal and that he did not appear to appreciate the seriousness of his situation.

On my return to Cambridge, Bailey had absented himself from the clinic in Naples and moved to Dubuisson's laboratory in Liege. He had had long contact with this group in the past for their interests in the muscle proteins corresponded with his own. Indeed there had been suggestions that he might spend sabbatical leave working with Dubuisson's group. His arrival, unheralded, in Liege and obviously not his normal self, caused some embarrassment. I was contacted and as I was soon to attend a conference organized at Liblice by the Czech Academy of Science, it was agreed that I should break my journey to Prague and see Bailey in Liege. This I did and was shocked by the state in which I found him, clearly acutely depressed, with bruises on his face from falling over, as his ability to orient himself on his feet was much impaired. It was not difficult to persuade him to join me on my return to the UK when I passed through Brussels a week later. I explained that he would not be arrested when he entered the country for we had contacted the Newcastle-under-Lyme police. The police had been informed that Bailey was a sick man and it was agreed that he should be allowed to enter the country without immediate arrest, if I would personally make myself responsible for surrendering him to them.

I broke my return journey from Prague on 24 September, to find Bailey waiting for me as we had arranged, sitting at the bar of Brussels airport. From Brussels we flew to London and travelled by train to Birmingham where he spent the night at my home. The following day, I drove to Newcastle-under-Lyme, arranged for him to see a psychiatrist, who confirmed that he was in need of treatment, and, in the company of his solicitor, surrendered him to the police. A magistrate was called out, his passport confiscated and he was put on bail, with my surety for £200, on condition that he went into hospital for treatment. After a few weeks in a hospital near Leek as a voluntary patient, he went to stay with his elderly father, who lived nearby, as he was not well enough to return to Cambridge.

As the day for his case to be dealt with at the Petty Sessions, 15 October, drew nearer, Bailey became very apprehensive and was most anxious that I should accompany him to the court. The counsel who had been instructed to defend him told us when we assembled at the courthouse, that the magistrates bench for his case was one of the most sympathetic that he could expect, including a well known professor of biology from the local university. The drunken driving charge was unassailable in view of the level of alcohol in his urine. I learnt, just before the proceedings started, that in certain circumstances the indecent assault charge could be dealt with summarily by the magistrates court. Counsel put to us that a quick way of dealing with the whole business without further court appearances was to plead guilty to both charges. After imparting this information, he left me alone with Bailey to decide which way we should plead. One of the aspects of his mental state was that he found it difficult to make a firm decision about anything and in effect, I was being left there to make a decision for him. From my knowledge of Kenneth's propensities, I felt that there was probably substance in the indecent assault charge as the law then applied; I got the impression that his solicitor certainly considered there was. My evidence of Bailey's homosexuality was, however, indirect and I felt that I could not decide how he should plead without direct confirmation. In response to my question, as we sat alone waiting for the court to start, he acknowledged that he was homosexual in a somewhat guarded reply. Knowing that I could get no decision from him, I said, 'Kenneth, you plead guilty'.

He was treated leniently by the court. Fined relatively modest sums for the two charges, he was put on probation for 2 years. He was not, however, fully recovered from the depression and spent the next few months living with his father in Newcastle-under-Lyme, visiting his psychiatrist from time to time and making occasional visits to stay with Maureen and I, in Birmingham. The problem was to rehabilitate him and I can remember endless discussions that I had with him on this matter. Return to Cambridge was possible, but difficult. Trinity was prepared to continue to pay his stipend but the Council decided that he should not be allowed to return to live in college and that if he did, he would be tactfully asked to move on. He was finding that life in Newcastle-under-Lyme lacked the mental stimulation that he needed and in May 1963, it was agreed with his psychiatrist that he had recovered enough to undertake a visit to Cambridge. This would allow him to meet old acquaintances and explore the possibilities of finding an institute where he might carry on working and still keep contact with his friends. He was very apprehensive of this trip and his own words were, 'I hope I have the strength to see it through'. After staying overnight in Birmingham, accompanied by Maureen, I drove him to Cambridge on 18 May and left him at Frank Young's house where he was to stay during his visit. Bailey visited a few friends and from all accounts seemed in quite good spirits. On the morning of 21 May, he was discovered dead at Young's house with the remnants of a mixture of gin and sodium veronal in a glass at his bedside.

This was a tragic end to the life of a very gifted scientist who had suffered from society's uncompromising attitude to homosexuality at that time. I suspect that when he was originally arrested for the drunken driving offence, the police put pressure on the young man who was with him to give evidence of indecent behaviour. Today, with the changed attitude in society in the United Kingdom, this would not happen and it is possible that Bailey would not have taken his own life. I often wonder when I think back over these events, whether I did the right thing in deciding that Bailey should plead guilty at the Petty Sessions. It still rests heavily on my conscience.

Move to Birmingham

In the late 1950s, my research was proceeding reasonably well and I was achieving some recognition by invitations to international meetings. Despite the excellent facilities for research in the biochemistry department and the stimulation of the Cambridge atmosphere, I was beginning to feel restricted so far as the resources available to me for my own personal research were concerned. The department was producing excellent graduates, many of whom wished to proceed to the Ph.D., but the competition for them was intense from the many senior distinguished research workers in the biochemistry and other departments in the faculty. After Chappell, I had been allocated, by 1957, only one other research student, Allison Newton, who worked on adenylic acid metabolism in muscle. I had always been interested in the fact that rabbit myosin possessed adenylic deaminase activity as great as its ATPase activity, which was very difficult to remove. This implied that the enzyme had a metabolic role and that there was probably a fairly active adenylic-inosinic acid cycle in muscle, although at that time there was little or no evidence for it. She showed that skeletal muscle had the ability to resynthesize adenylic from inosinic acid, using glutamine as a source of nitrogen for the amino group [30].

It was not the policy, at that time, for the department to encourage relatively junior staff, such as myself, to seek external grants that would enable them to build up their research teams. Thus, by 1957, I was looking around for a post that would enable me to expand my efforts in a way which did not seem possible in the Cambridge department. My personal research space consisted of a small laboratory bay which served as an office and which accommodated also my assistant. As there were few suitable academic posts available in the UK, I considered employment in the pharmaceutical industry. I was offered the post of research director at Evans Medical but after looking into the resources and prospects of the company, decided it was not the job for me. In retrospect, this decision was probably wise for very soon afterwards the company was taken over by another pharmaceutical group.

In the spring of 1959, I was offered the headship of the department of biochemistry, in the faculty of science at the University of Birmingham. Biochemical activity at Birmingham, at that time, was diffuse. Physiological chemistry was taught to the medical students by a group in the physiology department in the medical school. The biochemistry taught in the science faculty was largely of an applied nature, long influenced by an interest in fermentation which resulted from the fact that the department also contained the British School of Malting and Brewing, founded in 1899. This had been responsible for the training of a good fraction of the brewers and scientists engaged in the industry in the UK. Due to the rather applied nature of the course taught in the science faculty, Professor A.C. Frazer had established a department which taught an honours course in medical biochemistry in the faculty of medicine. My brief was to establish a course in general biochemistry in the faculty of science. As a consequence of these complications, Birmingham was considered a difficult place to build up a department of general biochemistry but I accepted the post for in the long term, prospects appeared good. A large new building to accommodate all the biology departments was planned, with the total space allocated to biochemistry greater than that available in any other university in the UK at that time.

First experience of the biochemistry department at Birmingham, to which I moved in September 1959, was something of a cultural shock. The department occupied part of the original redbrick building constructed in 1927, and the facilities available reflected that. To me, accustomed to then current state-of-the art facilities at Cambridge, it was extremely poorly equipped. There were no high speed centrifuges, only one spectrophotometer which was in the Brewing sub-department, and no cold room or animal house. Without substantial support from the faculty to purchase equipment and a generous temporary allocation of space in the department of chemistry, it would have been quite impossible for me to carry on my research. I was much indebted to Maurice Stacy, the head of the chemistry department, for research laboratory space, for the loan of equipment, the provision of a portable butcher's cold room which was essential for my work, and his strong general support. Despite the need to improvise at this stage, the prospects in Birmingham were very encouraging. The government was pouring vast sums of money into the expansion of higher education and Birmingham, as a major provincial university, was getting its share. In 1959, there was about £5 million worth of building work in progress on the campus and plans for the new biology block, costing well over £1 million, were being finalised. One could not fail to be impressed by the general vitality which pervaded the campus.

With the chair of biochemistry in the science faculty was associated the Headship of the British School of Malting and Brewing. Although my knowledge of brewing was minimal, I

held this position for a few years until the relationship between the general biochemical and the brewing activities
could be rationalized. The malting and brewing interests
were concentrated in a subdepartment, the head of which
became the director of the brewing school. When I arrived in
Birmingham, many of the brewing students were not candidates for a science degree but were simply awarded a diploma at the end of their course. This was reflection of the
fact that many of them were admitted to the university with
lower academic qualifications than were required for a degree course. It seemed to me that it was in the brewing industry's interest that their technical staff should be better
qualified. With this aim, I managed to persuade the industry
that malting and brewing training in Birmingham should be
at the post graduate level. It would not have been possible to
get the agreement of a somewhat conservative industry to
this change without the strong support of Norman Smiley,
the managing director of Guinness Ltd, and a powerful figure in the upper echelons of the brewing industry. From
1964, entrants to the brewing school joined with a first degree in an appropriate subject, and qualified after 1 year
with an M.Sc. in brewing science.

Just before I had agreed to move to Birmingham, Helmut
Mueller had joined me from the USA, hoping to spend a couple of years in Cambridge. Despite the upset to his plans, he
generously agreed to move with me. I well remember the
day, in the late summer of 1959, when Helmut and I hired a
van in Cambridge, loaded it up with equipment and drove to
Birmingham. By this time, David Baird and Lev Leadbetter
had been allocated to me as research students in Cambridge
but they would not complete their thesis work before I was
due to leave the department. Baird decided to stay in Cambridge, whilst Leadbetter joined me in Birmingham with
Steve Bondy, a Cambridge graduate who started research in
Birmingham. With Bondy, I made my first and only venture
into neurochemistry, driven by the conviction that, as in the

case of muscle, the brain must contain proteins that are specific to the tissue, which could be readily recognized as of functional importance [31]. Needless to say, we did not, with the methods available to us at the time, find any major fraction unique to the tissue which might be concerned with the specialized aspects of brain function, as is the case with muscle.

Myosin structure and function

Before I left Cambridge, I became interested in applying the newly developed ion exchange chromatography of proteins on cellulose derivatives to clean up myosin preparations from the deaminase and nucleoprotein impurities that they invariably possessed when prepared from rabbit skeletal muscle. There were certain difficulties in chromatographing myosin on DEAE cellulose because of the high salt concentrations required to keep it in solution. Nevertheless, two peaks of enzyme activity were consistently obtained [32]. Seen in retrospect, it is possible that these represent the two isoforms of rabbit skeletal myosin containing different essential light chains, although at the time, neither the knowledge for such an interpretation or the techniques to demonstrate it were available. Heavy meromyosin, which is soluble at low ionic strength and therefore more suitable for chromatography on DEAE cellulose, was shown to contain small amounts of a fragment of much lower molecular weight. By exploiting this observation, we were able to show that heavy meromyosin could be further degraded into subfragment 1 and subfragment 2 of molecular weights 130 and 100 kDa, respectively [33, 34]. Subfragment 1 was of particular interest as it was much less asymmetric than heavy meromyosin, exhibited all the ATPase activity present in the original heavy meromyosin and interacted with actin. As information about the structure of the myosin molecule progressed, it became clear that subfragment 1 represented a single head of the myosin molecule. Due to its size and shape, we considered that the molecule should crystallize but despite a number of attempts by me and my colleagues, we never succeeded. I derived some personal satisfaction when, over 30 years after its discovery, the crystallization of the reductively methylated form was announced, enabling the determination, for the first time, of the structure of a protein motor [35].

Encouraged by the preservation of biological activity after significant proteolysis of heavy meromyosin, we attempted to degrade the molecule further, in the hope that an even smaller active fragment could be obtained. Further proteolysis produced a fragment of 26 kDa apparent molecular mass containing 3-methyl histidine and methylated lysines, which was rather resistant to further degradation, and one of 53 kDa [36]. These inactive fragments correspond to two of the three proteolyic fragments shown later by other workers to be derived from the myosin head.

The interaction of actin and myosin

The ionic conditions in the cell are such that virtually all of the substrate for contraction exists as the MgATP complex. The influence that actin has upon the MgATPase activity of myosin is a striking property. In some way not understood, it changes its enzymic properties so that MgATP becomes an effective substrate which it is not for myosin alone. This effect of actin is fundamental for upon it depends the ability to regulate the contractile process. In vitro at least, this effect is very sensitive to ionic strength, in which respect it differs from the interaction of actin with myosin observed at high ionic strength. Under these conditions, when the two proteins are in solution they interact to produce a large increase in viscosity, implying that the interaction is stable to ionic strength. Studies on actoheavy meromyosin indicated that under conditions, when the complex was virtually completely

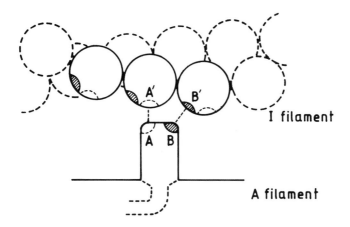

*Fig. 3. Scheme for the interaction of actin and myosin in the myofibril pro-
posed in 1965. A and B are the sites on the myosin head involved in the
ATPase activity and the formation of the strong, salt-stable, interaction
with actin, respectively. A' and B' are the sites on actin involved in activating the MgATPase of myosin and forming the salt-stable link with myosin,
respectively [40].*

dissociated, as indicated by viscometric measurements, the
MgATPase is high [37]. This suggested that the two types of
interaction are different, an aspect that has not been sufficiently emphasized in the past. Some 10 years later, Eisenberg and collaborators [38] reinvestigated this aspect and
confirmed the results but explained them by concluding that
the heavy meromyosin, and subfragment I, which was also
used in these studies, were in a 'refractory state'. A more
reasonable explanation, which was supported by treating
meromyosin and actin with sulphydryl inhibitors, indicated
that different interaction sites with different properties were
involved in the ATPase activation and the interaction resulting in the increase in viscosity [39]. In view of the evidence indicating different sites on the two proteins, it was
suggested that the myosin head could interact simultaneously with two actin monomers, with each interaction type

on a different monomer [40], and thus giving a polarity to the interaction (Fig. 3).

Biochemistry of muscle development and adaptation

The occupation, in 1962, of the space allocated to biochemistry in the new biology building, much increased the facilities for research. For the first time in Birmingham, the departments of biochemistry, physiological chemistry and medical biochemistry were accommodated in one building. The departments were administered independently but shared common services. With the much improved facilities and the availability of able students in the honours course who wished to carry out postgraduate research, this presented an opportunity to expand my activities. Although the main effort was directed at the biochemistry of the contractoregulatory process in skeletal muscle, I began to direct my attention also to the biochemistry of muscle development. At this time, I was joined by a number of able postgraduate students such as David Hartshorne, Jake Kendrick-Jones and Ian Trayer, all of whom are currently making substantial contributions to the muscle field.

Muscle tissue has the capacity to change its physiological properties during development and on maturity, to exist in a number of types, each with different properties specific for its particular contractile role. Interaction with nerve and the induced activity pattern are important factors in determining muscle properties, as had been indicated by the report describing how the properties of muscle could be changed by cross innervation [41]. At this time, starch gel electrophoresis of proteins had just been developed [42] and using it we were able to show that marked changes occurred in the soluble proteins of skeletal muscle sarcoplasm during neonatal life. These changes were a consequence of rapid increases in enzymes such as creatine kinase and 5' adenylic deaminase, which are normally of high activity in adult muscle. Increased contractile activity stimulated the production of these enzymes both in neonatal and adult muscle [43]. At that time, there were very few papers on foetal myosin and the evidence as to its nature was somewhat ambiguous. By isolation and purification of foetal myosin from the rabbit, we were able to show that as well as increasing in amount during development, it increased in specific activity [44, 45]. Our conclusion was that there was a foetal isoenzyme form of myosin which, during development, was replaced by the adult. Although this conclusion is fundamentally correct, we now know that there is a complex pattern of myosin isoform changes during development. It was a natural step to investigate changes in the sarcoplasmic reticulum during development for at that time there was only one rather preliminary study published. Unlike the other enzymes, the MgATPase peaked just before birth and thence fell sharply whilst the calcium pumping activity increased [46].

Methylated muscle proteins

It was essential for our work on muscle proteins to have good facilities for amino acid analysis. Ian Harris established these, and with Pete Johnson, noted during the analysis of actin and peptides derived from it, that the histidine peak often exhibited a trailing edge which was shown to be due to the presence of 3-methyl histidine [47]. At about the same time, Asatoor and Armstrong [48] independently reported a similar finding. Although 3-methyl histidine had previously been identified in urine and muscle, this was the first time it had been reported as a residue in a protein. Careful analysis of myosin and tropomyosin indicated that although absent from the latter protein, myosin from white skeletal muscle contained about 2 residues per molecule, compared with the single residue in actin. Myosin also differed from actin in that 3-methyl histidine was absent from the foetal form of the protein, providing further evidence for a unique foetal form that our enzymic studies had indicated earlier [49]. The discovery of methylated histidine prompted us to investigate the lysine fraction in amino acid hydrolysates of the myofibrillar proteins for methyl lysines had been reported to be present in flagellin. Mono and trimethyl lysines were shown to be present in both adult and foetal forms of myosin, unlike the case with 3-methyl histine where methylation occurred after birth. Madeline Hardy showed S-adenosyl methionine is a precursor for the methyl group on both amino acids and that the process was not inhibited by puromycin, suggesting that methylation of the myofibrillar proteins is a post synthetic modification [50, 51]. Even today the role of methylation of these two basic amino acids in the function of the muscle proteins is far from clear. Presumably, the increased hydrophobicity produced at specific points in the polypeptide chain by methylation is significant. Nevertheless, destruction of the methyl histidine of myosin by photo-oxidation has no marked effect on the ATPase activity or on its ability to combine with actin [52]. The fall in 3-methyl histidine content of myosin from the skeletal muscles of rabbits made dystrophic by vitamin E deficiency, probably reflects muscle regeneration and increased synthesis of foetal or slow muscle type of myosin [53].

Troponin system

When Ebashi [29] reported that 'natural tropomyosin' restored calcium sensitivity to the MgATPase of actomyosin that had been desensitized by treatment with trypsin, we immediately tested the tropomyosin-containing preparations isolated from myofibrils that we had studied some years earlier, the 'soluble' and 'extra protein' fractions. The latter is the protein fraction remaining in solution after actomyosin was precipitated from high ionic strength extracts of myofibrils. Both contained EGTA sensitizing factor, ESF, the name used for the calcium sensitizing system of the myofibril before troponin was adopted [54]. Extraction of natural actomyosin with low ionic strength buffers, pH7–8, to obtain the soluble fraction, removed the calcium sensitivity of the MgATPase. This provided a more convenient system on which to test ESF activity than the natural actomyosin desensitized by treatment with trypsin that was used by Ebashi. Reference to the two systems now available for assaying the ESF, desensitized actomyosin, DAM, and synthetic actomyosin, SAM, provided acronyms that often lightened the dull routine of enzymic assays.

It was noted that extra protein preparations, in addition to inhibiting the actomyosin MgATPase in the presence of EGTA, also had inhibitory activity in its absence [54]. As the preparations aged and the former activity decreased, the latter increased suggesting that the two properties might be related. The fact that in the absence of calcium, troponin inhibited the MgATPase activity of actomyosin in ionic conditions when the enzyme alone was active, indicated that the intact complex clearly has an inhibitory function. Thus it seemed likely that we were dealing with a modified form of troponin or a component derived from it. At this time, Hartshorne returned to Birmingham after spending 3 years in the USA. As I had accumulated several grams of the freeze-dried crude inhibitory protein, we examined its properties in more detail to convince ourselves that it was derived from the troponin complex. The fact that it inhibited the magnesium ATPase, i.e., that activated by actin, and not the calcium ATPase of desensitized actomyosin, strongly indicated that we were on the right track [55, 56]. At about the same time, Marcus Schaub joined us from Basle and the fractionation of the troponin complex became the major focus of our research. Hartshorne left to return to the USA, where he joined Mueller who had left me a few years earlier, and was now in Pittsburg. With Schaub, it was shown that the troponin complex could be separated by chromatography, on

Sephadex in 6 M urea, into the inhibitory protein and a fraction that we called the calcium sensitizing factor [57]. Remarkably, both factors were stable to the strong urea which my earlier experience with proteins suggested should have denatured them. Addition of the calcium sensitizing factor to the inhibitor restored its EGTA sensitivity. Independently, Hartshorne and Mueller [58] also separated troponin into similar factors which they named troponin B and A respectively.

The special relationship of calcium with the sensitizing factor was indicated by its marked change in electrophoretic mobility in 6 M urea, in the presence of EGTA [57], and its ability to bind calcium even on electrophoresis under these conditions [59]. In the initial fractionation experiments, there were indications that a component, in addition to tropomyosin, inhibitory and calcium sensitizing factors, was required to restore full calcium sensitivity to the MgATPase of actomyosin [57]. It was also noted that inhibitory factor preparations often contained a higher molecular weight component. This was later shown to due to a protein of apparent molecular weight 37 kDa, basic like the inhibitory protein, and which also contaminated tropomyosin preparations with which it formed a complex [60, 61]. In the late 1960s and early 1970s, a number of groups were engaged in fractionating troponin and isolating fractions with different activities, for which they used their own terminology. The result was that 3 or 4 different names were assigned to fractions that were clearly identical in properties. This made the field difficult to follow for those not engaged in it. It was a great step forward when, at an important meeting on muscle held in Cold Spring Harbour in 1972, the nomenclature suggested by the Gergely group, namely troponin I, C, and T, was adopted for general use.

Consolidation in Birmingham

By the mid 1960s, I was firmly established in Birmingham in good, excellently equipped accommodation. The department was well supported financially as money to expand the university facilities was still coming through from the government. The muscle group was beginning to build up a reputation and requests were coming in from foreign muscle scientists who wished to join us for a period. A very welcome visitor, in 1963, was Tony Martonosi, just after we had taken over laboratories in the new building.

Teaching schedules in the department had been revised to provide a widely based course in general biochemistry. The support for expansion had allowed more staff to be taken on to cope with the increase in student numbers and the need to have staff with interests not covered in the original department. Recruitment, at the time however, did present difficulties due to the national demand and a shortage of suitably qualified people. My personal research, conducted mainly by research students, had, up to that time, been supported by short term grants, usually for 3 years, and mainly from the Medical Research Council. I was very pleased when, in 1967, the Council provided me with long term support through a programme grant which was renewed until I was close to retirement. This enabled me to appoint more senior colleagues as post doctoral fellows in the muscle research group and thereby provide an element of continuity to the work.

Since occupying the new building, the medical faculty departments of physiological chemistry and medical biochemistry had operated completely independently. These departments taught different groups of students and provided, independently, lectures on topics similar to those being taught in the biochemistry department, based in the faculty of science. In 1968, things changed dramatically, for Professor A.C. Fraser, head of medical biochemistry, had resigned and the head of physiological chemistry, Professor W.V. Thorpe, was due to retire that year. The three departments were amalgamated and I was appointed chairman on a 5 year renewable basis, a position I held until I retired in 1985. Remaining in one post for 26 years does, perhaps, reflect a certain lack of initiative, but during this period I did have opportunities to move if I had wished to take advantage of them. During the 1960s, I was approached about a senior post in the CSIRO in Australia, the directorship of the National Institute for Dairying, Shinfield, and the chairmanship of the biochemistry department, Edmonton, Alberta. Research was going well, however, and I did not see the point of breaking out in a new direction at this time. The university had put a large investment into biochemistry in Birmingham, with the result that it was one of the best founded departments in the country, and future prospects appeared good. Problems might have been expected in merging the three departments of biochemistry but the whole process proceeded relatively smoothly and the medical bias preserved by providing an honours course in biochemistry, with options in medical and more general aspects of the subject.

Soon after arrival in Birmingham, I had developed another interest outside my academic work, which in due course turned out to be an important factor to be considered if I was to move outside the UK. After spending regular summer holidays with my family, which now included three children, Gillian, Jacqueline and Michael, on the Pembrokeshire coast in west Wales, I had become much attracted to this part of the country. In the summer of 1959, just before packing up in Cambridge for the move to Birmingham, I discovered a ruined water mill for sale close to the coastal village in which we had been spending our holidays. As a keen gardener, I had always wanted a garden with a stream in it. The property in question, Felin Werndew, welsh for 'the mill where the alders grow thick', had that feature but was completely derelict and overgrown with vegetation. Originally built in the 18th century, or possibly earlier, it had last worked as a corn mill in the early 1930s. It was an overshoot mill but the large steel wheel had been taken for scrap metal during the last war, the mill machinery had been removed and the roof had partially collapsed. The past 35 years have been spent in converting the ruins of the mill into a summer home, planting shrubs, adapting the stream, and building ponds and stone walls to produce a water garden that gives me great pleasure. The stream and ponds contain native brown trout which I have encouraged to proliferate. Nearby is the small bay, Cwm yr Eglwys, where I keep my boat and fish for mackerel and lobster. As an ongoing project, Felin Werndew has provided me with a different challenge and all important relaxation from my academic work. From time to time, unsuspecting research students and foreign visitors found themselves there for the weekend, their enjoyment of the Pembrokeshire countryside modulated by episodes of grass cutting and mixing concrete.

Given the opportunity, I would have enjoyed returning to Cambridge and although I was encouraged to apply for the chair in Cambridge when it became vacant on Young's retirement in 1974, I felt that chronologically, but not mentally, I was rather old for the post. In the event, H.L. Kornberg was appointed. Soon afterwards I was approached regarding the directorship of a research institute associated with the department of physiology at the University of Los Angeles. This was intended for muscle research and funded by the Muscular Dystrophy Associations of America, with the longer term intention of extending the building for cardiac research, also an active interest of the department of physiology. The concept of concentrating muscle research in this way very much appealed to me and the post had many attractions, not the least of which, was the more liberal attitude to the age of retirement in the USA. Also, at that time, it was becoming clear that conditions in Birmingham were changing with strong pressures to economize and cut back staff posts. Finally, after much heart searching, but largely because of family considerations, I decided to stay in Birmingham. A factor, although minor, was the thought of having to undertake a round trip of 10 000 miles to cut the grass at Felin Werndew!

Components of the troponin complex

The remarkable stability of biological properties of the components of the troponin complex much facilitated their study. Troponin C obviously complexed with troponin I, for its role in the complex was to neutralize the inhibitory activity of the latter protein. The stability and strength of the affinity between the two proteins was such that the complex was not dissociated by 8 M urea, unless EGTA was present [62, 63]. Jim Head studied this highly specific reaction which could be followed by gel electrophoresis, presenting a very convenient way of detecting either protein in mixtures. The property was employed to devise an affinity chromatographic method for the isolation of troponin I from muscle in one step using troponin C covalently linked to sepharose [64]. This useful property was discovered because at that time we frequently monitored protein samples by electrophoresis in buffered strong urea solutions, under which conditions separation was due to charge. If we had used gel electrophoresis in SDS, the conditions more commonly used at that time, we would not have made the observation for the complex was dissociated under those conditions.

The stability of the activity of troponin I to dissociating conditions suggested tertiary structure was probably not particularly important for its inhibitory activity and that it might be possible to produce a small biologically active fragment from it. Preliminary experiments with my research assistant at the time, Heather Cole, indicated that the inhibitory activity of troponin I was not destroyed by treatment with cyanogen bromide. Further study by Henry Syska and Mike Wilkinson led to the isolation of the cyanogen bromide peptide of rabbit skeletal troponin I, representing residues 96 to 116, which possessed the number of the properties of the whole molecule [65]. The inhibitory activity of this fragment, the inhibitory peptide, on the MgATPase of actomyosin was much increased by tropomyosin and neutralized by troponin C. The intact molecule of troponin I inhibits the actomyosin MgATPase at a molar ratio of actin to troponin I of 1:1 but in the presence of tropomyosin, the inhibitory activity is much stimulated so that this ratio approaches 7:1 [62]. This is a functionally important property and upon it depends the fact that in the myofibril, one molecule of the troponin complex controls the interaction with myosin of seven actin monomers in the thin filament. The fact that a small peptide of troponin I also exhibits this property is remarkable and must reflect, in some way, the mechanism involved. It was noted that the inhibitory peptide, as well as the cyanogen bromide peptide corresponding to residues 1–47, bound to a troponin C affinity column, suggesting that these two regions are involved in the interaction with troponin C [65]. Similar studies on the latter protein enabled the region important for the calcium-dependent interaction with troponin I to be defined [66]. Emboldened by the success of this approach, we turned our attention to troponin T, and Pete Jackson and Godfrey Amphlett were able to demonstrate that tropomyosin interacts with the region represented by residues 70–160 [67].

The fact that many of the individual biological properties of the troponin complex are retained by relatively small regions of the polypeptide chains of its components indicates that compact, well defined interaction sites are involved in its function. Relatively small peptides that are biologically active are particularly suitable for nuclear magnetic resonance (NMR) studies that enable the amino acid residues involved in the interactions to be defined with precision.

This was first pointed out to me early in the 1970s, during a visit to the University of Pittsburg by the NMR group there. Although there was some discussion of possible collaboration at the time, this did not come to anything. It was not until the late 1970s, that we started a collaboration with Barry Levine, then in the department of chemistry at the University of Oxford, to use this technique to study the interaction between I, C and actin [68–70]. From these studies, it was possible to define, in considerable detail, the nature of the interactions between troponin I, actin and troponin C, and the effect of calcium on them (Fig. 4).

Calmodulin and calmodulin-binding proteins

As the nature of the regulatory system of striated muscle was beginning to be defined, attention was directed to smooth muscle, in which tissue it was clear, by 1976, that myosin light chain kinase had a more important regulatory role than in striated muscle. Although it was not possible to isolate troponin I from smooth muscle, we had evidence in the mid 1970s that there was a troponin C-like protein in smooth muscle and indeed in many other tissues [71, 72]. This protein was identified by the formation of a calcium-dependent complex, which was stable to high urea concentrations, when skeletal muscle troponin I was added to the tissue homogenates. Although it behaved in a similar manner to troponin C, in this respect it clearly was not identical with it from its amino acid analysis. About this time we were joined by Tom Vanaman from Duke University who had been working on the bovine brain modulator protein and had shown its similarity to troponin C. Roger Grand's investigations indicated that the troponin C-like protein of uterus smooth muscle was identical with the bovine brain modulator protein, now known as calmodulin [73]. A surprising property of calmodulin is that its complex with skeletal troponin I can replace the whole troponin complex in the cal-

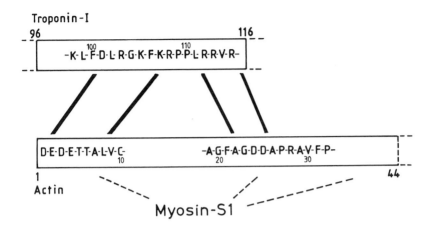

Fig. 4. Schematic representation of the interaction between the N-terminal residues of actin and the basic segment of troponin I that possesses inhibitory activity. The corresponding acidic residues of actin that jointly contribute to binding the inhibitory segment derive from different regions of the N-terminus of actin. An extended trans pro-pro linkage separates the two portions of the inhibitory segment that interacts with actin [70].

cium sensitive regulation of the actomyosin ATPase [74]. This implies that when calmodulin interacts with troponin I, troponin T is not essential, as is the case with the troponin complex, for calcium regulation of the actomyosin ATPase.

These observations prompted us to employ a calmodulin-sepharose affinity column to isolate proteins that formed calcium dependent complexes from brain and other tissues. Proteins of 22, 61, 77 and 140 kDa were isolated in relatively high yield [75]. The 22 kDa protein was purified and identified as myelin basic protein [76] and the other proteins were partially purified. Similar complexes were observed in other tissues and it is likely that the 140 kDa protein is identical with the later identified calmodulin-binding protein, caldesmon [77].

Phosphorylation of the myofibrillar proteins

Troponin complex

A brief report that troponin B could be phosphorylated by cAMP dependent protein kinase [78], appeared at about the time when we had characterized the three components of the troponin complex. This stimulated us to examine this aspect for there was much current interest in phosphorylation as a regulatory process. Variable reports in the literature as to the precise site of phosphorylation arose because, in the original studies, it was not appreciated that the troponin B preparations contained both troponin I and T. With the availability of the purified proteins, we were able to show that most of the phosphate in troponin preparations was associated with troponin T which, in vitro, was a much more effective substrate for phosphorylase kinase than cAMP dependent protein kinase [79, 80]. On the other hand, troponin I was phosphorylated in vitro by both kinases, the former being specific for an N-terminal site and the latter for a site adjacent to the inhibitory peptide [81]. Phosphorylation by both enzymes was inhibited by troponin C providing further evidence that these regions of troponin I were involved in troponin I–C interaction [80]. As yet, no clear biological role for the phosphorylation of troponin I of skeletal muscle is obvious. In vivo, the protein is partially phosphorylated at levels which are not significantly changed by contractile activity. Phosphorylated troponin I will still interact with troponin C but the covalent modification introducing negative charges at, or close to, the interaction sites would be expected to change the binding constant for the complex.

The situation with cardiac troponin I is rather different. This was shown to be phosphorylated by cAMP dependent protein kinase, 30 times faster than the skeletal isoform [82]. Aided by the sequence studies of my colleagues in the muscle group [83], we were able to show that although cardiac troponin I contained analogous potential phosphorylation sites to the skeletal isoform, it also had an additional one that was the preferred by cAMP dependent protein kinase. This was located in the 30 residue polypeptide at the N-terminus that was present in the cardiac but not the skeletal isoform. This site was identified as serine 20 in the amino acid sequence then available for rabbit cardiac troponin I. At this time, Arthur Moir, who was principally concerned with the phosphorylation work, was joined by John Solaro from the Medical College of Virginia and we started to relate our studies on the isolated protein to the perfused rabbit heart. These were much facilitated by the affinity chromatographic procedure with troponin C columns that enabled the isolation of troponin I from whole hearts in one simple step. Troponin I from the normally functioning perfused rabbit heart contained about 1 mole of phosphate per mole, which rose to about two under influence of adrenaline. In both cases, all the phosphate was located in an N-terminal cyanogen bromide peptide, but although serine 20 was phosphorylated, we could not detect phosphate in the other serine and threonine residues in the peptide [84]. This was a puzzling problem that we were unable to explain at the time. It was resolved some 13 years later when it was pointed out by Mitmann et al [85], that there was an error in the sequence that we had used. They showed that the N-terminal peptide of cardiac troponin I contains a double serine site, residues 22 and 23 in the rabbit, both of which can be phosphorylated to account for the 2 moles of phosphate per mole after intervention with adrenaline. An important consequence of the change from the monophosphorylated to the diphosphorylated form is a decrease in the calcium sensitivity of the troponin regulated MgATPase of the cardiac myofibril [86, 87].

It seems likely that basic similarities in function exist between cardiac troponin I and the phosphorylation sites in other regulatory systems consisting of a serine flanked by another serine or a threonine residue [88]. Where dual phosphorylation occurs in adjacent sites, this raises the question of the relative roles of the two sites and whether ordered phosphorylation and dephosphorylation occurs. Aided by NMR, we were able to show in the human cardiac troponin I serine 24 is phosphorylated first by cAMP dependent kinase but that the major conformational change in the peptide occurred when serine 23 is subsequently phosphorylated [89]. It is this latter stage of phosphorylation that we believe is responsible for the change in calcium sensitivity of the troponin complex.

Myosin

In the late 1960s, Bill Perrie started to investigate the light chains of myosin, the nature of which, at that time, had not been decisively resolved. In our hands, fresh myosin preparations invariably migrated with four bands corresponding to the light chains, whereas other groups of investigators were reporting three on electrophoresis in SDS [90]. This was another example of the value of electrophoresis in 8 M urea, under which conditions the light chains were separating by charge rather than by molecular weight. For a number of reasons, we suspected that we were observing a phosphorylated form of the light chain migrating with intermediate velocity. Larrie Smillie, on sabbatical leave from Edmonton, then joined us and in due course we were able to show that the additional light chain band was in fact the phosphorylated derivative of the so-called regulatory light chain, and to identify the site of phosphorylation [91, 92]. The regulatory light chains of all vertebrate muscle types were shown to be phosphorylated [93, 94] and new enzymes, myosin light chain kinase [95], which was later purified by Euclides Pires [96], and myosin light chain phosphatase [97], were identified. In our earlier studies of the enzyme, we used a light chain preparation isolated from myosin as substrate, unaware that it contained a trace of calmodulin [98] which was shown by others to be a requirement for activity of the enzyme [99, 100] and confirmed by Angus Nairn [98]. In striated muscle, the current view is that only one of the two serine residues of the regulatory light chain is phosphorylated by the kinase. The serine of the light chain is the preferred site in vertebrate smooth muscle but in the presence of excess enzyme, the adjacent threonine residue is also phosphorylated [101, 102]. It is uncertain whether this has any physiological significance.

Whereas in vertebrate smooth muscle, myosin light chain kinase has a role in activating the actomyosin ATPase, its function in striated muscle is less well defined. In the latter tissue, phosphorylation of the light chain is not essential for the MgATPase of actomyosin. In our original studies, which were carried out in the presence of saturating amounts of actin, we were unable to demonstrate any activation of the enzyme. Phosphorylation of the light chain does presumably modulate the ATPase activity in some way for it decreases the K_m for actin [103], and active phosphorylation followed by dephosphorylation is associated with contractile activity, particularly in fast skeletal muscle. It has been suggested that light chain phosphorylation is responsible for post tetanic potentiation of muscle but in our studies with rabbit muscle in vivo, the correlation between the two effects is not complete [104].

Muscular dystrophy

My first real contact with muscular dystrophy came late in 1960s, when I attended an annual general meeting of the Muscular Dystrophy Group of Great Britain. For the first time, I shared a conference hall which contained several hundred parents and close relatives of young boys suffering from progressive muscle weakness which is a feature of Duchenne muscular dystrophy. All present knew that these boys, many of whom attended the meeting in their wheel chairs, would be dead by the end of their second decade, or soon afterwards. As a parent, one could sense the anguish of relatives of boys suffering from this condition which provided the bulk of the patients with which the group was concerned. As a muscle scientist, it was disturbing that, despite the great advances that had been made in the understanding of muscle function, research had thrown little light on the nature of the lesion in this distressing condition. It brought home to me, very forcibly, how isolated most of us working the muscle field were from the real human problems related to muscle function. I later joined the medical research committee of the Muscular Dystrophy Group, an association I consider was one of the most rewarding of my professional experiences. I was member of this committee for 20 years but rather selfishly turned down the opportunity to become its chairman. I did not wish to increase the demands on my time already fully occupied with the demands of the departmental chairmanship and felt a strong desire to keep in close contact with my research.

At that time, a major interest of the group was Duchenne muscular dystrophy, for it is a well defined condition with a relatively large number of cases, whereas many of the other types of dystrophy, with equally distressing consequences, are rarer and less well-defined. For some time it had been obvious that a defect in the muscle membrane was associated with the dystrophy but there was no clear evidence of its origin. A popular view, at the time, was that there was possibly a defect in the process of muscle development. This suggestion had risen from the fact that the foetal forms of many of the proteins in muscle, persisted in the adult dystrophic muscle. We now know that the presence of the foetal forms is a reflection of the fact that the dystrophic muscle attempts to regenerate and therefore some of the biochemical changes observed are a consequence of the condition rather than the cause. The uncertainty as to the cause of Duchenne muscular dystrophy resulted in a wide range of approaches to be supported by the British and American

Muscular Dystrophy Groups, the two major sources of research funds. Muscle science, world-wide, benefited very much from this broad approach and I have reason to be grateful to both bodies for their financial support in the past, for my personal research.

Before the genetic lesion was identified, precise diagnosis of Duchenne patients, and the female carriers, was far from perfect. A widely used procedure depended on the elevation of the serum creatine kinase level but was not entirely satisfactory because it did not give clear cut results, particularly for a significant fraction of the female carriers of the condition. This test depended on the increased permeability of the muscle membrane allowing a normal component of muscle to leak out into the circulation. As the condition was clearly genetic, it was reasonable to expect that some protein was missing or present in a modified form, either in the muscle or possibly in some of the body fluids. I had long been impressed with the resolving power of two-dimensional protein electrophoresis as a means of defining the protein composition of a tissue. When Anderson and Anderson [105] described the IsoDalt apparatus to carry out, in a reproducible manner, high resolution electrophoresis of many samples simultaneously, this seemed to be the equipment to be applied to the problem. I had the IsoDalt apparatus made, and Neil Frearson soon had it functioning to examine the proteins present in the urine from Duchenne boys. Although there is not much protein in normal urine, that present is a complex mixture, fairly constant in composition under normal circumstances. The composition of the urinary protein fraction can reflect an individual's clinical condition and it seemed likely that it might exhibit changes specific to dystrophy. Unlike tissues, to obtain samples of which require biopsy procedures, urine can easily be obtained non-invasively from relatively large numbers of patients. Several proteins were detected on the electropherograms of the urine of Duchenne boys which were not present in normal samples, and the most conspicuous and consistent of these was named spot C [106]. It was concluded that protein responsible for Spot C was not a mutant gene product but its presence in urine was an index of skeletal muscle damage and, as such, potentially valuable as a diagnostic tool. Although a consistent constituent of Duchenne urine, the amounts were extremely small and many litres of urine had to be processed to get enough for analysis and study. So far it has not been possible to identify Spot C.

The discovery that the dystrophin gene was the locus for mutations responsible for Duchenne muscular dystrophy has completely revolutionized research on the muscular dystrophies. With our ongoing interest in muscle proteins, we carried out an early study into the role of dystrophin, and by using synthetic peptide fragments corresponding to regions of the molecule, were able to use NMR to identify sites in the N-terminus involved in binding to actin [107].

Tropomyosin

The isolation and crystallization of tropomyosin was introduced as an exercise in the practical class on protein chemistry that Kenneth Bailey ran for the final year of the Cambridge undergraduate biochemistry course, in 1946, the year he reported its discovery. As his assistant in teaching this class, I had to make sure I could handle the protein myself and it became the first muscle protein I crystallized. As a consequence of this early association, I retained a special affinity for tropomyosin and frequently used it in my research. I had always considered the standard Bailey preparation to be homogenous for it satisfied the criteria used at that time. This assumption was a reflection of the lack of resolution of the methods then available for detecting the heterogeneity, such as that due to the isoform composition of many proteins. On application of the newly introduced gel electrophoresis under dissociating conditions to our, so-called, homogenous tropomyosin from fast skeletal muscle, we were surprised to find that it migrated as two bands. Pete Cummins, who had joined me as a postgraduate student, was able to separate the rabbit skeletal tropomyosin into two components corresponding to these bands. As he was unable to demonstrate any significant difference in biological properties of the two fractions and only slight differences in amino composition, they were considered to be isoforms and named α and β tropomyosins [108]. The ratio of α and β isoforms, which could be distinguished immunochemically [109], is characteristic of the muscle type [108, 110]. Dave Heeley was able to show that traces of other isoforms of tropomyosin in skeletal muscle can be detected under electrophoretic conditions of high resolution [110]. It was noted that the isoforms were also partially phosphorylated but as originally reported [111], this was not a dynamic process that changed with muscle activity. Marked changes in phosphorylation level did occur during development, particularly in cardiac muscle [112]. As was the case with other myofibrillar proteins, it was shown that the pattern of tropomyosin isoform expression in the muscle cell is under nervous control. Cross innervation changes the isoform composition to that characteristic of the muscle which is normally innervated by the crossed nerve [110].

Isoforms and antibodies to the components of the troponin complex

As is the case with tropomyosin, the isoform pattern of the troponin complex is characteristic of the muscle type. Many investigators have reported similar findings with myosin but with this protein, the situation is much more complicated for more than one gene is involved in the synthesis of the molecule. As we had shown that unique isoforms of the muscle regulatory proteins were associated with each muscle cell type, it was suggested that this fact could be utilized both as

a means of accurate typing of cells and diagnostically for detecting changes associated with disease [113]. Using an an-
tibody to troponin C, Tamio Hirabayashi was able to clearly
locate the calcium binding protein in the I filament [114] and
show that it contributed to the 38.5 nm meridional X-ray reflection of vertebrate striated muscle [115].

Troponin components, particularly troponin I and T,
which existed in forms unique to the muscle type, seemed to
offer particular promise for typing muscle cells. Initial
studies with polyclonal antibodies to troponin I indicated
that the different isoforms, e.g., those from cardiac and
skeletal muscle, could readily be distinguished immunochemically, which would be expected from what was known
of their amino-acid sequences [116]. In view of the considerable homology of the troponin I isoforms, the possibility of
cross reaction existed with the polyclonal antibodies and it
was decided to use the, then newly developed, monoclonal
technique to prepare them. Tej Dhoot, who was very skilled
in this technique, joined me and soon built up a library of
antibodies to the individual isoforms of the troponin components, the first of its kind. With these it was possible to
show that with mature skeletal muscles, unlike is often the
case with myosin, only one isoform, of troponin I in particular, corresponding to the muscle type existed in a mature
normal cell. Thus, providing a simple and elegant method of
muscle cell typing [117–120]. Use of these antibodies enabled the changes in gene expression occurring during development [121], regeneration [122], denervation [123, 124],
hormonal intervention [125], and disease, etc., to be followed. Cross innervation experiments using the antibodies
demonstrated changes occurring in the troponin components
in parallel with those of the tropomyosin isoforms [110, 126].

It is now clear that certain fixed combinations of the isoforms of the myofibrillar proteins correspond to the activity
pattern or stage of development of a particular muscle. The
differences between the structures of the isoforms of a par-

ticular protein are relatively small. Presumably, they are
such as to modulate the function of the complex as a whole,
in a manner appropriate for the changed activity pattern.
Much has to be learn about the mode of regulation of complex changes in gene expression which occur when muscle
responds to its activity pattern and external factors, the so-
called plasticity of muscle.

The Italian connection

My first sight of Italy, in February 1942, was not under very
auspicious circumstances. After a night battened down in
the hold of an Italian cargo boat, I and my fellow prisoners
were allowed on deck briefly, to get some air before docking
at Naples. In the dull grey of the early morning, Capri and
Ischia looked rather foreboding. Despite the fact that I spent
the next 18 months confined to POW compounds in various
regions, it was apparent that Italy was a beautiful country
and I resolved to revisit it when life in Europe returned to
normality. It was not until some 12 years after the war, that
I renewed my association with the country, when Andrea
Corsi, then working at the University of Padua, joined me in
Cambridge. Since that time, I have maintained contacts with
Corsi, the muscle scientists in Padua, particularly Alfredo
Margreth and Massimo Aloisi, and Tony Raggi from Pisa,
who spent some time with the muscle group in Birmingham.
Visits have included lecture tours and assistance in organising meetings on various aspects of muscle research. I remember with particular pleasure, meetings in the late
summer, held in the Tyrol at Bressione, at the summer campus of the University of Padua. I revisited Mantua several
times and retraced my wartime meanderings through the
town. Little did I imagine during my first involuntary visit,
that long after the war I would become a member of the Accademia Virgiliana, based in the city and named after its
famous citizen of long ago. I was greatly honoured and felt

much pleasure and personal satisfaction when in 1990, I was elected Foreign Member of the Accademia Nazionale dei Lincei, the national academy.

My last, and I hope not final, visit was to the European Muscle Congress in Florence, in September 1995. After the meeting, I joined up with two old army friends, Jim Bourn and George White, for a sentimental journey. We had last been together in Italy just before we escaped, at the time of the Armistice in 1943. Two of us had escaped from the train taking us to Germany whilst it stopped in Mantua but were both recaptured. We retraced together the routes we took through the town 52 years ago. Remarkably, these were easily recognisable for little had changed apart from the bridge where I had been recaptured (Fig. 5). It was a particularly emotive experience when we visited the house high

Fig. 5. Standing in September 1995, by the bridge over the Largo Superiore, Mantua, where I was recaptured by a German guard 52 years earlier. In 1943, there was an ancient bridge which was later destroyed by Allied bombing and subsequently rebuilt after the war.

up in the mountains, south of Rimini, and met the surviving daughters of the family which had befriended Jim, the only one of our group who had managed to avoid recapture by the Germans. At great personal risk to him and his family, their father, a minor landowner, had hidden Jim in his house until his farm was overrun by the advancing American forces, some 14 months later. We were interested, and I may say pleased, to find that the old Carthusian monastery, near Salerno, in which we had languished for many months from 1942 to 1943, had been splendidly restored from its semi-derelict state. It is now a national monument and museum for the display of ecclesiastical art treasures from various parts of Italy. Usage which was much more to our liking than that of which we had had personal experience.

Committees and miscellaneous responsibilities

My major aim and interest at Birmingham was to build up a first class department and maintain a research group on muscle, of international standing. To do this, and carry out my share of the teaching duties, as well as the administrative responsibilities that went with the chairmanship of the department, was demanding. This meant that opportunities for sabbatical leave abroad that were offered to me from time to time, where not taken up. I was reluctant to lose the close contact with the research in my group which would inevitably result. In fact, during the whole of my time at the department of biochemistry at Birmingham, I only spent 1 month away on sabbatical leave. That was in 1964, with Al Stracher at the department of biochemistry in the Downstate Medical Centre of the State University of New York, Brooklyn. In some ways I did not feel the need for the mental rejuvenation which sometimes comes from working in a new environment for a substantial period. The ongoing research in the department provided its own stimulation, as did attendance at most of the major meetings on muscle occurring world-wide. The regular Gordon Conferences on muscle, held at various venues on the east coast of the USA were particularly valuable in this respect. Meetings and exchange of ideas and results on muscle tended to be more frequent in the USA. Indeed, I often learnt about new work carried out in the UK at meetings in the USA.

In recent years, the European Muscle Conference that meets annually has become an important forum for the presentation and discussion of muscle research. I feel a tenuous personal responsibility for this organisation which was started largely by the efforts of Marcus Schaub. When Marcus was working in Birmingham, I took him along to a rather loose organisation known as the Muscle Club, that met occasionally in Britain. About once a year, a dinner would be held, usually in turn at one of the colleges in Oxford or Cambridge, and at the Universities of Birmingham and London, where there was active interest in muscle. The aim was to encourage contact and the exchange of ideas between muscle scientists in the UK. Occasionally, but not always, a formal lecture was given, but an important factor on these occasions was the food and the wine served. Marcus observed these occasions with interest and was fascinated by the apparently reserved and low-key expression of interest, by what seemed to him to be, the British muscle mafia. On his return to Switzerland, he set about organising the European Muscle Club, a much more dynamic organisation than the British equivalent and open to all 'continental' Europeans. Since 1971, it has held annual meetings in the countries of Europe in turn. In the initial years, British scientists were not eligible to be members for it was felt that the club might be dominated by the relatively large number of people working on muscle in the UK at that time. I am glad to say this restriction soon disappeared as did the somewhat effete British Muscle Club. It is now known as the European Society for Muscle Research and continues to meet annually, attracting a large number of contributors. Marcus Schaub must take a great deal of credit for the success of this organisation. But for his energy and enthusiasm, it would never have got off the ground. Almost single-handed as secretary and with the support of Caspar Ruegg, Gabriel Hamoir, and later Witold Drabikowski, he has been largely responsible for its growth to what is a major annual international muscle meeting. His recent retirement from that post has left an enormous gap that will not be easily filled.

From my early days in Birmingham, I was increasingly drawn into the work of the committees of the Science (now Biology and Biotechnology) and Medical Research Councils. I particularly enjoyed my membership of the Animals Committee of the Agricultural Research Council. This body was responsible for the allocation of funds for research related to farm animals and the assessment of the work of institutes concerned with scientific background to agriculture. Often we met at these institutes and were shown their work. I learnt a great deal about the animal side of farming, in which I had always had an interest. Indeed, I believe that in an alternative life, I could have taken up the role of a prosperous farmer with few regrets. Towards the end of my tenure of the chair at Birmingham, I became chairman of the Advisory Board to the Meat Research Institute, one of the Agricultural Research Council Institutes, at Langford, Bristol. In addition to my work with the Muscular Dystrophy Group, I was involved in the allocation of medical research funds through membership of the British Heart Research Committee and the Systems Board of the Medical Research Council. After my election to the Royal Society in 1974, an event which gave me much satisfaction, I was drawn into service on a number of the society's committees including the council which, although demanding, was interesting.

Editorial work, I studiously avoided, for I did not enjoy it and felt enough of my time was taken up by the frequent requests from a variety of journals to referee papers. Despite my shortcomings in this direction, I devoted considerable time to the Biochemical Society, which as an energetic and prosperous organisation, has done much to foster the advance of biochemistry in the UK and indeed, in Europe. At various times between 1955 and 1983, I spent in all, 11 years on the committee, the last three as chairman. In the latter capacity, it was a special pleasure to participate in a joint meeting of the Japanese and British Biochemical Societies, at Osaka in 1982, and meet many old friends. At the international level, I became chairman of the National Committee for Biochemistry and of the Nominations Committee of the International Union of Biochemistry and Molecular Biology.

Retirement

My retirement from the department of biochemistry in 1985, was marked by a special meeting on muscle, to which many of my former students and colleagues from the UK and abroad contributed. Efficiently organised by Ian Trayer, it was an event I appreciated and I was touched by the support it received from friends and colleagues whom I much respected. Occasions such as this are designed to mark the end of a way of life, which in my case I had much enjoyed. It is an event that, so far as I can judge from some current attitudes, cannot come too soon for a number of my younger colleagues. For my part, I did not feel physically or mentally ready to discard completely, the scientific life style. It would not have been appropriate, or fair, to my successor, for me to stay on in the department, of which I had been head for 26 years. A move to Felin Werndew, built up over the years as a potential retirement home, would have been easy for Maureen and I. This, however, would have meant complete abandonment of any possibility of carrying on with research and access to library facilities would have been very difficult. Also, I was still involved in external committee work that required me to maintain contact with science to be effective.

There were possibilities of moving to the biophysics department at Kings College London or the biochemistry department at the University of Aberystwyth. These would have involved moving house and I was relieved and grateful when John Coote, head of the physiology department in the medical school in Birmingham, generously offered to provide me with accommodation. Since that time, with the help of Val Patchell, I have managed to maintain limited research activity on dystrophin and particularly on phosphorylation and its role in the calcium regulation of contractile activity in muscle. A substantial part of this work has been carried out in collaboration with Barry Levine, from whom I have learnt much about the potential of NMR for the illumination of molecular processes in biochemistry. A stimulating and imaginative scientist, Barry has, in our frequent discussions, done much to retard the progressive mortification of the central nervous system which a septuagenarian has to endure.

Acknowledgements

I cannot thank my parents enough for providing me with a reasonable set of genes. Despite my shortcomings, I have been supported strongly throughout my scientific life by my wife Maureen, and children, Gillian, Jacqueline and Michael; to them I am especially grateful. In any career there may be unknown individuals who have had a profound influence. My most sincere thanks go to three persons who, without their knowledge, played vital roles. The first is to the Arab who, with great skill, removed the revolver from my Sam Brown belt, without me being aware at the time, as I walked through the back streets of Cairo, in 1941. The chances are that if it had been in my possession at the moment of capture in the western Desert I would not have been taken alive. The second is to the soldier of the Panzer Armee Afrika, who on 2 February 1942, had me fixed in the sights of his machine gun, with the result that the bullets were close enough to crack as they passed my head, for failing to hit me. The third is to some person in the legal department of the Wehrmacht who arranged for me to be court-martialed in Hildesheim on the day, in 1944, my camp in Brunswick was heavily bombed by the American air force.

It goes without saying that a career in biochemistry depends vitally upon one's associates. Over the years I have been fortunate to work with a group of able colleagues at all levels, who have been principally responsible for what has been achieved. Most have been mentioned in the text or in the references. If they have not, this is an oversight on my part. I am particularly grateful to Val Patchell who as colleague and assistant has managed to stay with me for far longer than any anyone else. Her loyalty, cheerfulness, energy and practical skill were particularly important in enabling me to keep close to research after I was retired from the biochemistry department in 1985. To all these colleagues I express my heartfelt thanks.

References

1   A. Szent-Gyorgyi, Studies from the Institute of Medical Chemistry, University Szeged, A. Szent-Gyorgyi (Ed.), Vols. I–III, S. Karger, Basel, 1941–1943.
2   R.S. Bear, J. Am. Chem. Soc. 67 (1945) 1625.
3   W.T. Astbury, S.V. Perry, R. Reed and L.C. Spark, Biochem. Biophys. Acta 1 (1947) 379–392.
4   K. Bailey and S.V. Perry, Biochem. Biophys. Acta 1 (1947) 506–616.
5   F.B. Straub and G. Feuer, Biochem. Biophys. Acta 4 (1950) 455–470.
6   F. Sanger, Annu. Rev. Biochem. 57 (1988) 1–28.
7   A. Szent-Gyorgyi, Acta Phys. Skand. 9 (Supp. XXV) (1945) 1–115.
8   S.V. Perry, Physiol. Rev. 36 (1956) 1–76.
9   S.V. Perry, J. Gen. Physiol. 33 (1950) 563–577.
10 M. Heidenhain, in: Plasma und Zelle, Gustav Fischer, Jena, 1911, pp. 507–686.
11 A.F. Schick and G.M. Hass, Science 109 (1949) 486–487.
12 S.V. Perry, Biochem. J. 48 (1951) 257–265.
13 A.G. Szent-Gyorgyi, Arch. Biochem. Biophys. 42 (1953) 305–320.

14 S.V. Perry, Biochem. J. 51 (1952) 495–499.
15 S.V. Perry and A. Corsi, Biochem. J. 68 (1958) 5–12.
16 S.V. Perry, Biochim. Biophys. Acta 8 (1952) 499–509.
17 H. Kumagai, S. Ebashi and F. Takeda, Nature 176 (1955) 166.
18 D.J. Baird and S.V. Perry, Biochem. J. 77 (1960) 262–271.
19 J.B. Chappell and S.V. Perry, Biochem. J. 55 (1953) 586–595.
20 J.B. Chappell and S.V. Perry, Nature 173 (1954) 1094–1095.
21 I. Banga and A. Szent-Gyorgyi, Stud. Inst. Med. Chem. Inst. Szeged 3 (1943) 72–75.
22 E. Bozler, Am. J. Physiol. 167 (1951) 276–283.
23 S.V. Perry and T.C. Grey, Biochem. J. 64 (1956) 184–193.
24 J. Raaflaub, Helv. Chim. Acta 30 (1955) 1798.
25 H. Senn, Diss Universitat, Zurich (1954).
26 S.V. Perry and T.C. Grey, Biochem. J. 64 (1956) P6.
27 A. Weber, J. Biol. Chem. 234 (1959) 2764–2769.
28 A. Corsi and S.V. Perry, Biochem. J. 68 (1958) 12–17.
29 S. Ebashi, Nature 200 (1963) 1010.
30 A.A. Newton and S.V. Perry, Biochem. J. 74 (1960) 127–136.
31 S.C. Bondy and S.V. Perry, J. Neurochem. 10 (1963) 593–601; 603–609.
32 S.V. Perry, Biochem. J. 74 (1960) 94–101.
33 H. Mueller and S.V. Perry, Biochem. J. 85 (1962) 431–439.
34 J.M. Jones and S.V. Perry, Biochem. J. 100 (1966) 120–130.
35 I. Rayment, W.R. Rypniewski, K. Schmidt-Base, R. Smith, D.R. Tomchick, M.M. Benning, D.A. Winkelmann, G. Weisenberg and H.M. Holden, Science 261 (1993) 50–58.
36 D. Stone and S.V. Perry, Biochem. J. 131 (1972) 127–137.
37 L. Leadbeater and S.V. Perry, Biochem. J. 87 (1963) 233–238.
38 A.B. Fraser, E. Eisenberg, W.W. Kielley and F.D. Carlson, Biochemistry 14 (1975) 2207–2214.
39 S.V. Perry and J. Cotterill, Biochem. J. 92 (1964) 603–608.
40 S.V. Perry and J. Cotterill, Nature 206 (1965) 161–163.
41 A. Buller, J. Eccles and R. Eccles, J. Physiol. 150 (1960) 417–437.
42 O. Smithies, Biochem. J. 61 (1955) 629–641.
43 J. Kendrick-Jones and S.V. Perry, Nature 208 (1965) 1068–1070.
44 S.V. Perry and D.J. Hartshorne, in: G. Gutmann and P. Hnik (Eds.), The Effect of Use and Disuse in Neuromuscular Function, Czechoslovak Academy of Science, Prague, 1963, pp. 491–498.
45 I.P. Trayer and S.V. Perry, Biochem. Z. 345 (1966) 87–100.
46 D.L. Holland and S.V. Perry, Biochem. J. 114 (1969) 161–170.
47 P. Johnson, C.I. Harris and S.V. Perry, Biochem. J. 105 (1967) 361–369.

48   A.M. Asatoor and M.D. Armstrong, Biochem. Biophys. Res. Commun. 26 (1967) 168–174.
49   I.P. Trayer, C.I. Harris and S.V. Perry, Nature 217 (1968) 452–453.
50   M.F. Hardy, C.I. Harris, S.V. Perry and D. Stone, Biochem. J. 120 (1970) 653–660.
51   M.F. Hardy and S.V. Perry, Nature 223 (1969) 300–302.
52   P. Johnson and S.V. Perry, Biochem. J. 119 (1970) 293–298.
53   G.E. Lobley, S.V. Perry and D. Stone, Nature 3 231 (1969) 317–318.
54   S.V. Perry, V. Davies and D. Hayter, Biochem. J. 99 (1966) 1c.
55   D.J. Hartshorne, S.V. Perry and V. Davies, Nature 209 (1966) 1352–1353.
56   D.J. Hartshorne, S.V. Perry and M.C. Schaub, Biochem. J. 104 (1967) 907–913.
57   M.C. Schaub and S.V. Perry, Biochem. J. 115 (1969) 993–1004.
58   D.J. Hartshorne and H. Mueller, Biochem. Biophys. Res. Commun. 31 (1968) 647–653.
59   M.C. Schaub, S.V. Perry and W. Hacker, Biochem. J. 126 (1972) 237–249.
60   J.M. Wilkinson, S.V. Perry, H. Cole and I.P. Trayer, Biochem. J. 124 (1971) 55P.
61   J.M. Wilkinson, S.V. Perry, H. Cole and I.P. Trayer, Biochem. J. 127 (1972) 215–225.
62   S.V. Perry, H.A. Cole, J.F. Head and F.J. Wilson, Cold Spring Harbor Symp. Quant. Biol. 37 (1972) 251–262.
63   J.F. Head and S.V. Perry, Biochem. J. 137 (1974) 145–154.
64   H. Syska, S.V. Perry and I.P. Trayer, FEBS Lett. 40 (1974) 253–257.
65   H. Syska, J.M. Wilkinson, R.J.A. Grand and S.V. Perry, Biochem. J. 153 (1976) 375–387.
66   R.A. Weeks and S.V. Perry, Biochem. J. 173 (1978) 449–457.
67   P. Jackson, G.W. Amphlett and S.V. Perry, Biochem. J. 151 (1975) 85–97.
68   R.J.A. Grand, B.A. Levine and S.V. Perry, Biochem. J. 203 (1982) 61–68.
69   D.C. Dalgarno, R.J.A. Grand, B.A. Levine, A.J.G. Moir, G.M.M. Scott and S.V. Perry, FEBS Lett. 150 (1983) 54–58.
70   B.A. Levine, A.J.G. Moir and S.V. Perry, Eur. J. Biochem. 172 (1988) 389–397.
71   J.F. Head, R.A. Weeks and S.V. Perry, Biochem. J. 161 (1977) 465–471.
72   R.J.A. Grand, S.V. Perry and R.A. Weeks, Biochem. J. 177 (1979) 521–529.
73   R.J.A. Grand and S.V. Perry, FEBS Lett. 92 (1978) 137–142.

74   G.W. Amphlett, T. Vanaman and S.V. Perry, FEBS Lett. 72 (1976) 163–168.
75   R.J.A. Grand and S.V. Perry, Biochem. J. 183 (1979) 285–295.
76   R.J.A. Grand and S.V. Perry, Biochem. J. 189 (1980) 227–240.
77   K. Sobue, Y. Muramoto, M. Fujita and S. Kakiuchi, Proc. Natl. Acad. Sci. USA 78 (1981) 5652–5655.
78   C. Bailey and C. Villar-Palasi, Fed. Proc. 30 (1971) 1147.
79   S.V. Perry and H.A. Cole, Biochem. J. 131 (1973) 425–428.
80   S.V. Perry and H.A. Cole, Biochem. J. 141 (1974) 733–743.
81   A.J.G. Moir, J.M. Wilkinson and S.V. Perry, FEBS Lett. 42 (1974) 253–256.
82   H.A. Cole and S.V. Perry, Biochem. J. 149 (1975) 525–533.
83   J.M. Wilkinson and R.J.A. Grand, Nature 271 (1978) 31–35.
84   A.G.J. Moir, R.J. Solaro and S.V. Perry, Biochem. J. 185 (1980) 505–513.
85   K. Mittman, K. Jaquet and L.M.G. Heilmeyer, Jr., FEBS Lett. 273 (1990) 41–45.
86   R.J. Solaro, A.J.G. Moir and S.V. Perry, Nature 262 (1976) 615–617.
87   K.P. Ray and P.J. England, FEBS Lett. 70 (1976) 11–16.
88   S.V. Perry, in: K. Maruyama, Y. Nonomura and K. Kodama (Eds.), Calcium as Cell Signal, Igaku-Shoin, Tokyo, 1995, pp. 49–59.
89   P.G. Quirk, V.B. Patchell, Y. Gao, B.A. Levine and S.V. Perry, FEBS Lett. 370 (1995) 175–178.
90   W.T. Perrie and S.V. Perry, Biochem. J. 119 (1970) 31–39.
91   W.T. Perrie, L.B. Smillie and S.V. Perry, Cold Spring Harbor Symp. Quant. Biol. 37 (1972) 17–18.
92   W.T. Perrie, L.B. Smillie and S.V. Perry, Biochem. J. 135 (1973) 151–164.
93   W.T. Perrie, M.A.W. Thomas and S.V. Perry, Biochem. Soc. Trans. 1 (1973) 860–861.
94   N. Frearson, B.W.W. Focant and S.V. Perry, FEBS Lett. 63 (1976) 27–32.
95   E. Pires, M.A.W. Thomas and S.V. Perry, Febs Lett. 41 (1974) 292–296.
96   E. Pires and S.V. Perry, Biochem. J. 167 (1977) 137–146.
97   M. Morgan, S.V. Perry and J. Ottaway, Biochem. J. 157 (1976) 687–697.
98   A.C. Nairn and S.V. Perry, Biochem. J. 179 (1979) 89–97.
99   M. Yazawa and K. Yagi, J. Biochem. (Tokyo) 82 (1977) 287–289.
100 R. Dabrowska, D. Aromatorio, J.M.F. Sherry and D.J. Hartshorne, Biochem. Biophys. Res. Commun. 78 (1977) 1263–1272.

101 H.A. Cole, H.S. Griffiths, V.B. Patchell and S.V. Perry, FEBS Lett. 180 (1985) 165–169.
102 B.A. Levine, H.S. Griffiths, V.B. Patchell and S.V. Perry, Biochem. J. 254 (1988) 277–286.
103 S.M. Pembrick, J. Biol. Chem. 255 (1980) 8836–8841.
104 S.A. Westwood, O. Hudlicka and S.V. Perry, Biochem. J. 218 (1984) 841–847.
105 N.G. Anderson and N.L. Anderson, Anal. Biochem. 85 (1978) 331–340; 341–354.
106 N. Frearson, R.D. Taylor and S.V. Perry, Clin. Sci. 61 (1981) 141–149.
107 B.A. Levine, A.J.G. Moir, V.B. Patchell and S.V. Perry, FEBS Lett. 298 (1992) 44–48.
108 P. Cummins and S.V. Perry, Biochem. J. 133 (1973) 765–777.
109 P. Cummins and S.V. Perry, Biochem. J. 141 (1974) 43–49.
110 D. Heeley, G.K. Dhoot, N. Frearson, S.V. Perry and G. Vrbova, FEBS Lett. 152 (1983) 282–286.
111 H. Ribulow and M. Barany, Arch. Biochem. Biophys. 179 (1977) 718–720.
112 D.H. Heeley, A.J.G. Moir and S.V. Perry, FEBS Lett. 146 (1982) 115–118.
113 S.V. Perry, in: A.T. Milhorat (Ed.), Exploratory Concepts in Muscular Dystrophy II, Exerpta Medica, Amsterdam, 1974, pp. 319–328.
114 T. Hirabayashi and S.V. Perry, Biochem. Biophys. Acta 351 (1973) 273–289.
115 E. Rome, T. Hirabayashi and S.V. Perry, Nature New Biol. 244 (1973) 154–155.
116 P. Cummins and S.V. Perry, Biochem. J. 171 (1978) 251–259.
117 G.K. Dhoot, P.G.H. Gell and S.V. Perry, Exp. Cell Res. 117 (1978) 357–370.
118 G.K. Dhoot and S.V. Perry, Nature 278 (1979) 714–718.
119 G.K. Dhoot, N. Frearson and S.V. Perry, Exp. Cell Res. 122 (1979) 339–350.
120 G.K. Dhoot and S.V. Perry, in: S. Ebashi (Ed.), Muscular Dystrophy, University of Tokyo Press, Tokyo, 1982, pp. 89–102.
121 G.K. Dhoot and S.V. Perry, Exp. Cell Res. 127 (1980) 75–87.
122 G.K. Dhoot and S.V. Perry, Muscle Nerve 5 (1981) 39–47.
123 G.K. Dhoot and S.V. Perry, Cell Tissue Res. 225 (1982) 201–215.
124 G.K. Dhoot and S.V. Perry, Exp. Neurol. 82 (1983) 131–142.
125 G.K. Dhoot and S.V. Perry, FEBS Lett. 133 (1981) 225–229.
126 G.K. Dhoot, S.V. Perry and G. Vrbova, Exp. Neurol. 72 (1981) 513–530.

Name Index

Yocum, C.F., 270
Young, F.G., 410, 418, 437
Young, W.J., 83

Zalkin, H., 258

Zeidler, D., 77
Zhu, Q.-S., 188
Zimm, B.H., 28, 30, 34
Zuckerman, S., 117